THE 100+ SERIES™

GEOMETRY

Essential Practice for Advanced Math Topics

Carson-Dellosa Publishing LLC
Greensboro, North Carolina

Visit carsondellosa.com for correlations to Common Core, state, national, and Canadian provincial standards.

Carson-Dellosa Publishing LLC
PO Box 35665
Greensboro, NC 27425 USA
carsondellosa.com

ISBN 978-1-48380-080-6

08-174207784

Table of Contents

Introduction

What are the Common Core State Standards for Middle School Mathematics?

In grades 6–8, the standards are a shared set of expectations for the development of mathematical understanding in the areas of ratios and proportional relationships, the number system, expressions and equations, functions, geometry, and statistics and probability. These rigorous standards encourage students to justify their thinking. They reflect the knowledge that is necessary for success in college and beyond.

Students who master the Common Core standards in mathematics as they advance in school will exhibit the following capabilities:

1. Make sense of problems and persevere in solving them.

Proficient students can explain the meaning of a problem and try different strategies to find a solution. Students check their answers and ask, "Does this make sense?"

2. Reason abstractly and quantitatively.

Proficient students are able to move back and forth smoothly between working with abstract symbols and thinking about real-world quantities that symbols represent.

3. Construct viable arguments and critique the reasoning of others.

Proficient students analyze problems by breaking them into stages and deciding whether each step is logical. They justify solutions using examples and solid arguments.

4. Model with mathematics.

Proficient students use diagrams, graphs, and formulas to model complex, real-world problems. They consider whether their results make sense and adjust their models as needed.

5. Use appropriate tools strategically.

Proficient students use tools such as models, protractors, and calculators appropriately. They use technological resources such as Web sites, software, and graphing calculators to explore and deepen their understanding of concepts.

6. Attend to precision.

Proficient students demonstrate clear and logical thinking. They choose appropriate units of measurement, use symbols correctly, and label graphs carefully. They calculate with accuracy and efficiency.

7. Look for and make use of structure.

Proficient students look closely to find patterns and structures. They can also step back to get the big picture. They think about complicated problems as single objects or break them into parts.

8. Look for and express regularity in repeated reasoning.

Proficient students notice when calculations are repeated and look for alternate methods and shortcuts. They maintain oversight of the process while attending to the details. They continually evaluate their results.

How to Use This Book

In this book, you will find a collection of 100+ reproducible practice pages to help students review, reinforce, and enhance Common Core mathematics skills. Use the chart provided on the next page to identify practice activities that meet the standards for learners at different levels of proficiency in your classroom.

Common Core State Standards* Alignment: Geometry

Domain: Expressions and Equations	
Standard	**Aligned Practice Pages**
8.EE.B.5	78–82
8.EE.C.7b	25
Domain: Functions	
Standard	**Aligned Practice Pages**
8.F.A.1	64, 65
Domain: Geometry	
Standard	**Aligned Practice Pages**
8.G.A.1	7–10, 35
8.G.A.2	9, 10, 13, 15, 26
8.G.A.3	7, 8, 11
8.G.A.5	13–15, 21, 27–34, 36–53, 55, 56, 99, 100
8.G.C.9	94–97
Domain: Trigonometric Functions	
Standard	**Aligned Practice Pages**
HSF-TF.A.3	64, 65
Domain: Congruence	
Standard	**Aligned Practice Pages**
HSG-CO.A.1	26–29
HSG-CO.A.2	11, 87, 88
HSG-CO.A.3	10, 42, 45, 46, 88
HSG-CO.A.5	7–9
HSG-CO.B.7	15
HSG-CO.B.8	16–20
HSG-CO.C.9	22, 23, 26, 28, 29, 30–33, 47, 48, 51
HSG-CO.C.10	22, 23, 34, 47–50
HSG-CO.C.11	23, 43, 44, 50
HSG-CO.D.12	24, 35–41, 77, 98

Domain: Similarity, Right Triangles, and Trigonometry	
Standard	**Aligned Practice Pages**
HSG-SRT.A.2	52–56
HSG-SRT.A.3	52, 53, 55, 56
HSG-SRT.B.4	57, 58
HSG-SRT.C.6	59–61
HSG-SRT.C.7	59–61
HSG-SRT.C.8	62, 63
HSG-SRT.D.10	64, 65
Domain: Circles	
Standard	**Aligned Practice Pages**
HSG-C.A.2	66–70, 72–74
HSG-C.A.3	74
HSG-C.A.4	67, 68
HSG-C.B.5	69–72, 74–76
Domain: Expressing Geometric Properties with Equations	
Standard	**Aligned Practice Pages**
HSG-GPE.A.1	83
HSG-GPE.A.2	84
HSG-GPE.A.3	85, 86
HSG-GPE.B.4	78
HSG-GPE.B.5	79–82
HSG-GPE.B.6	25
HSG-GPE.B.7	88–90, 100
Domain: Geometric Measurement and Dimension	
Standard	**Aligned Practice Pages**
HSG-GMD.A.1	93
HSG-GMD.A.3	94–97
HSG-GMD.B.4	101

Name________________________________ 8.G.A.1, 8.G.A.3, HSG-CO.A.5

Rotations

A **rotation** is a turn about a point. The original figure and its image have the same orientation.

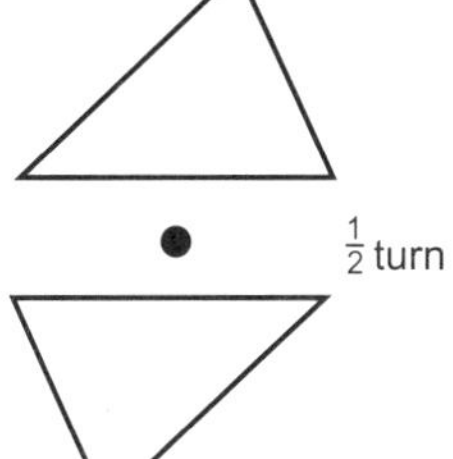

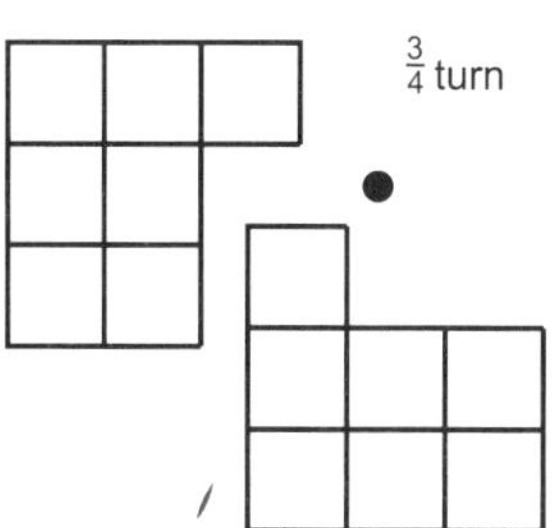

Name the image of the figure named under a half turn about point O.

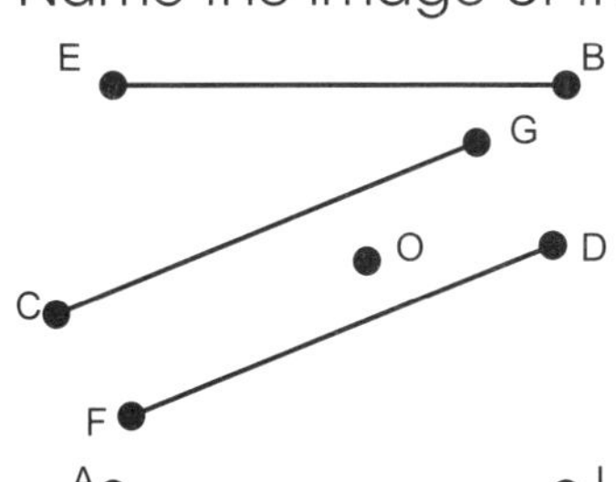

1. $\overline{EB}$
2. $\overleftrightarrow{DF}$
3. $\overline{AI}$
4. Point B
5. Point G
6. Point O

9. Which of these figures represent rotations?

a.

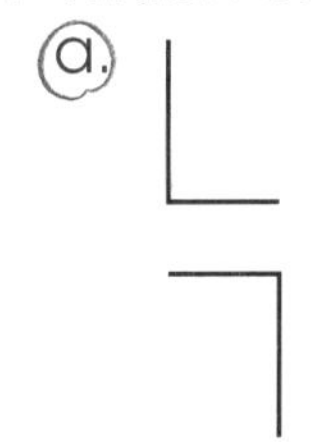

b.

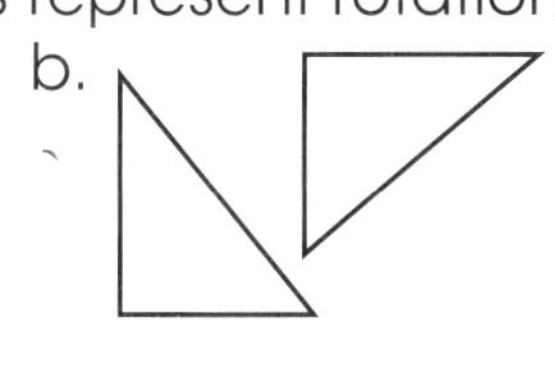

c.

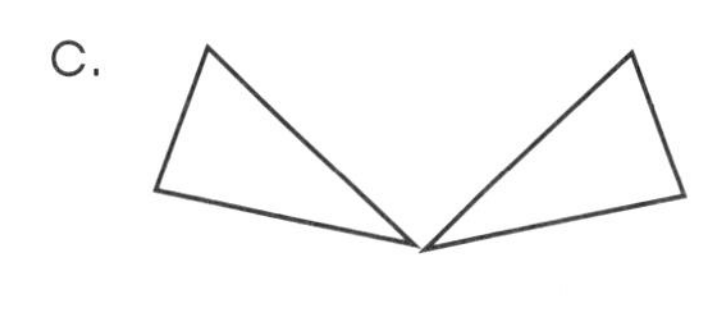

Sketch the image using the indicated rotation about point O.

10.

$\frac{1}{4}$ turn

11.

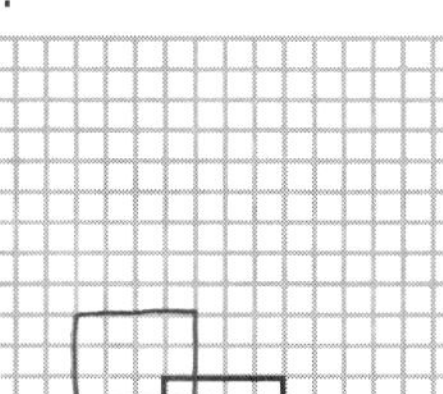

$\frac{1}{2}$ turn

12.

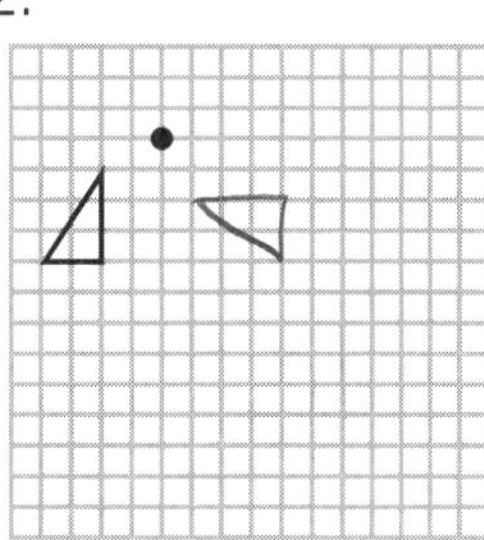

$\frac{3}{4}$ turn

13.

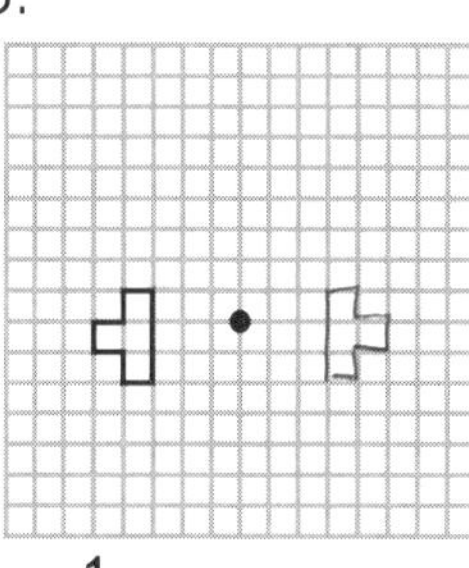

$\frac{1}{2}$ turn

14.

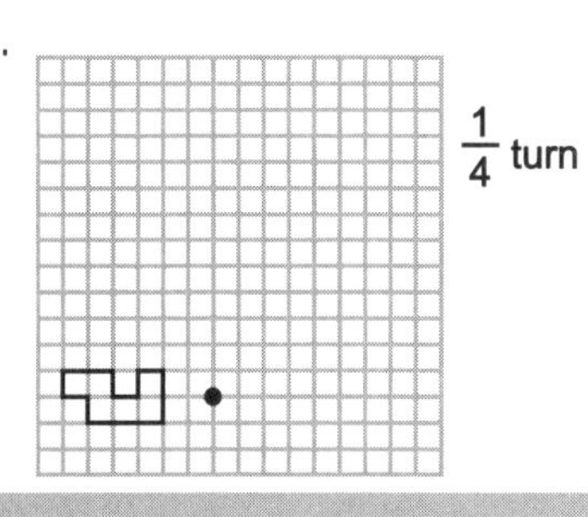

$\frac{1}{4}$ turn

15.

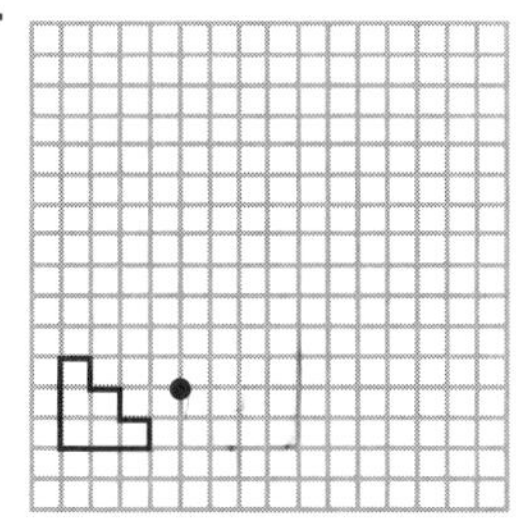

$\frac{1}{2}$ turn

16.

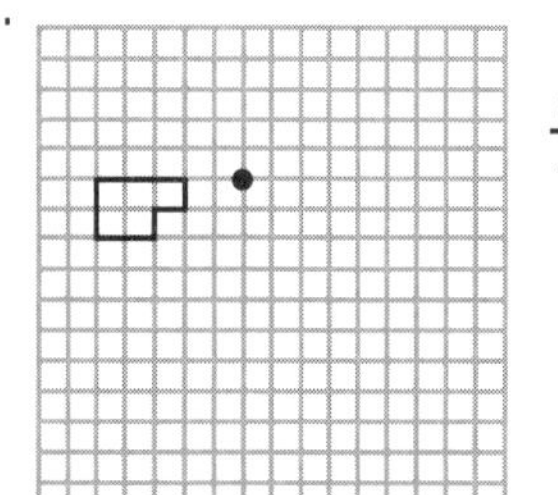

$\frac{3}{4}$ turn

Name________________________________ 8.G.A.1, 8.G.A.3, HSG-CO.A.5

Reflections

A **reflection** requires a flip. The original figure and its image have opposite orientations.

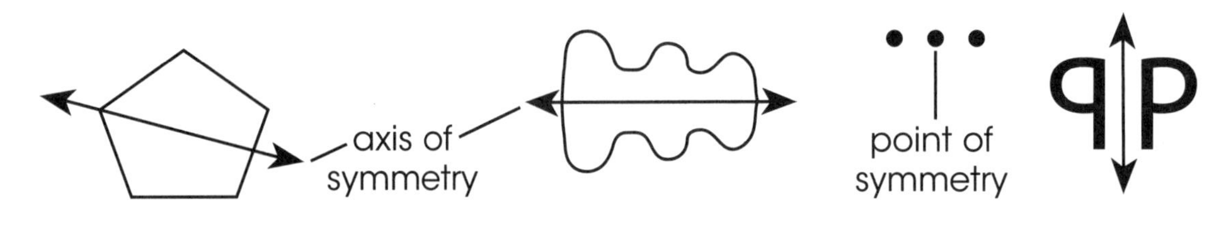

1. Which of these figures represent reflections?
   a. (b.) (c.)

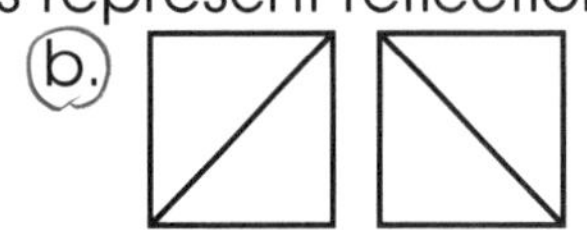

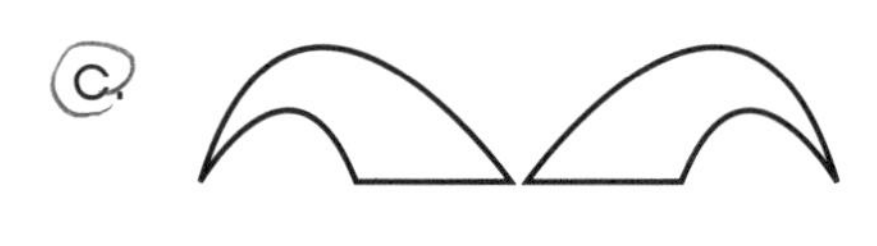

Sketch the axis (or axes) of symmetry for each figure.

2. 3. 4. 5. 6.

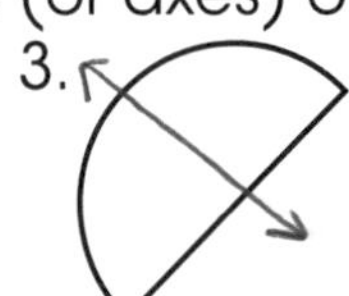

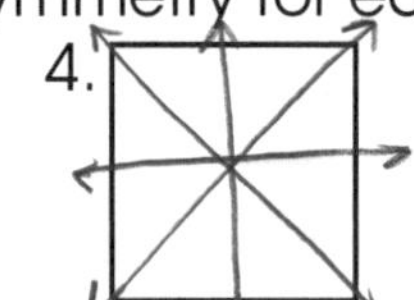

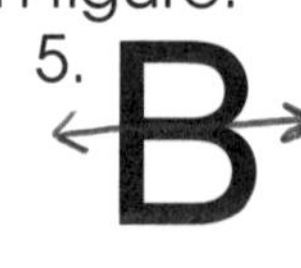

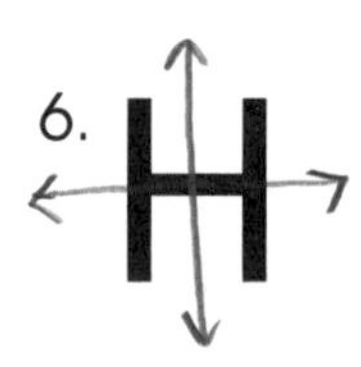

Sketch the reflection of the given figures across the line.

7. 8. 9.

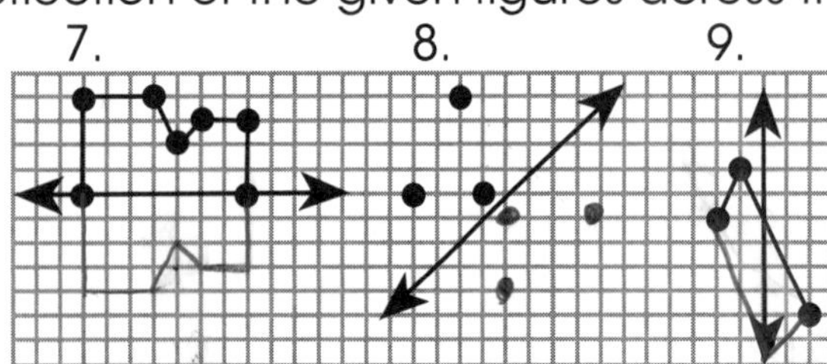

Name the reflection of these points:

10.

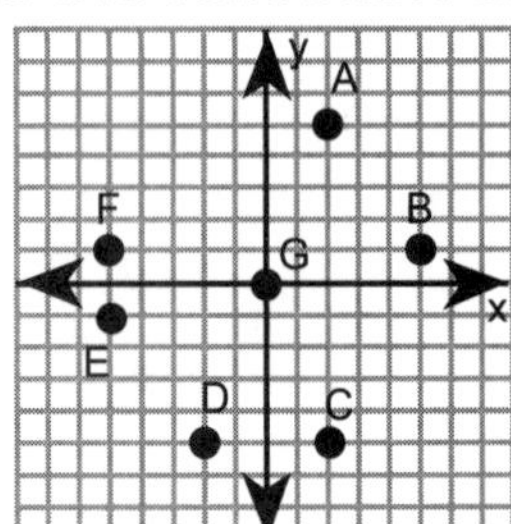

a. B across line y F
b. F across line x E
c. A across point G D
d. D across line y C

11.

a. T across line x K
b. M across line y A
c. S across point V L
d. R across line x L
e. Y across line x N
f. R across point V D

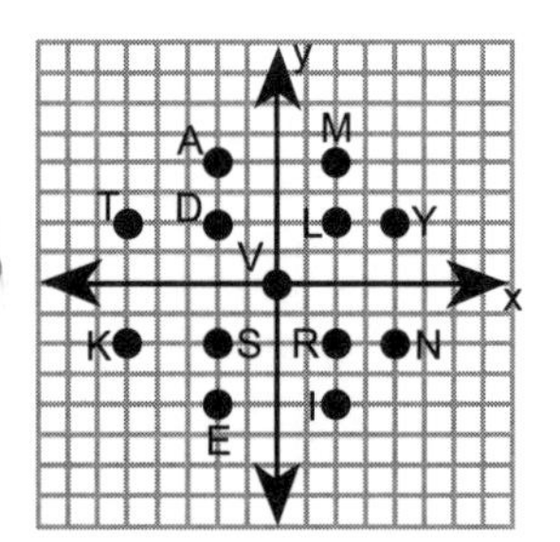

Name________________________________

8.G.A.1, 8.G.A.2, HSG-CO.A.5

Translations

A **translation** is a slide. The original figure and its image have the same orientation.

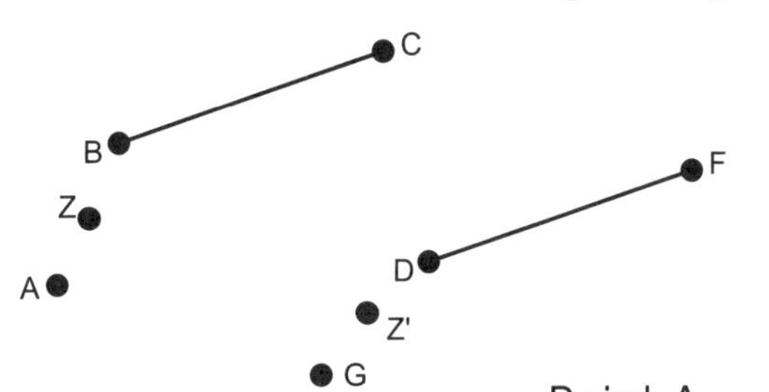

Name the image of each figure under the translation ZZ′.

Point A → Point G; $\overline{BC} \rightarrow \overline{DF}$

1. Which of these figures represent translations?

a. 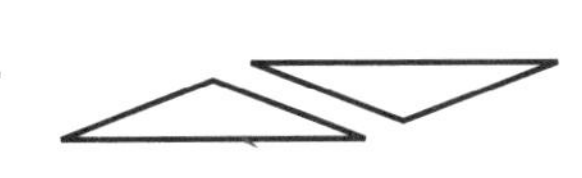

(b.) 

(c.) 

d. 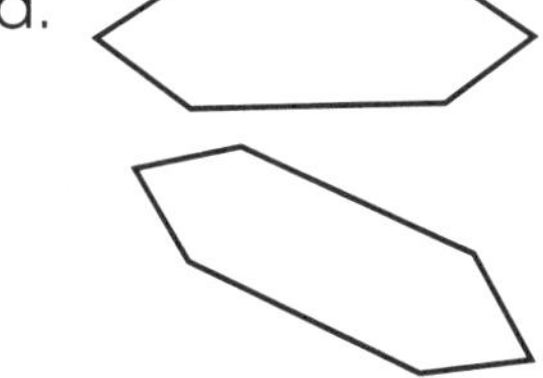

(e.) 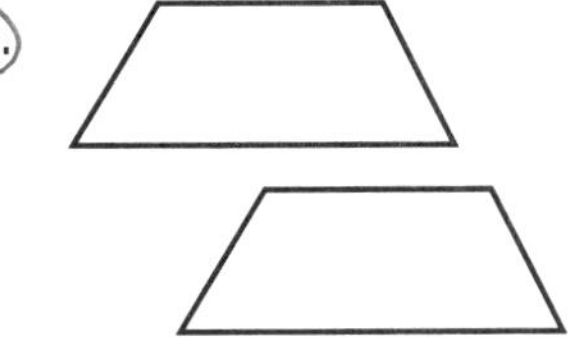

f. 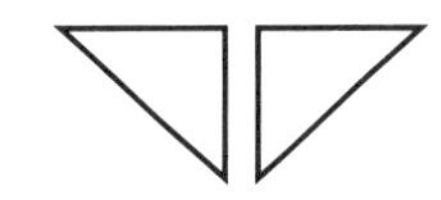

Name the image of each figure under the translation EE′.

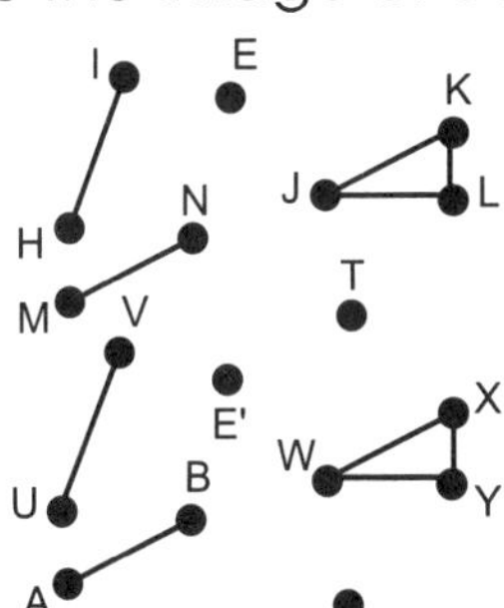

2. Point T Z
3. $\overline{MN}$ $\overline{AB}$
4. ΔJKL ΔWXY
5. Point N B
6. $\overline{HI}$ $\overline{UV}$
7. Point H U

Name________________________________ 8.G.A.1, 8.G.A.2, HSG-CO.A.3

Mixed Practice with Transformations

1-5. Sketch each letter which is point symmetric and mark the point of symmetry.

6-13. Sketch each letter that is line symmetric and draw **all** lines of symmetry.

Each of these figures has been moved in a series of basic motions. Name the motion indicated by the lettered arrow.

14.

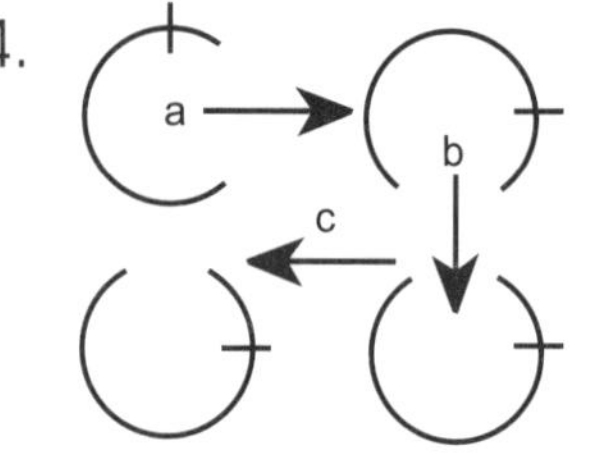

a. rot
b. ref
c. trans

15.

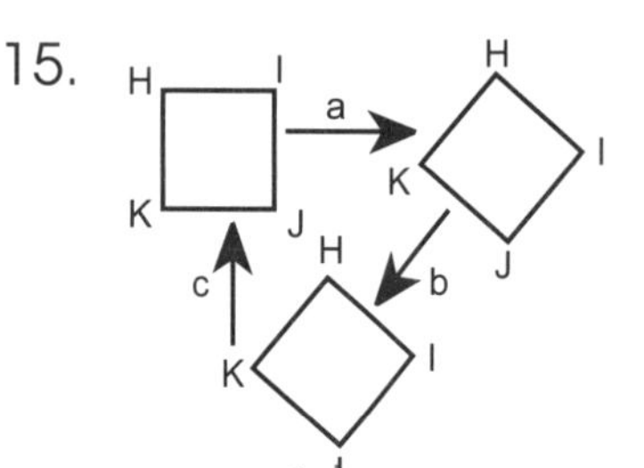

a. rot
b. trans
c. rot

16.

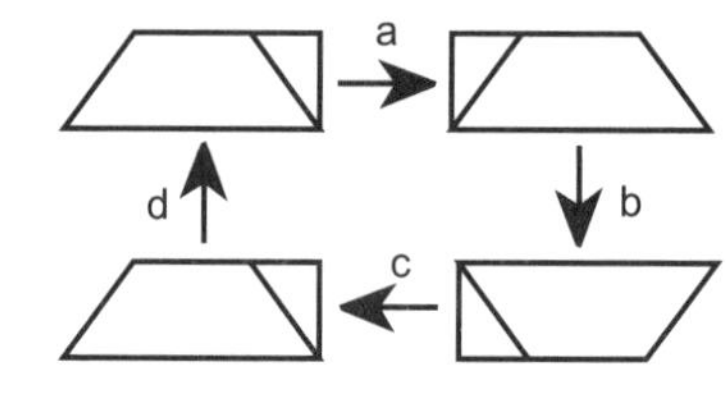

a. ref
b. ref
c. rot
d. trans

Tell which single basic motion will make these figures coincide.

17.

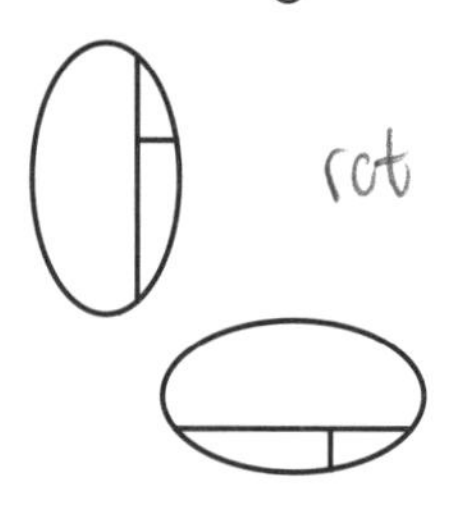

rot

19.

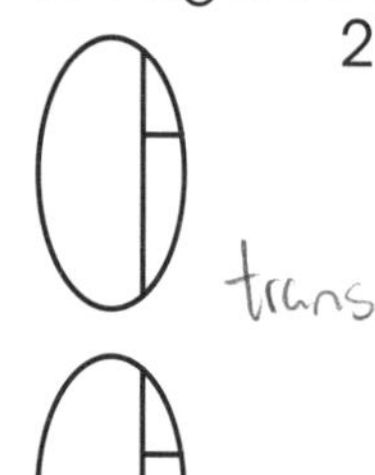

trans

21.

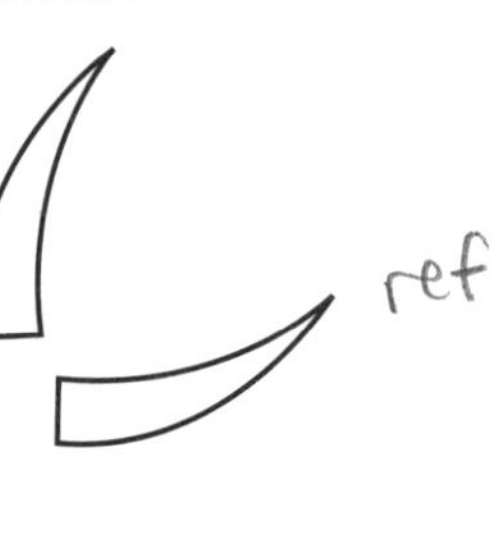

ref

18.

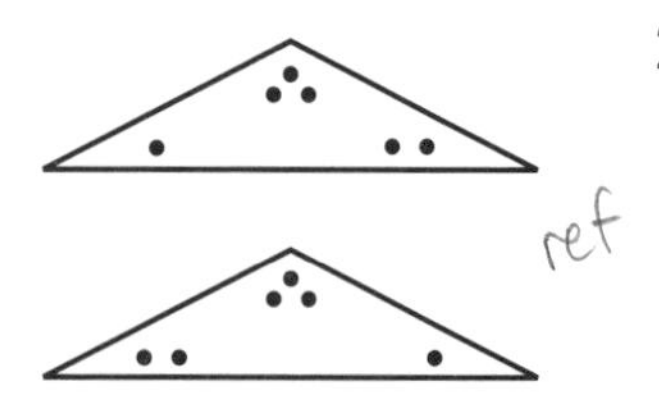

ref

20.

trans

22.

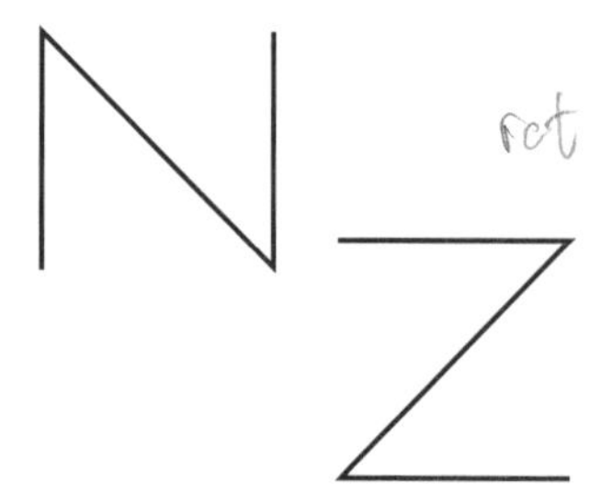

rot

Name______________________________ 8.G.A.3, HSG-CO.A.2

Transformations with Dots and Graphs

For each point named, give its reflection across the

a. x-axis b. origin c. y-axis

1. $(2, {}^{-}3)$
2. $({}^{-}4, {}^{-}1)$
3. $(5, 5)$
4. $({}^{-}1, 2)$
5. (a, b)

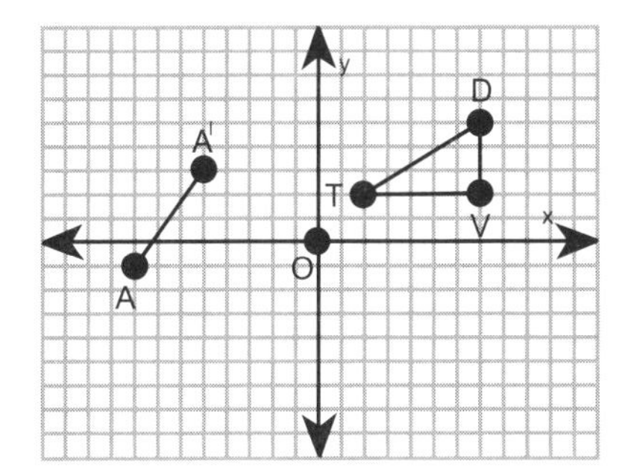

6. Find the image of ΔTDV:

   a. for the rotation of a $\frac{1}{4}$ turn counterclockwise.

   b. for the translation AA'.

   c. for the reflection across O.

7. Find the image of QRST:

   a. for the reflection across the x-axis.

   b. for the rotation about O of a $\frac{1}{2}$ turn clockwise.

   c. for the translation of BB'.

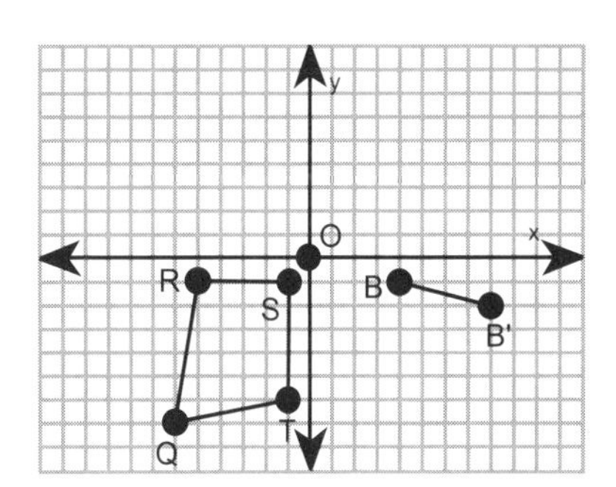

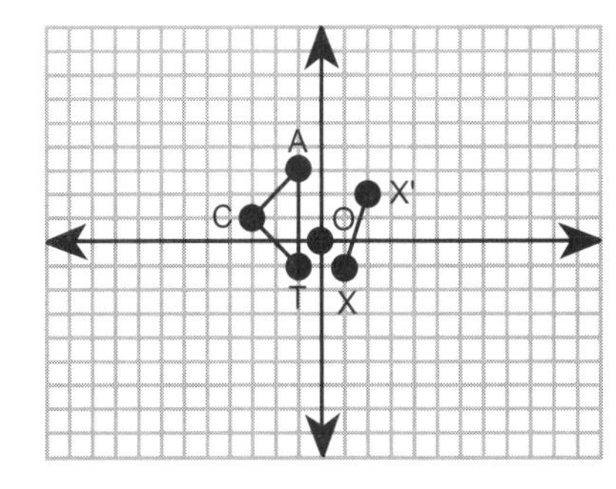

8. Find the image of ΔCAT:

   a. for the reflection across the y-axis.

   b. for the translation XX'.

   c. for the rotation about O of a $\frac{3}{4}$ turn clockwise.

Name____________________

Just for Fun

Draw each figure without lifting your pencil from the paper and without tracing any line more than once.

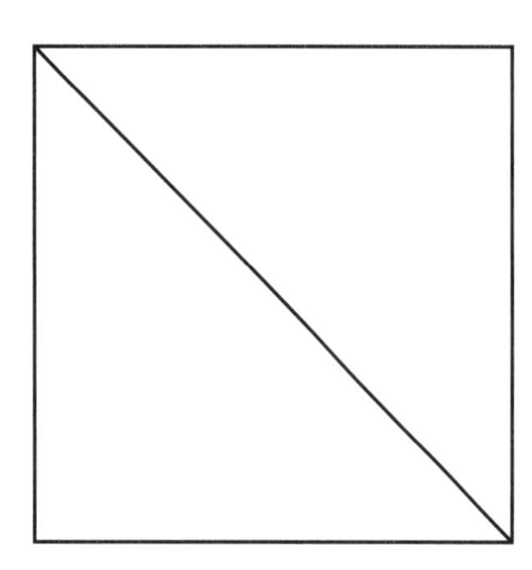

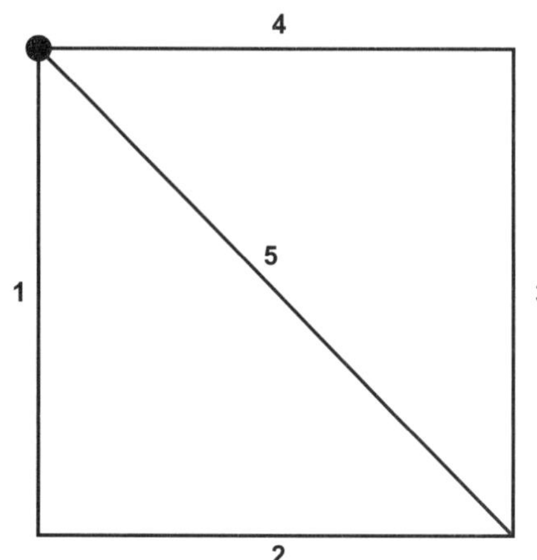

1. 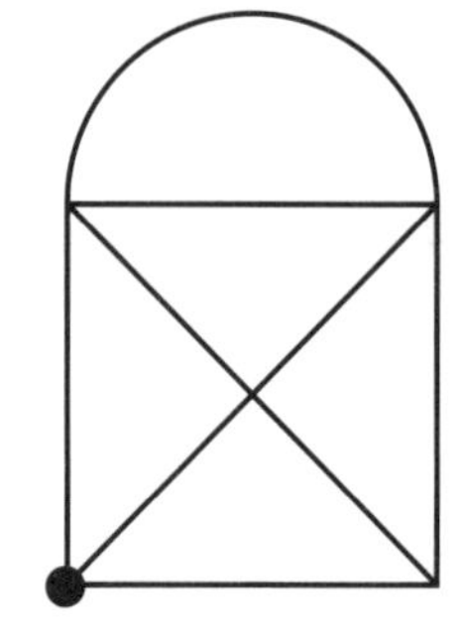

2. 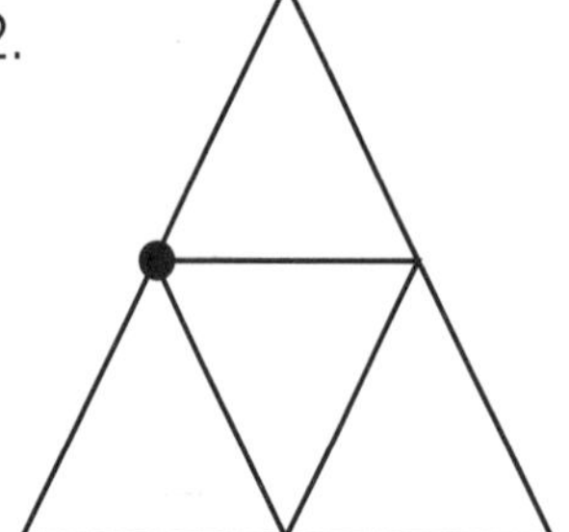

3. 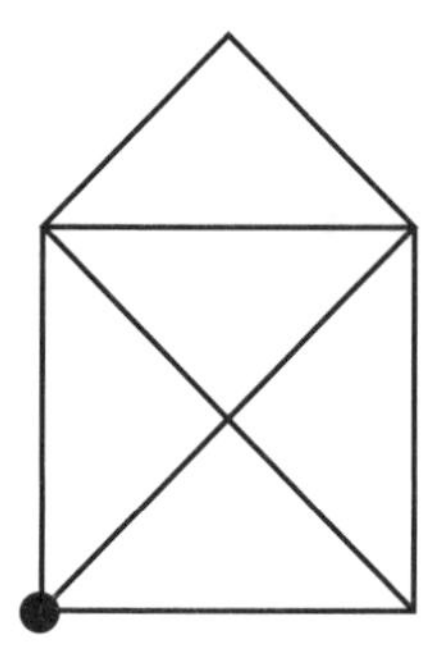

4. 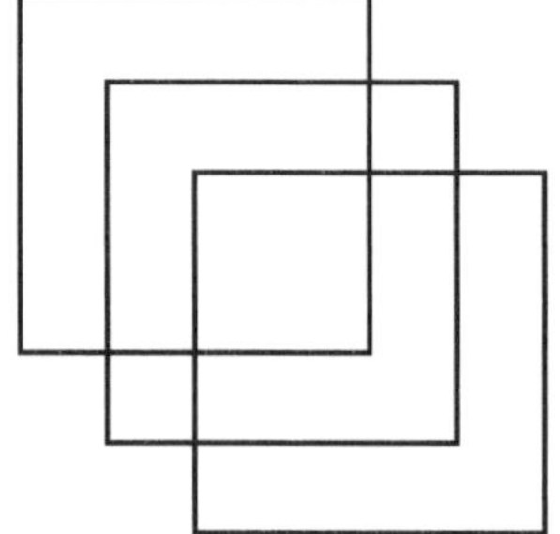

5. 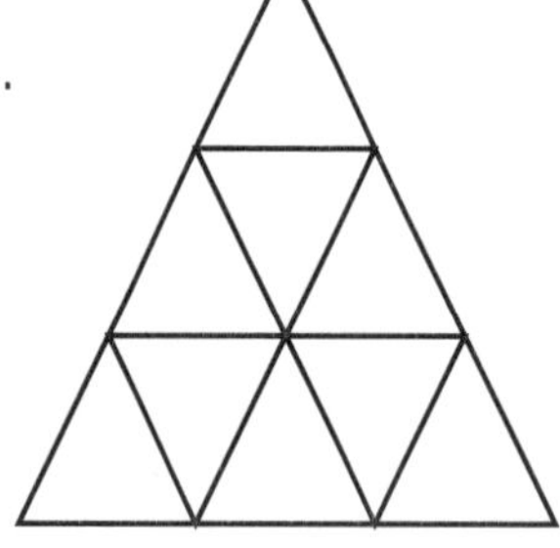

6. 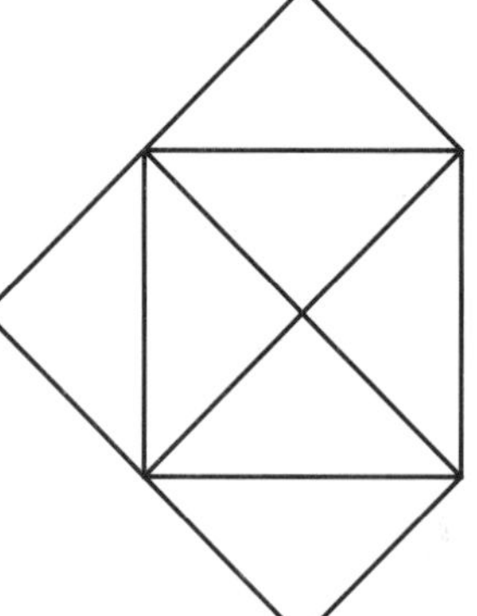

7. 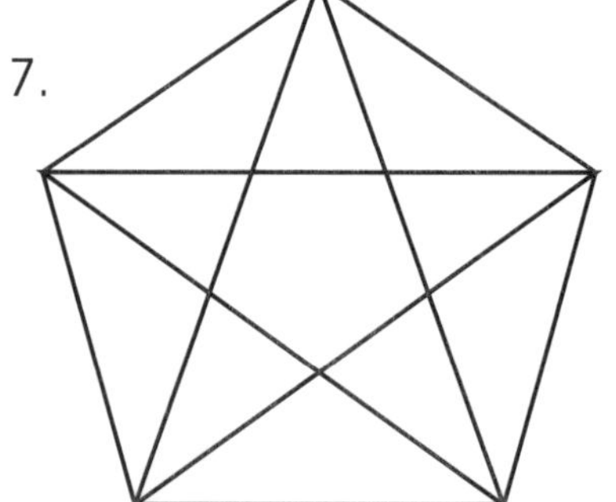

8. 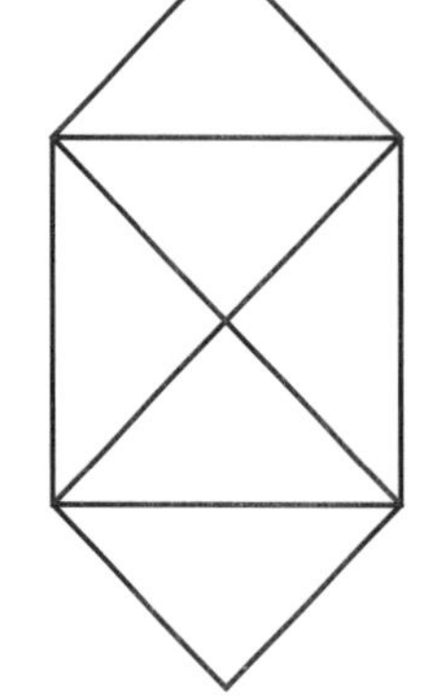

9. 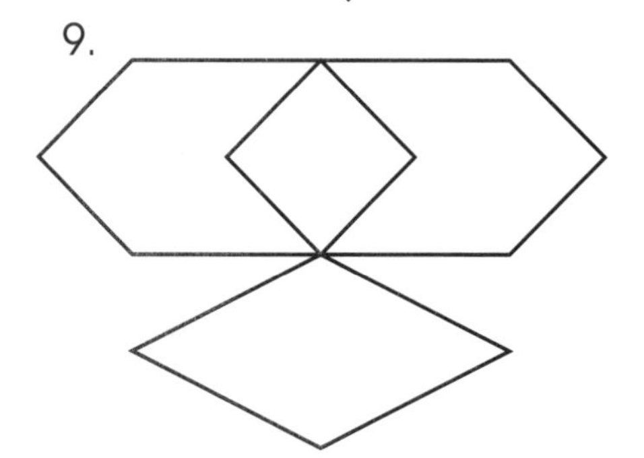

Name________________________________

8.G.A.2, 8.G.A.5

Triangles

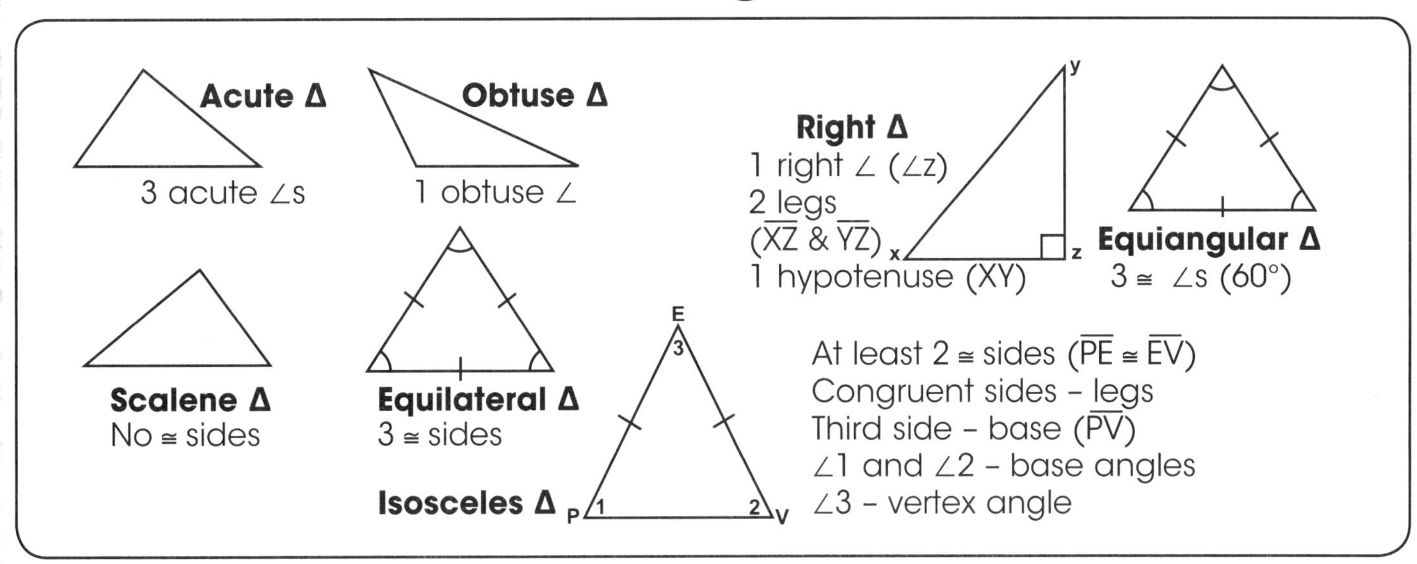

Classify each triangle by its angles and by its sides.

1. 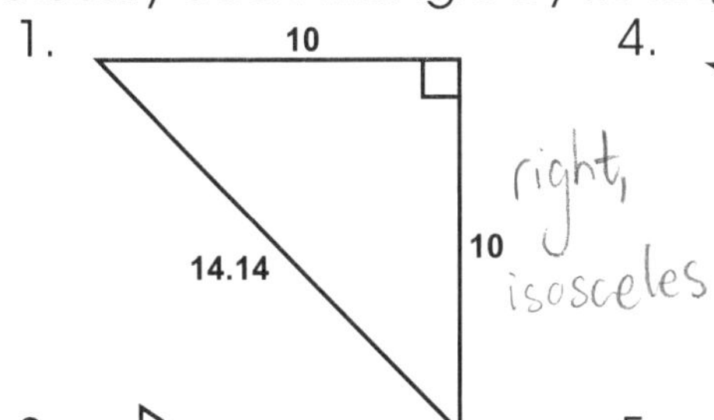

right, isosceles

2. 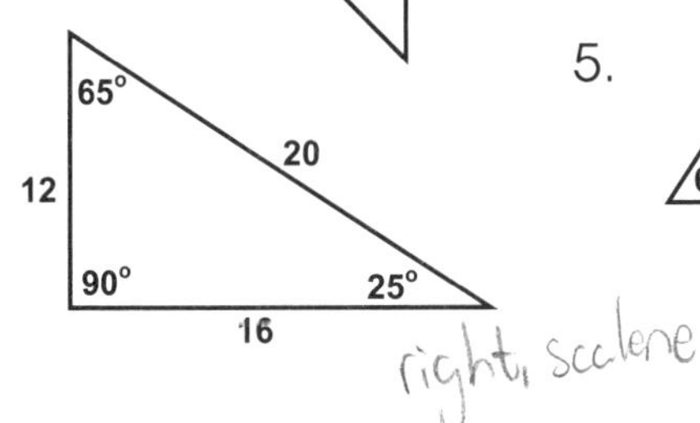

right, scalene

3. 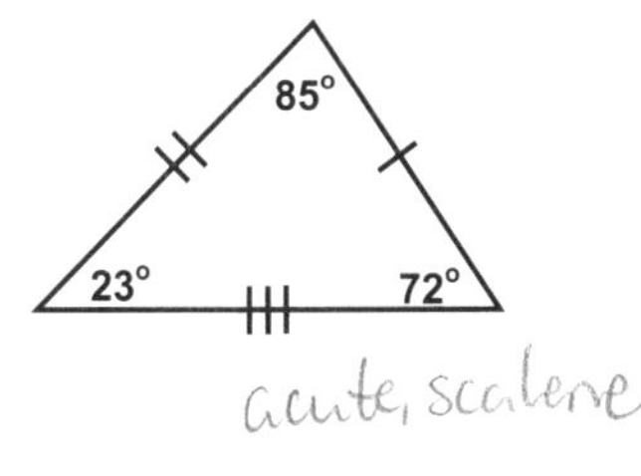

acute, scalene

4. 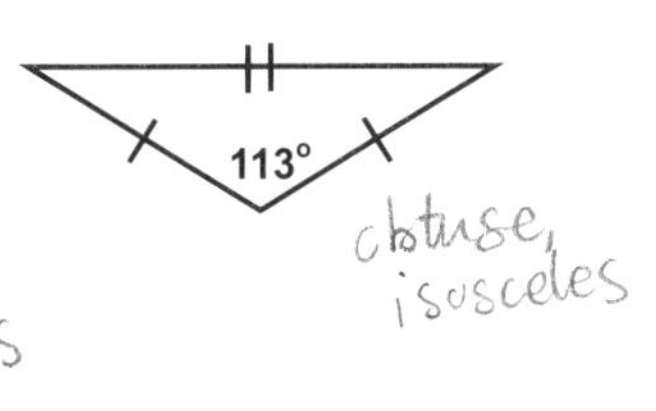

obtuse, isosceles

5. 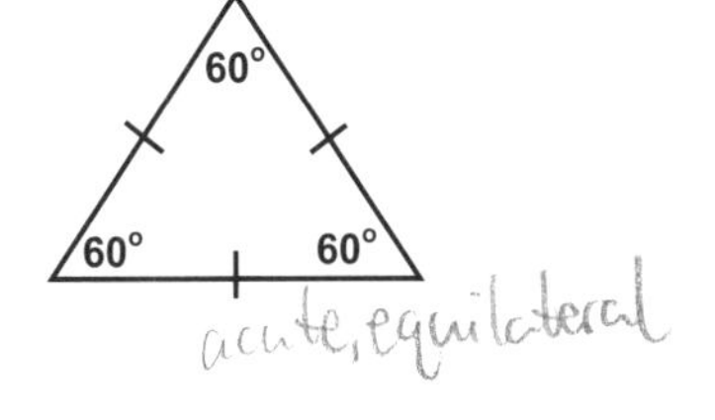

acute, equilateral

6. 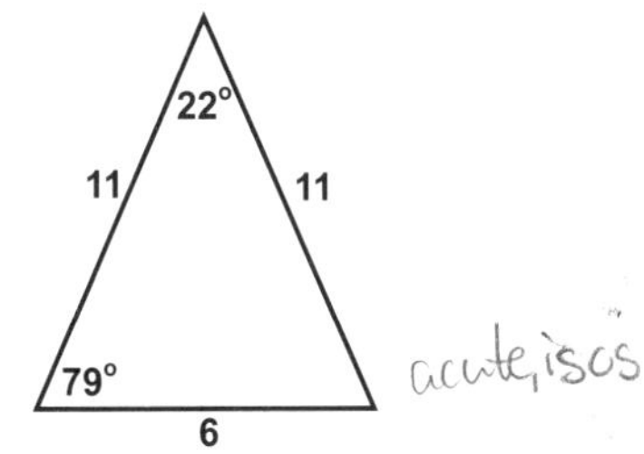

acute, isos

7. 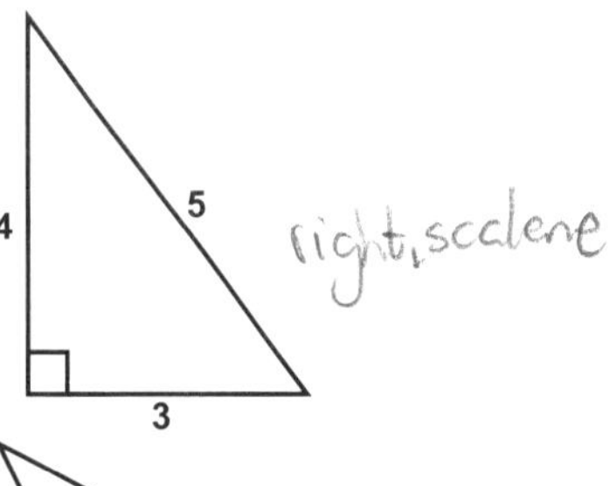

right, scalene

8. 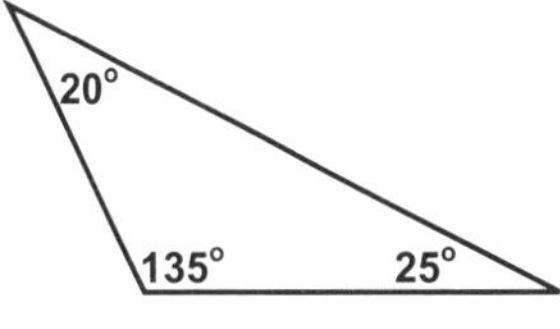

9. 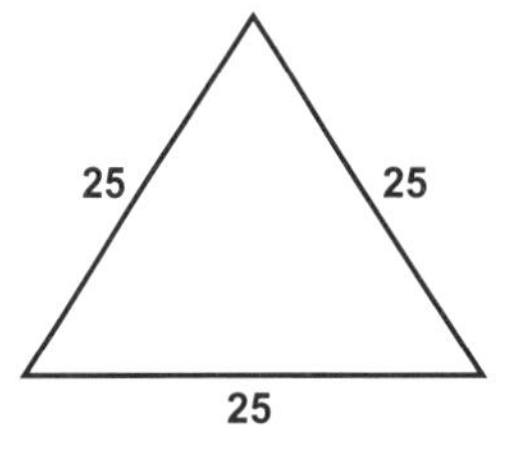

acute, equil

10. 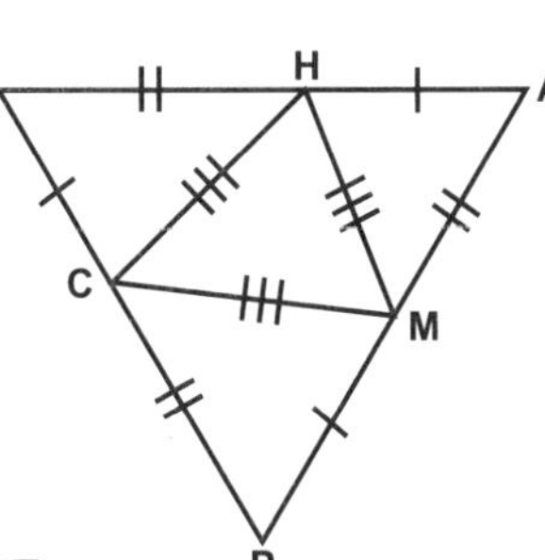

a. Name all equilateral triangles. ΔCHM, ΔPAR
b. Name all isosceles triangles. ΔPAR, CHM
c. Name all scalene triangles. ΔCRH, ΔAHM, ΔCPM

11. True or false: an equilateral triangle is always isosceles.

Name________________________________

8.G.A.5

Included Sides and Angles

L

S B

Side LB is included by ∠L and ∠B, the angles whose vertices are the endpoints of the segment. ∠S is included by $\overline{LS}$ and $\overline{BS}$, the segments which form the sides of the angle. ∠B lies opposite $\overline{LS}$. $\overline{BS}$ lies opposite ∠L.

1. The side opposite ∠P is $\overline{AL}$.

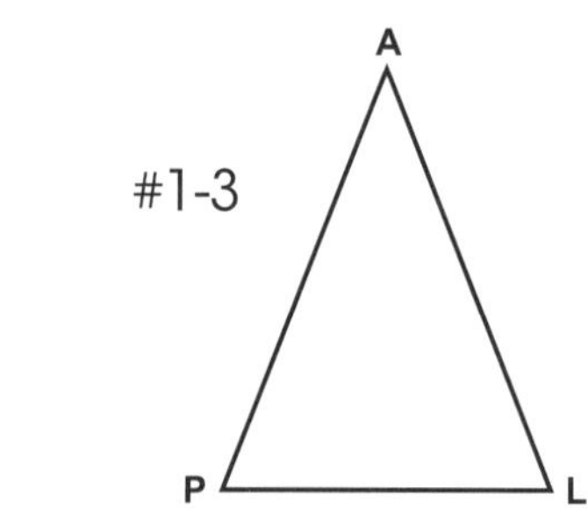

2. The angle included by $\overline{AP}$ and $\overline{LA}$ is ∠A.
3. The side included by ∠P and ∠L is $\overline{LP}$.

4. The side included by ∠A and ∠ARB is $\overline{AR}$.

5. The angles opposite $\overline{BR}$ are ∠K and ∠A.

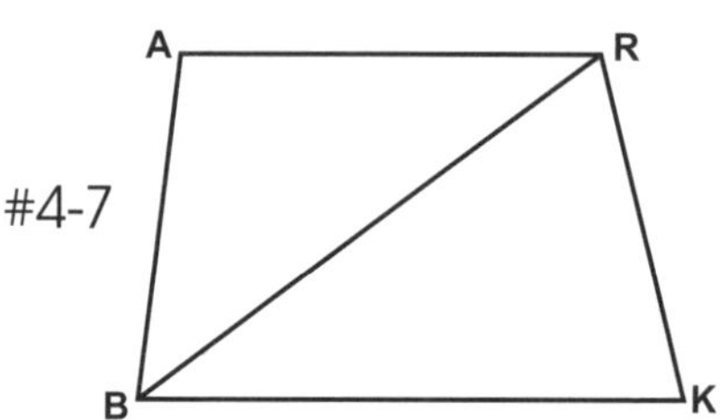

6. The angle included by $\overline{RB}$ and $\overline{KB}$ is ∠B.

7. The side opposite ∠KRB is $\overline{BK}$.

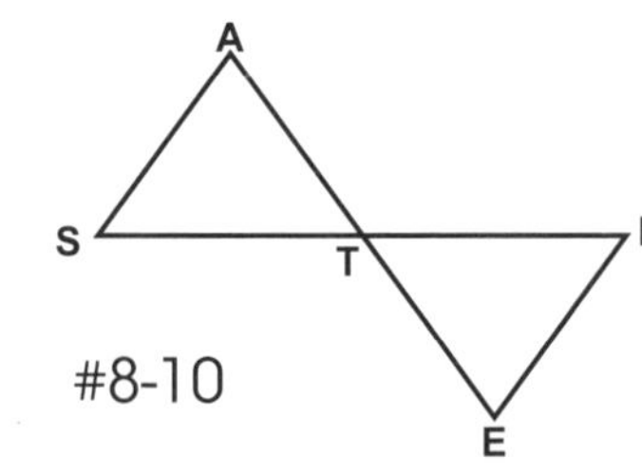

8. In ΔBET, the side opposite ∠E is $\overline{BT}$.
9. In ΔSAT, the angle included by $\overline{AT}$ and $\overline{TS}$ is ∠ATS.
10. The segment included by ∠A and ∠STA is $\overline{AT}$.
11. The side opposite ∠MTH is $\overline{MH}$.

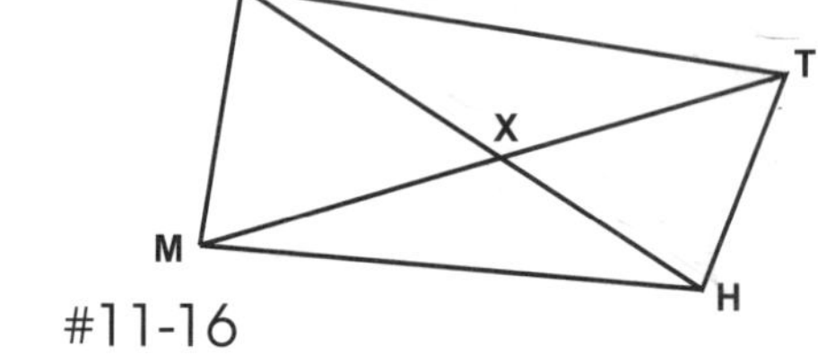

12. The side included by ∠AHT and ∠HAT is $\overline{AH}$.
13. The angle included by $\overline{AX}$ and $\overline{TX}$ is ∠AXT.
14. The angles opposite $\overline{AT}$ are ______, ______, and ______.
15. In ΔAXM, the side opposite ∠M is ______.
16. The segment included by ∠MXH and ∠MHX is ______.

Name______________________________ 8.G.A.2, 8.G.A.5, HSG-CO.B.7

Congruence of Triangles

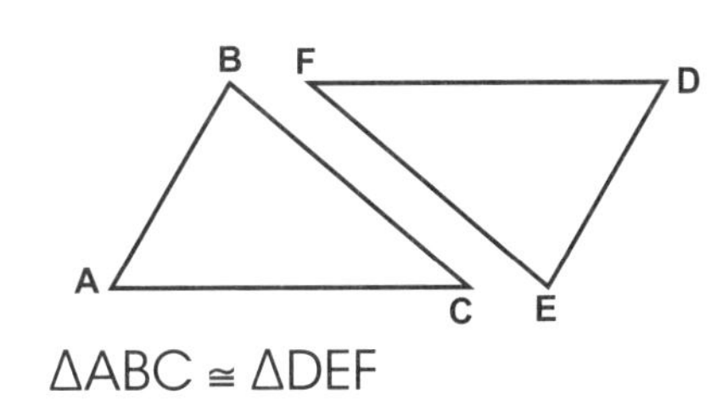

$\triangle ABC \cong \triangle DEF$

Corresponding Parts ≅

$\overline{AB} \cong \overline{DE}$	$\angle A \cong \angle D$
$\overline{BC} \cong \overline{EF}$	$\angle B \cong \angle E$
$\overline{AC} \cong \overline{DF}$	$\angle C \cong \angle F$

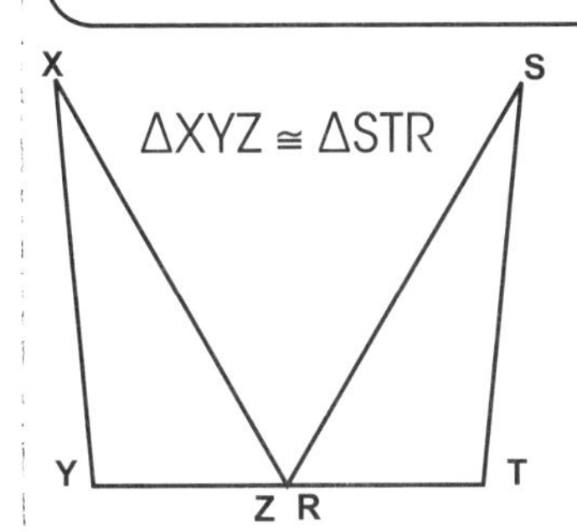

$\triangle XYZ \cong \triangle STR$

1. Name 3 pairs of congruent angles.
2. Name 3 pairs of congruent sides.

Which statements appear to be true?

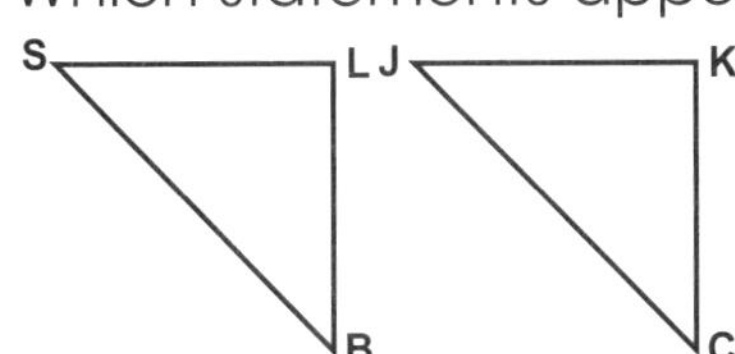

3. $\angle SLB \cong \angle JKC$
4. $\angle LBS \cong \angle JCK$
5. $\angle SLB \cong \angle CKJ$
6. $\angle SLB \cong \angle CJK$
7. $\angle SBL \cong \angle JCK$
8. $\angle BLS \cong \angle JKC$

For the given congruence, list the six pairs of congruent parts.

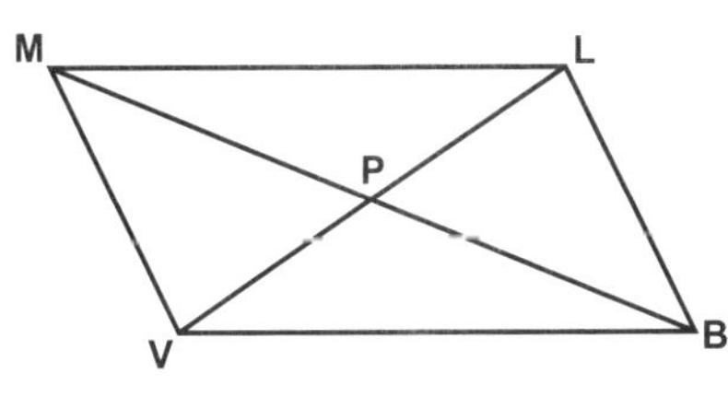

#9-11

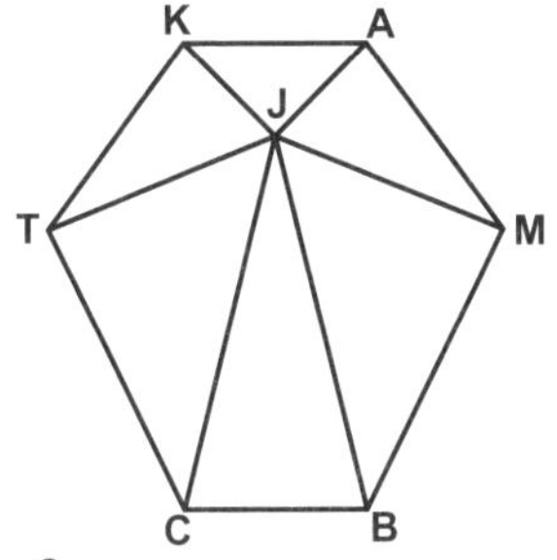

#12-13

#14-16

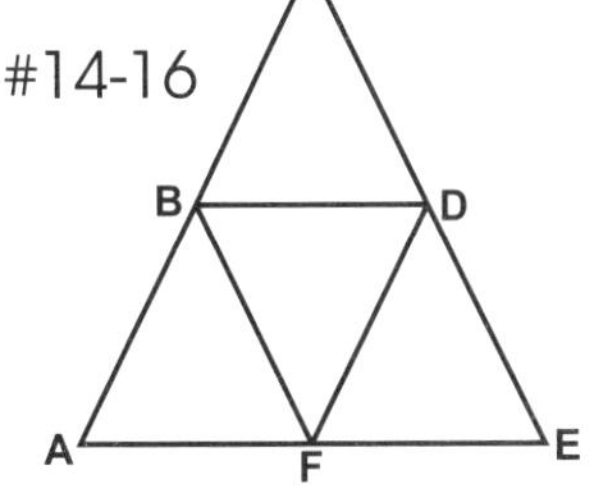

9. $\triangle MLB \cong \triangle BVM$
10. $\triangle LMP \cong \triangle VBP$
11. $\triangle LPB \cong \triangle VPM$
12. $\triangle MJB \cong \triangle TJC$
13. $\triangle TKJ \cong \triangle MAJ$
14. $\triangle CBD \cong \triangle DFE$
15. $\triangle BAF \cong \triangle BDF$
16. $\triangle CBD \cong \triangle FBD$

Name________________________________

HSG-CO.B.8

Ways to Prove Triangles Congruent

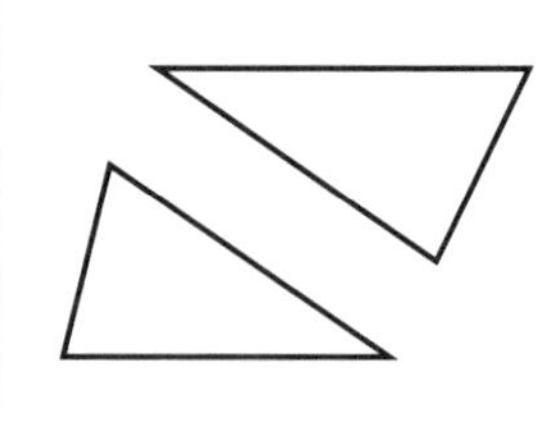

SSS (side, side, side) = three sides of one triangle congruent to the corresponding parts of another triangle ⇒ ≅ Δs.

SAS (side, angle, side) = two sides and the included angle of one triangle congruent to the corresponding parts of another triangle ⇒ ≅ Δs.

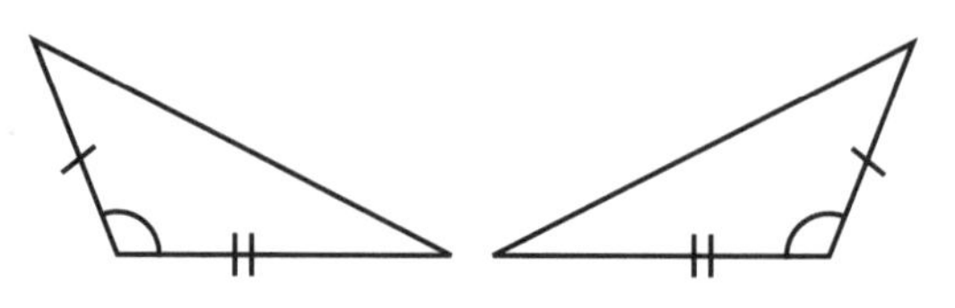

Identify which property will prove these triangles congruent (SSS, SAS, or none).

1.

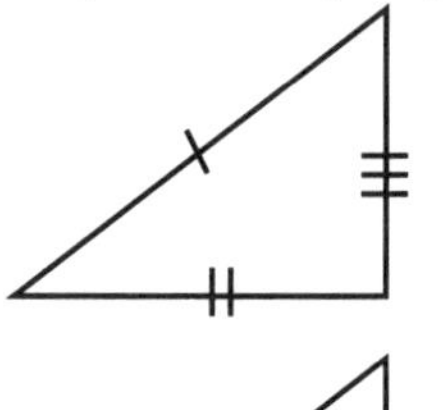

2.

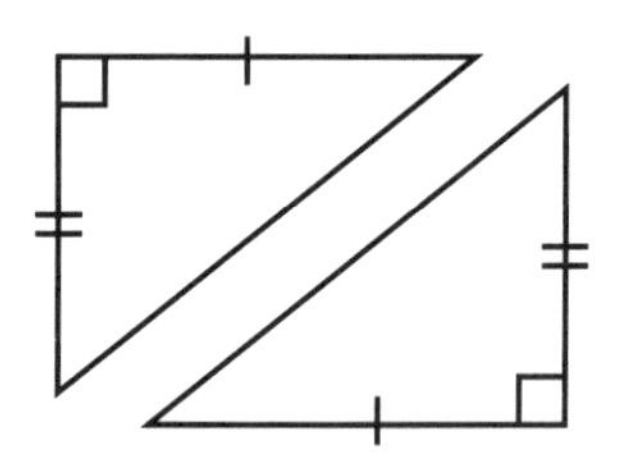

3.

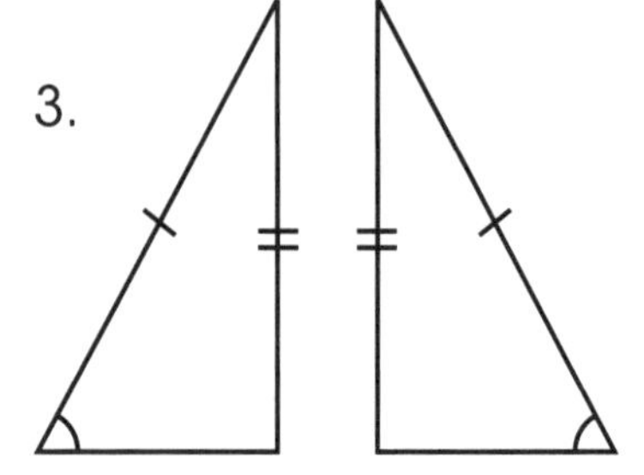

4.

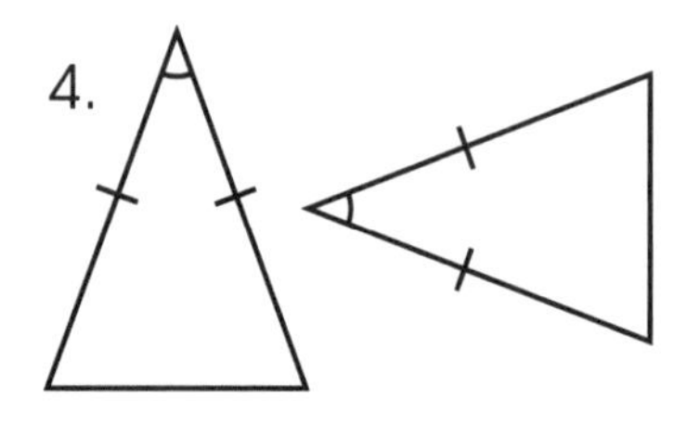

5.

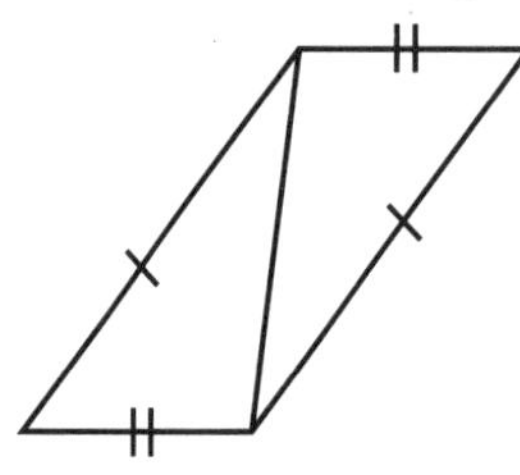

6.

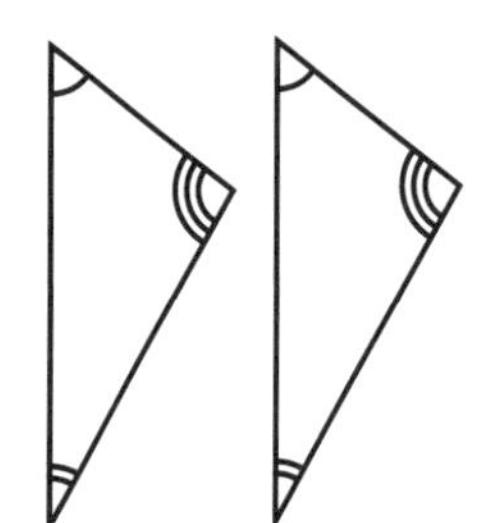

7.

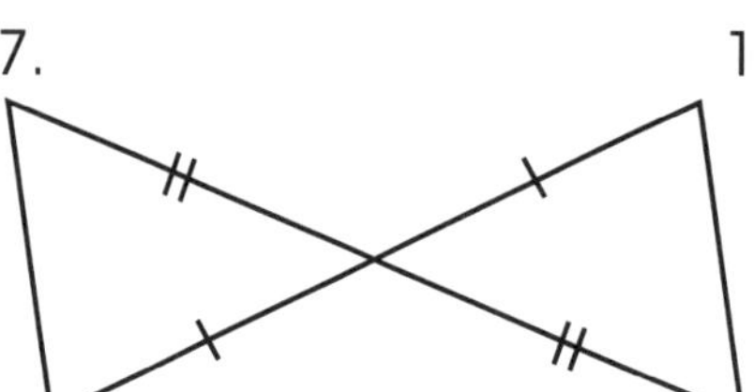

8.

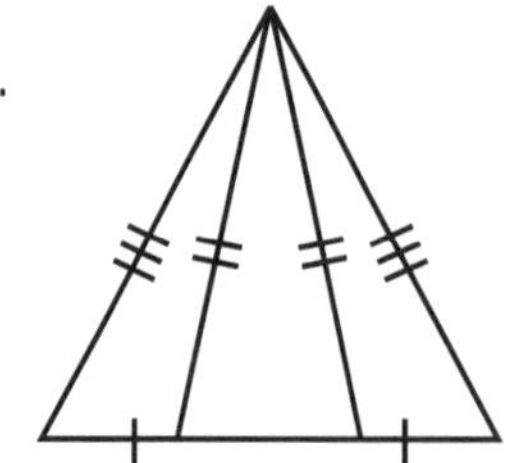

9.

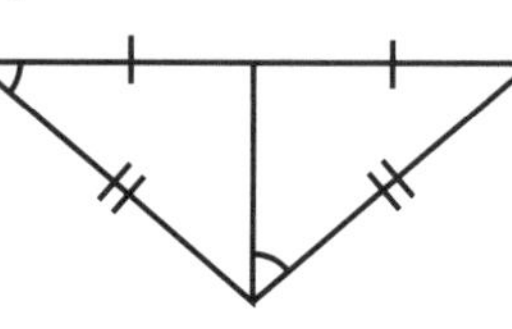

10.

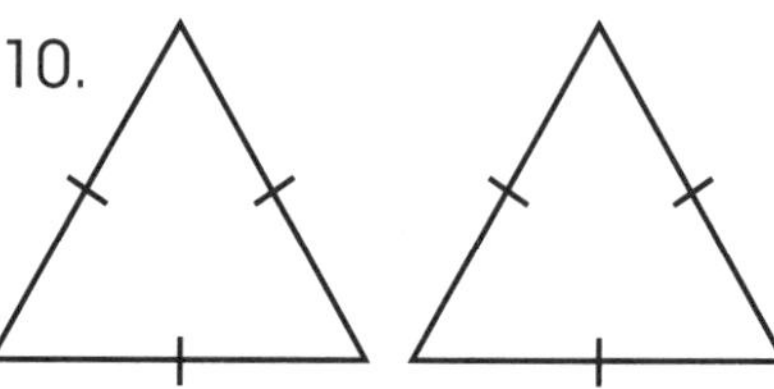

11.

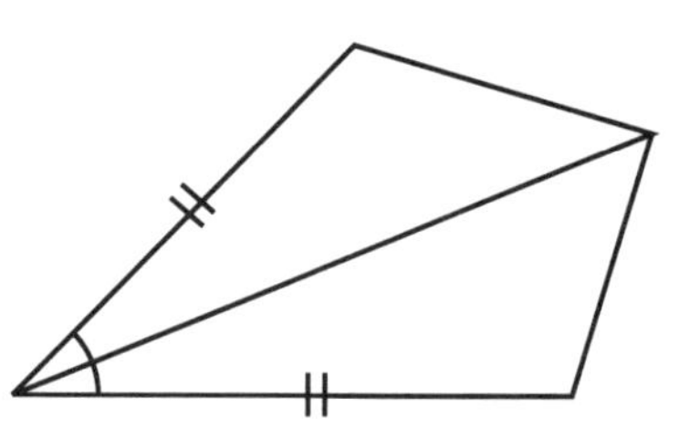

12.

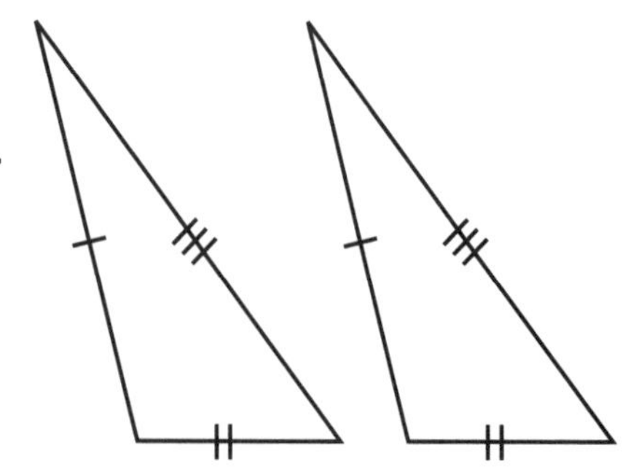

Name____________________________________ HSG-CO.B.8

More Ways to Prove Triangles Congruent

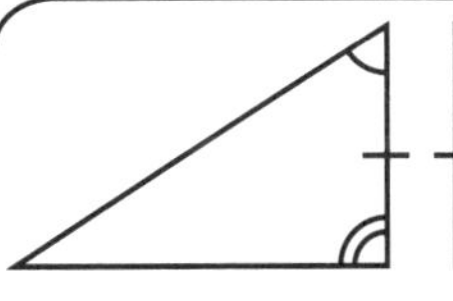

ASA (angle, side, angle) = two angles and the included side of one triangle congruent to the corresponding parts of another triangle ⇒ ≅ Δs.

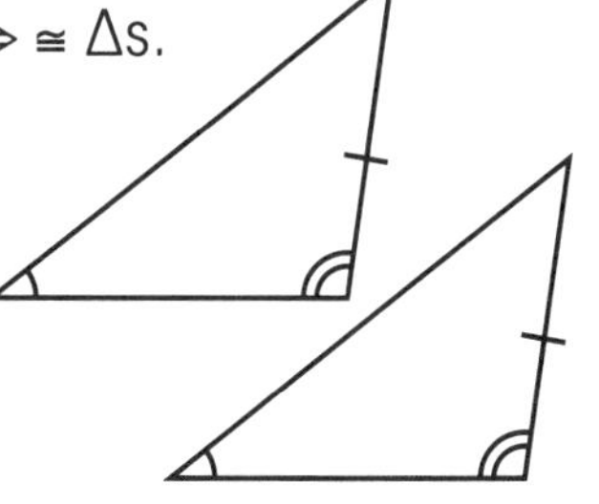

AAS (angle, angle, side) = two angles and the non-included side of one triangle congruent to the corresponding parts of another triangle ⇒ ≅ Δs.

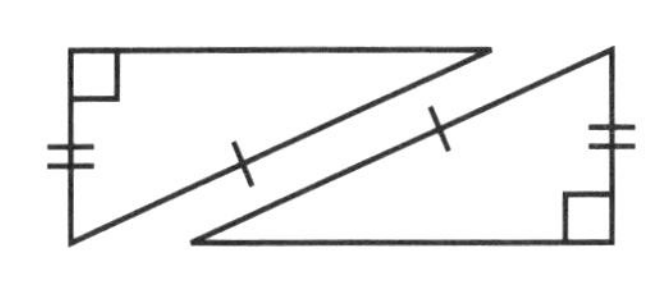

HL (hypotenuse, leg) = the hypotenuse and a leg of one right triangle congruent to the corresponding parts of another triangle ⇒ ≅ Δs.

Identify which property will prove these triangles congruent (ASA, AAS, HL, or none).

1. 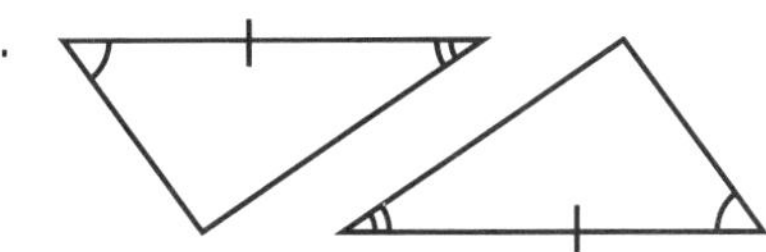

2. 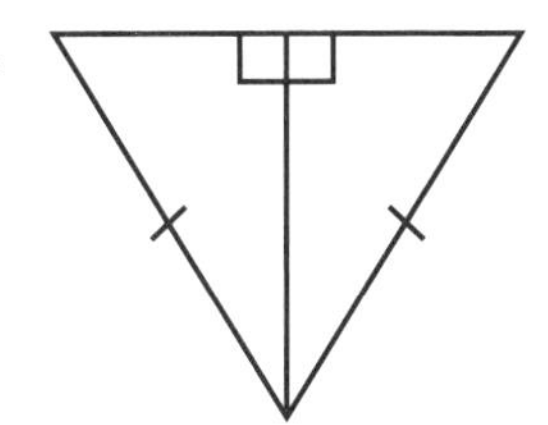

3. 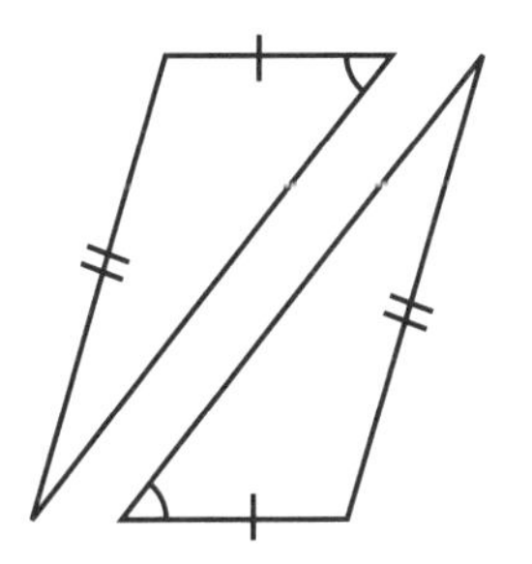

4. 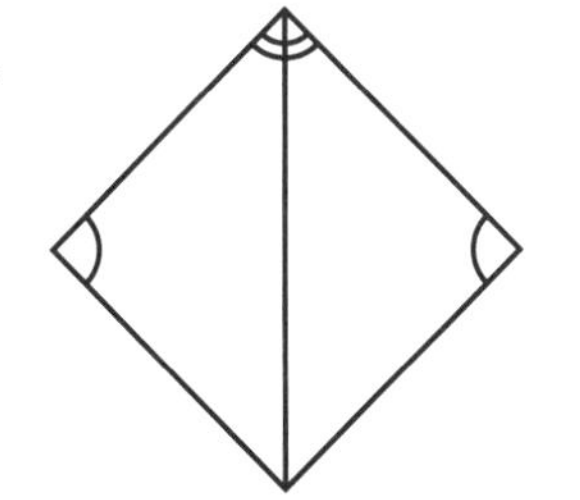

5. 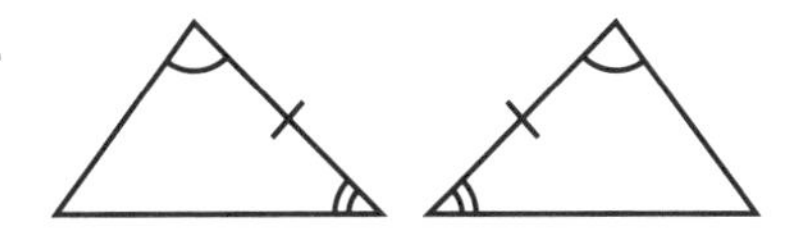

6. 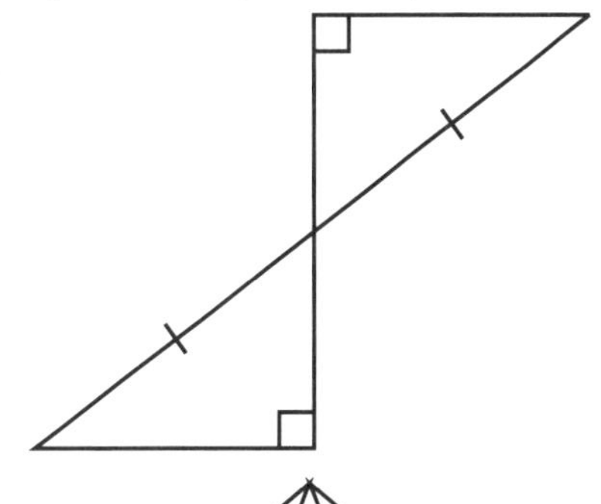

7. 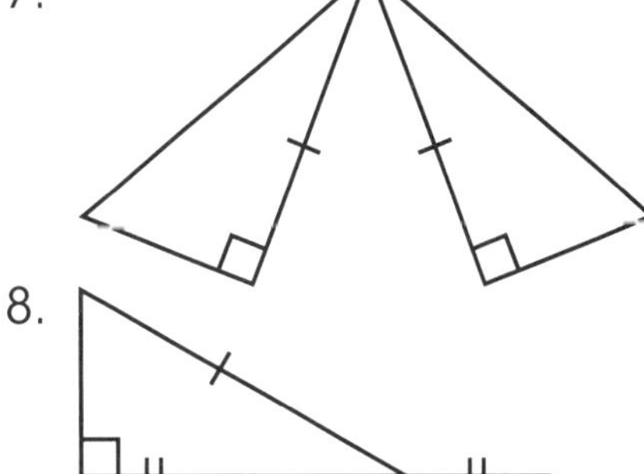

8.

9. 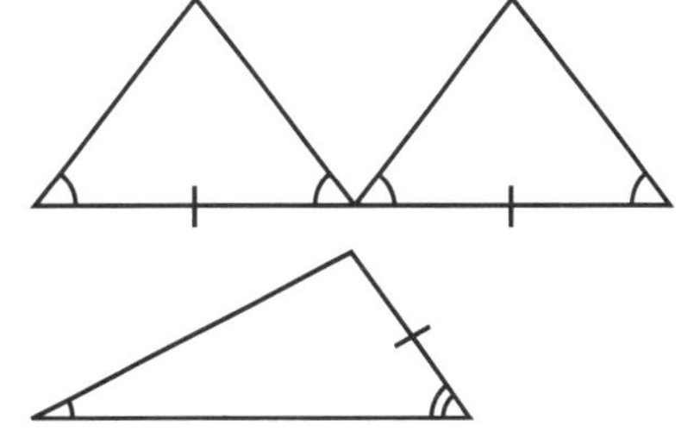

10.

Name____________________________________

HSG-CO.B.8

Congruent Triangles Review

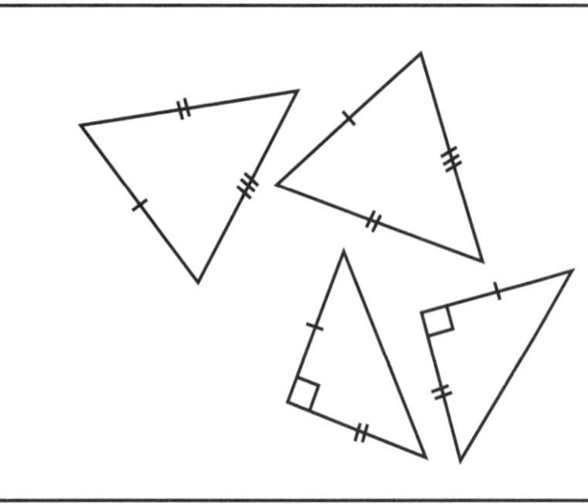

SSS — Side, Side, Side
three sides of one triangle are congruent to the corresponding sides of another triangle → ≅ △s

SAS — Side, Angle, Side
two sides and the included angle of one triangle are congruent to the corresponding parts of another triangle → ≅ △s

State whether these pairs of triangles are congruent by SSS or SAS. If neither method works, write N.

1. 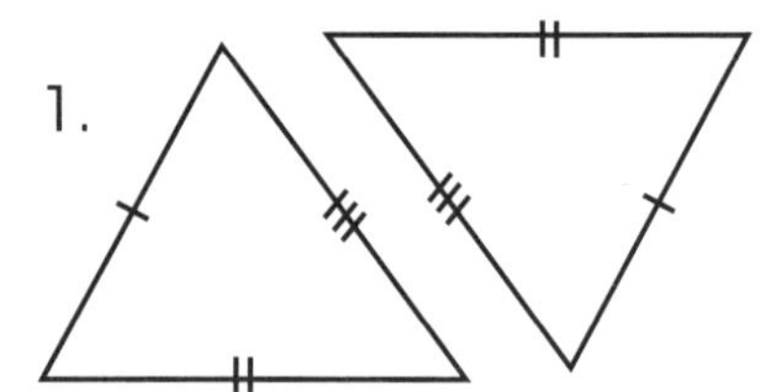

2. 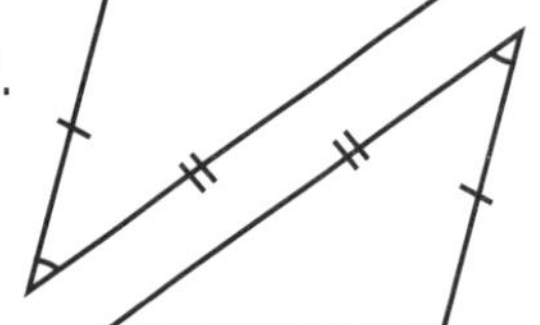

3. 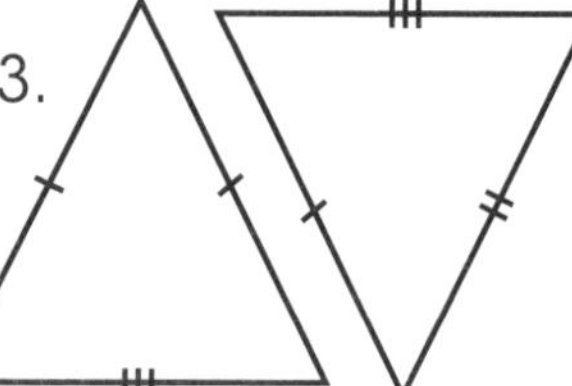

4. 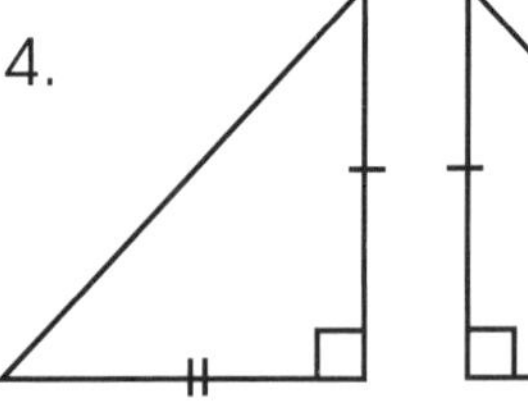

5. 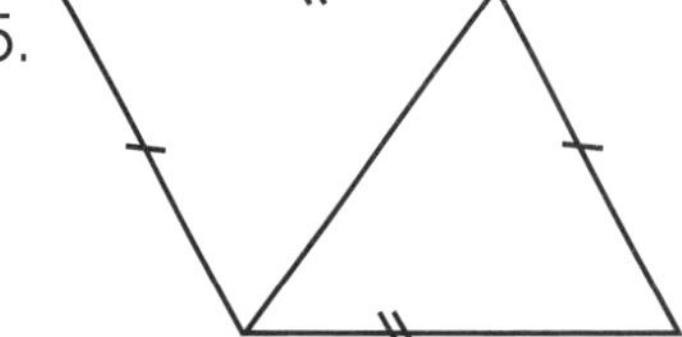

6. 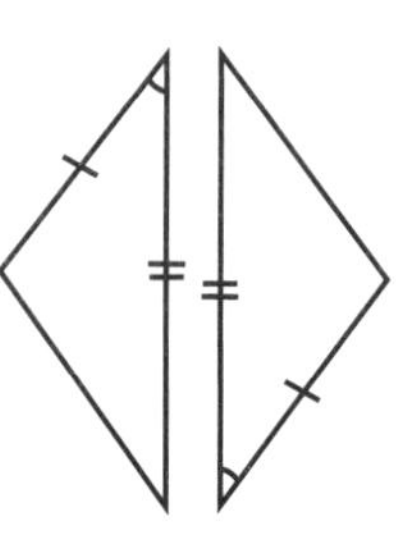

7. 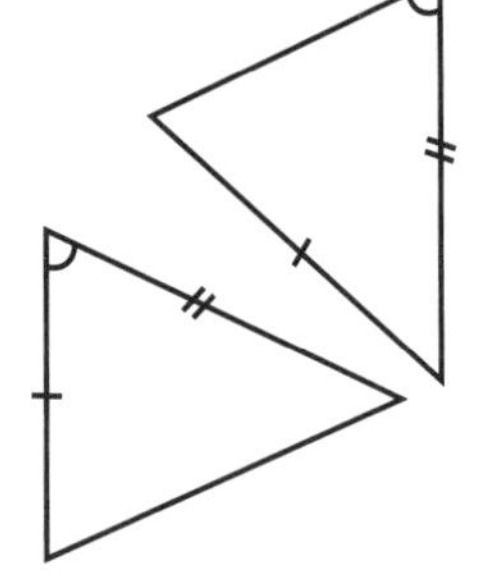

8. 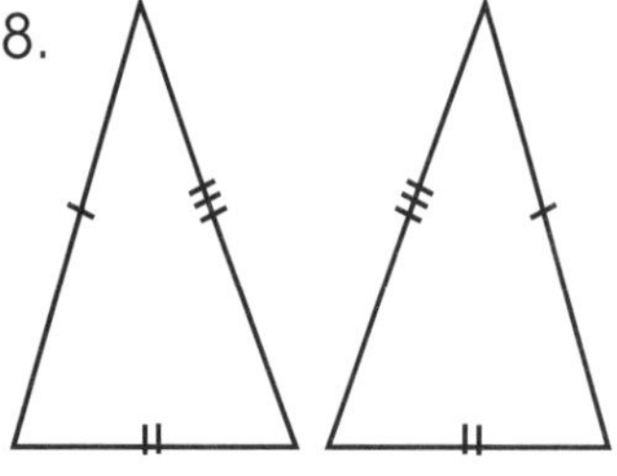

9. 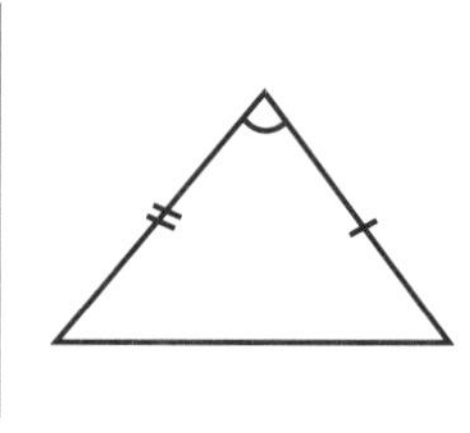

10. 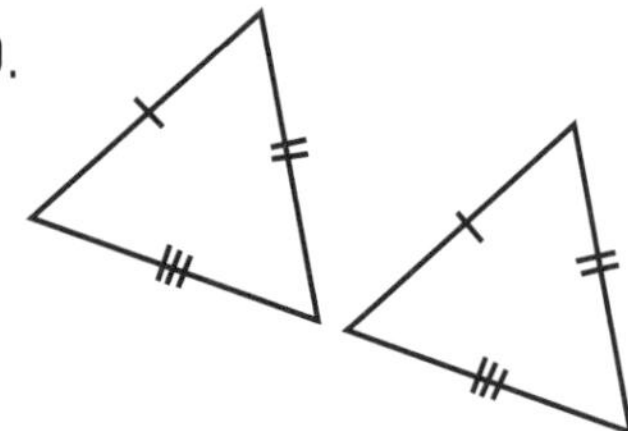

11. 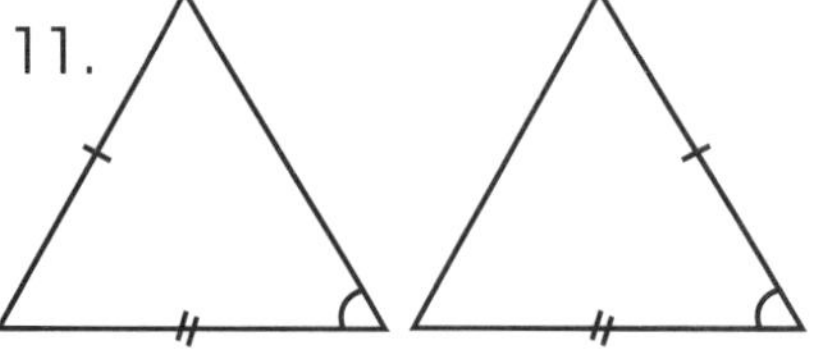

12. 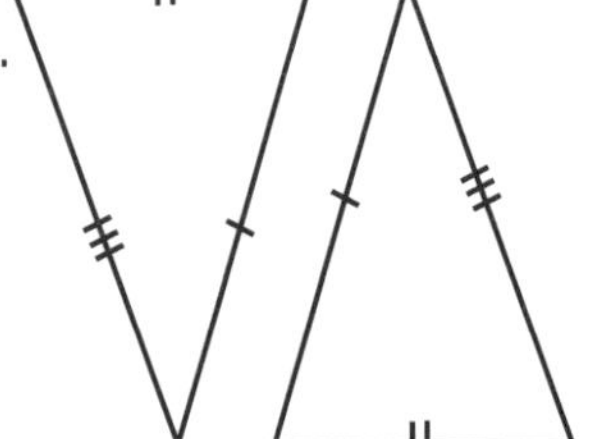

Name________________________________ HSG-CO.B.8

More Congruent Triangles Review

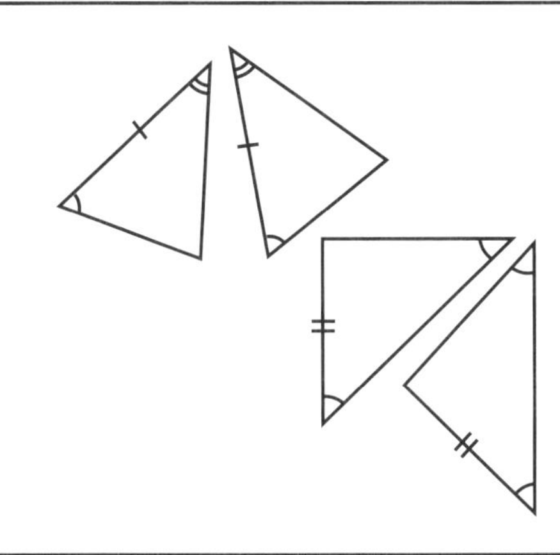

ASA — Angle, Side, Angle
two angles and the included side of one triangle are congruent to the corresponding parts of another triangle → ≅ △s

AAS — Angle, Angle, Side
two angles and the non-included side of one triangle are congruent to the corresponding parts of another triangle → ≅ △s

State whether these pairs of triangles are congruent by ASA or AAS. If neither method works, write N.

1. 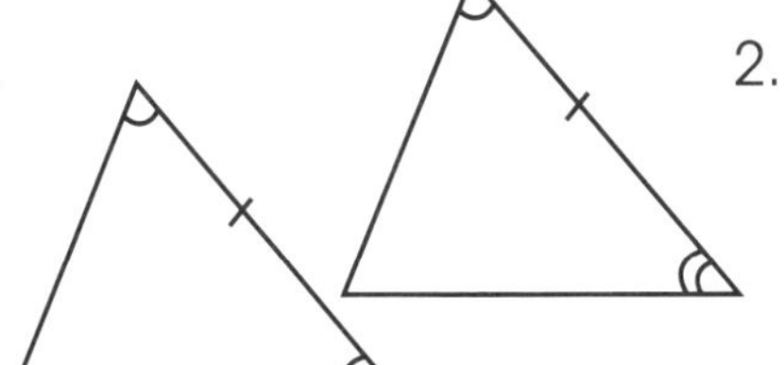

2. 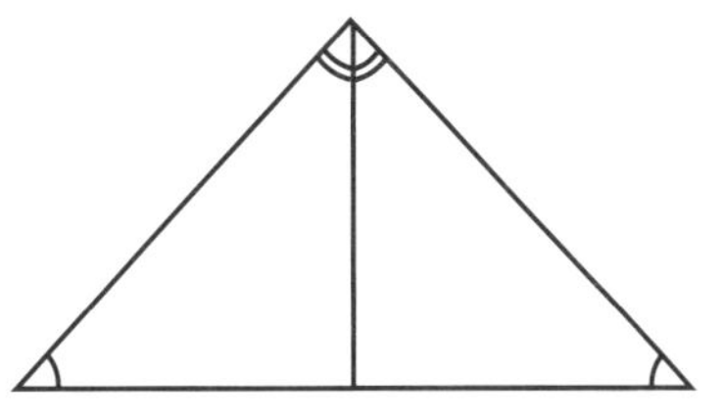

3. 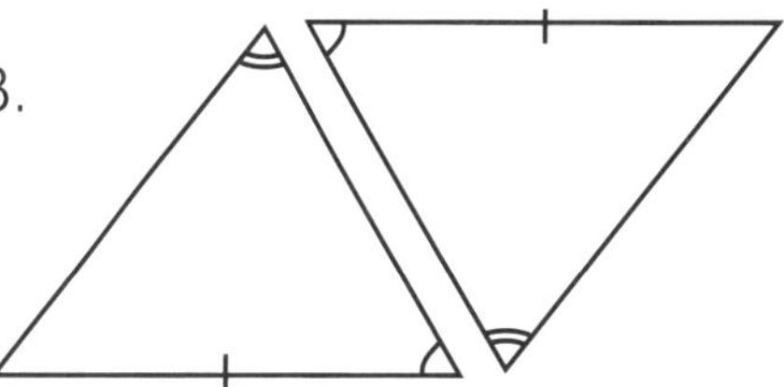

4. 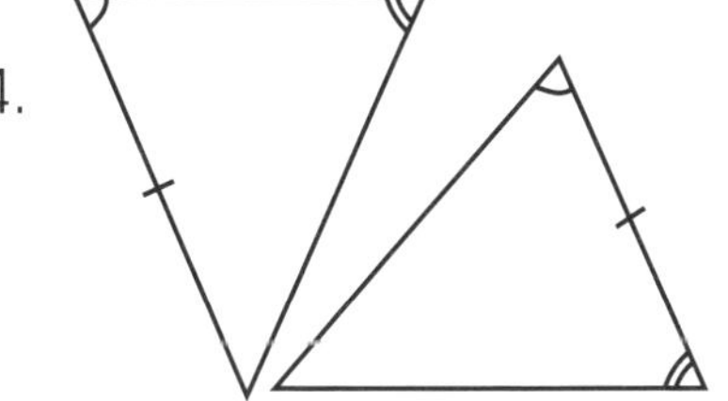

5. 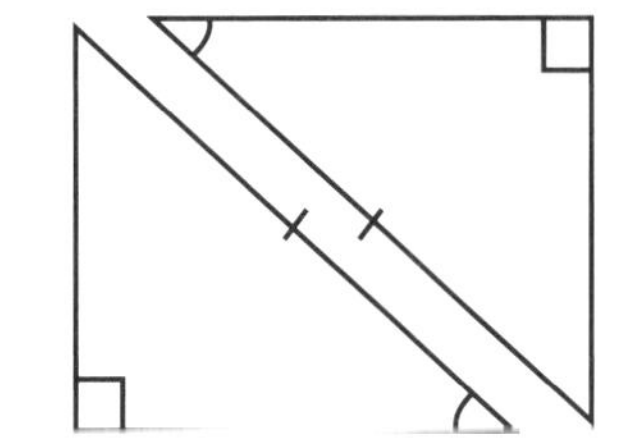

6. 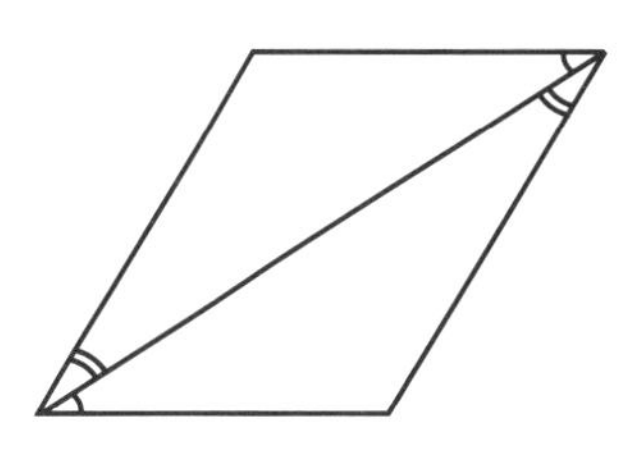

7. 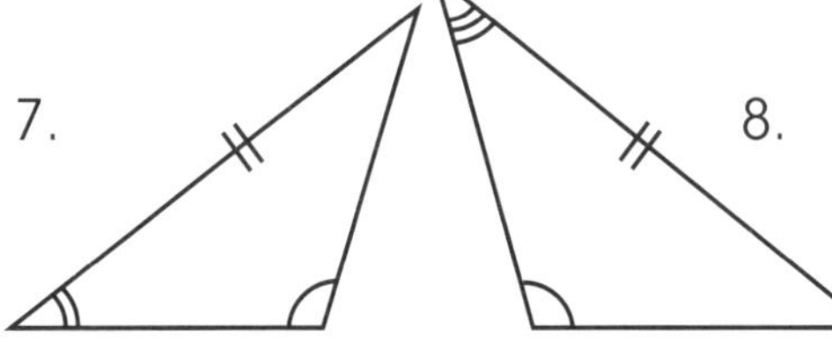

8. 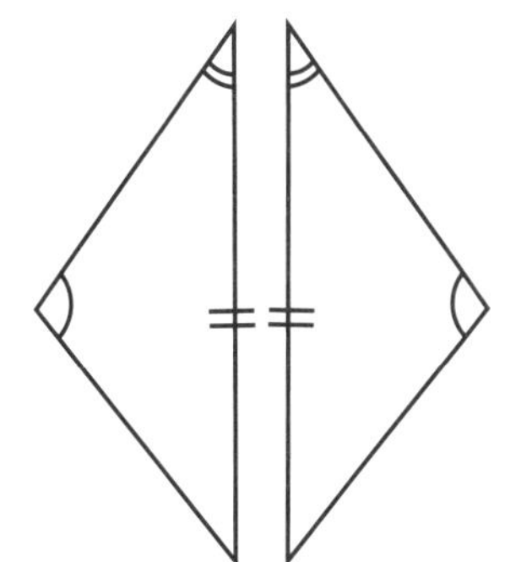

9. 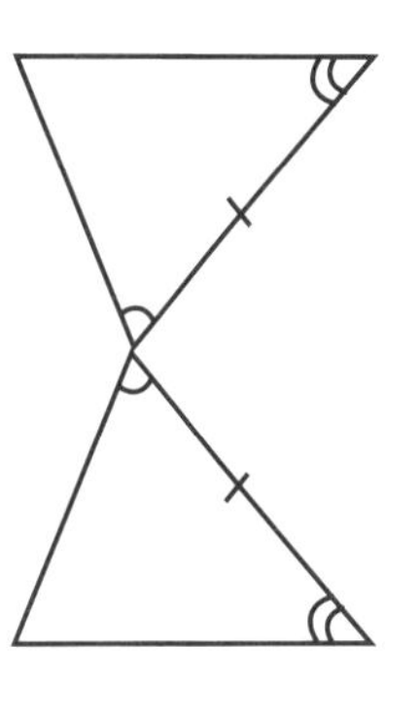

10. 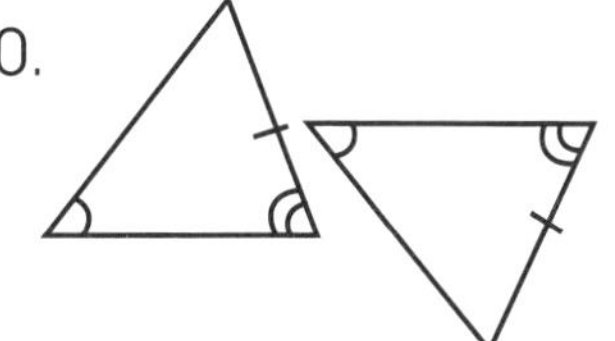

11. 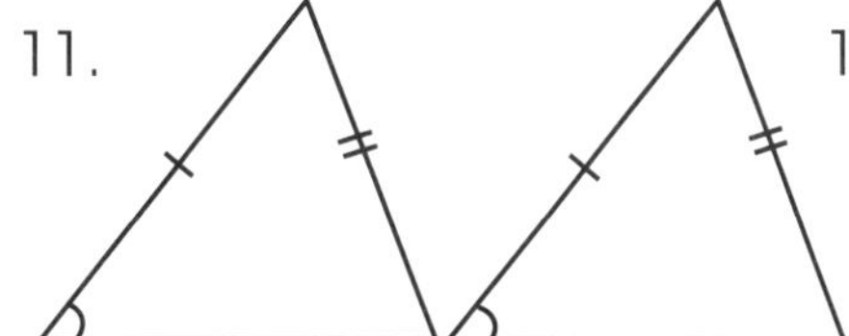

12. 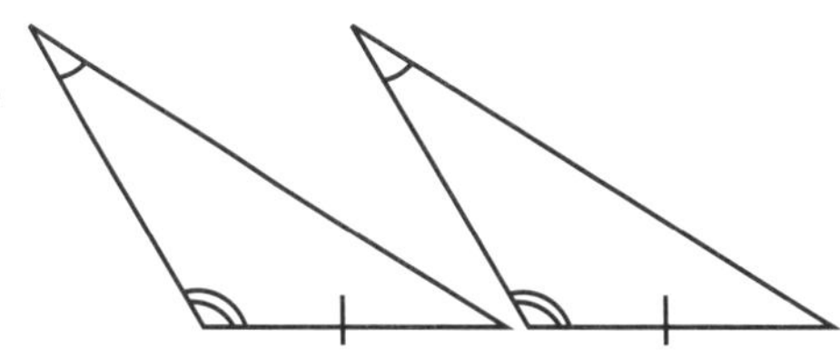

Name________________________________

HSG-CO.B.8

Triangle Congruence Practice

Identify which property will prove these triangles congruent (SSS, SAS, ASA, AAS, HL, or none).

1. 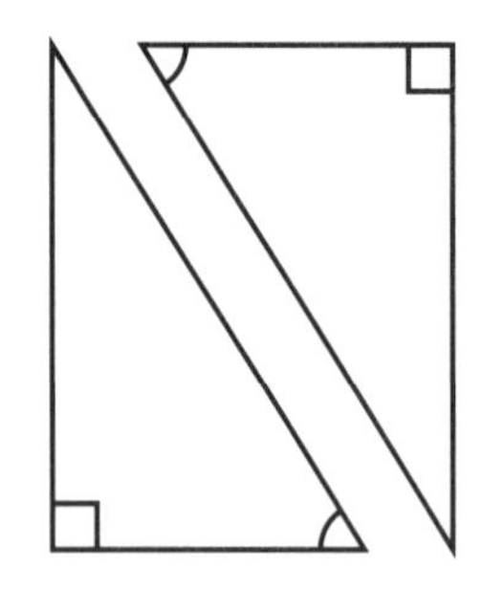

2. 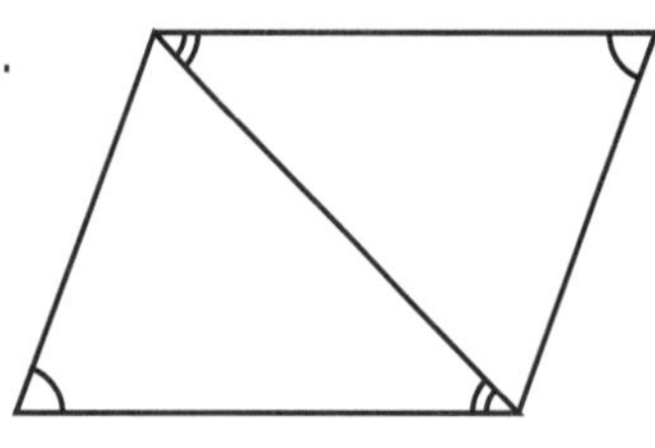

3. 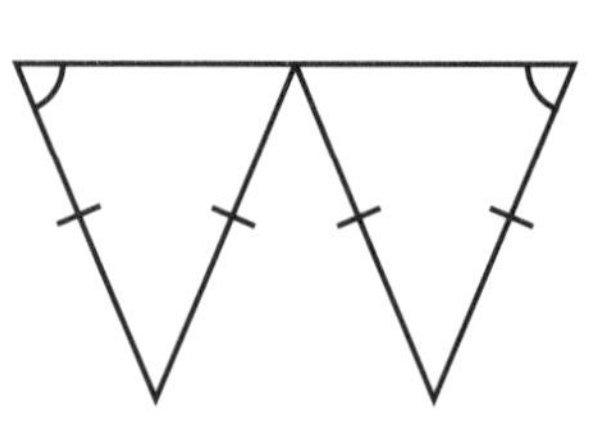

4. 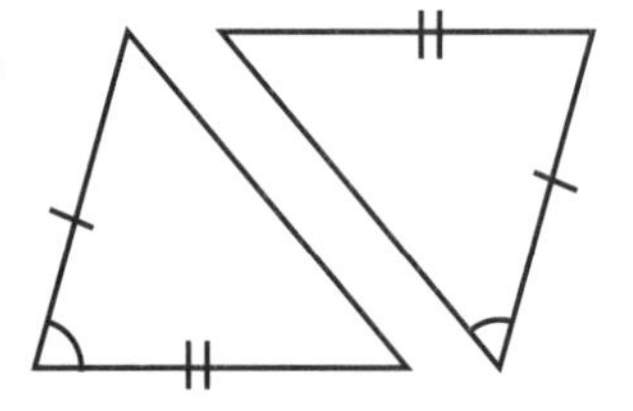

5. 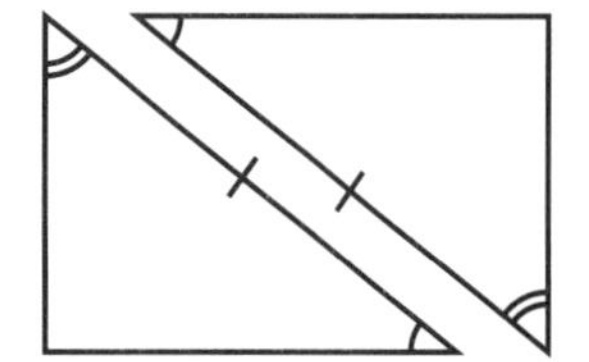

6. 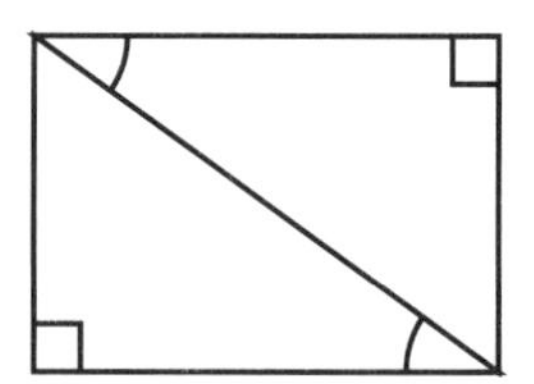

7.

8. 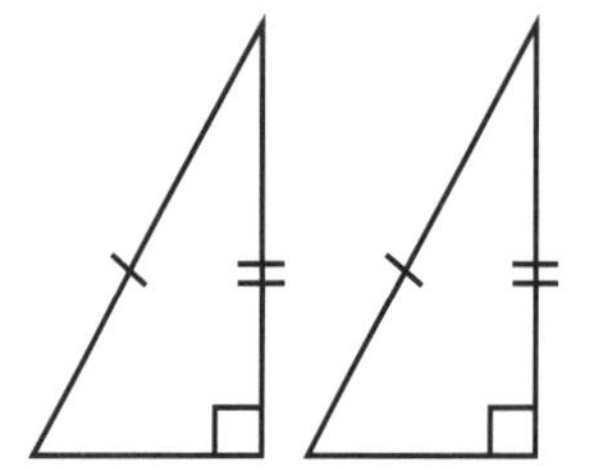

9. 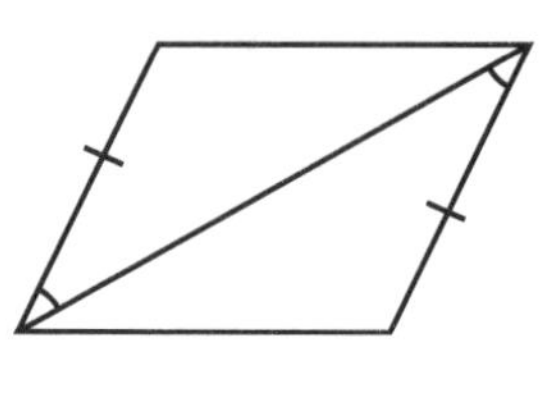

10. 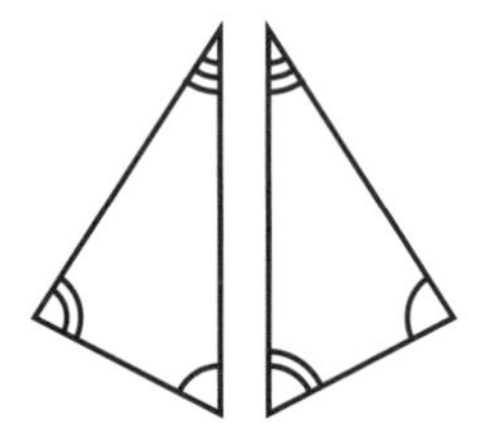

11. 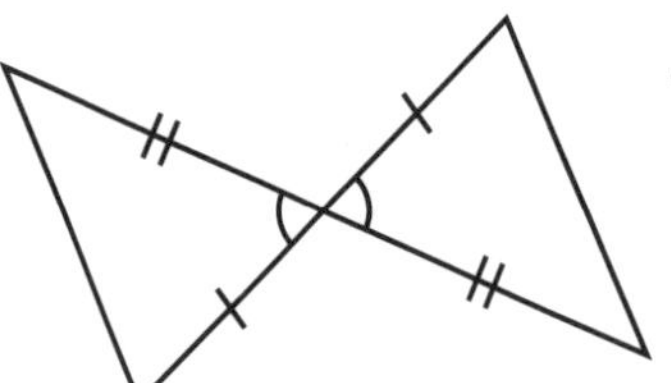

12. 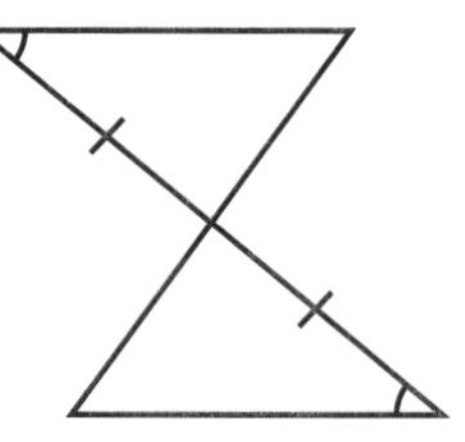

13. 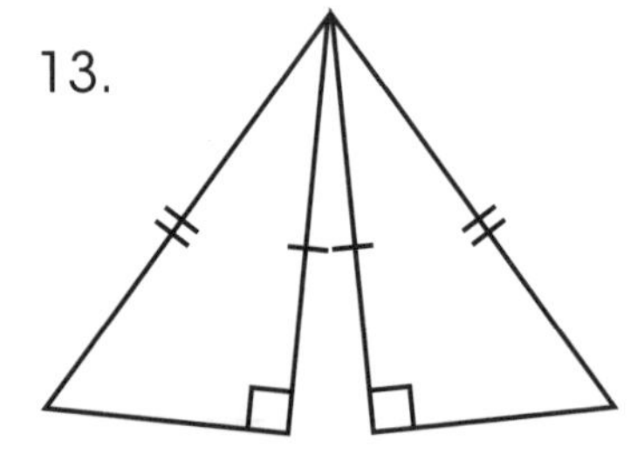

14. 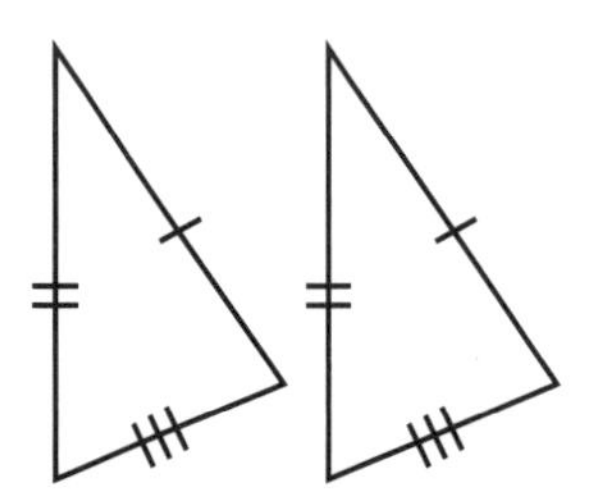

15. 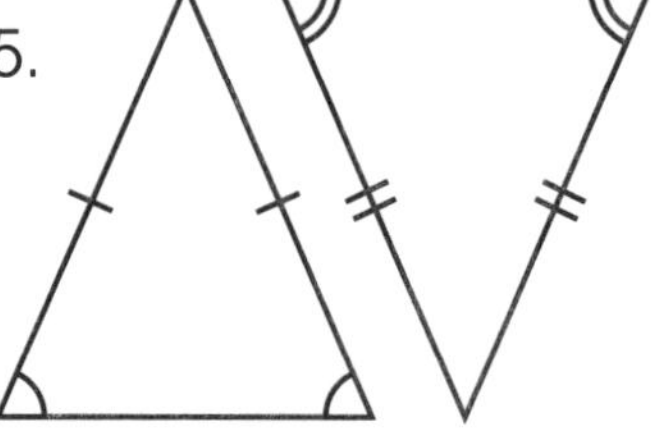

Name________________________________

8.G.A.5

Triangle Inequality Properties

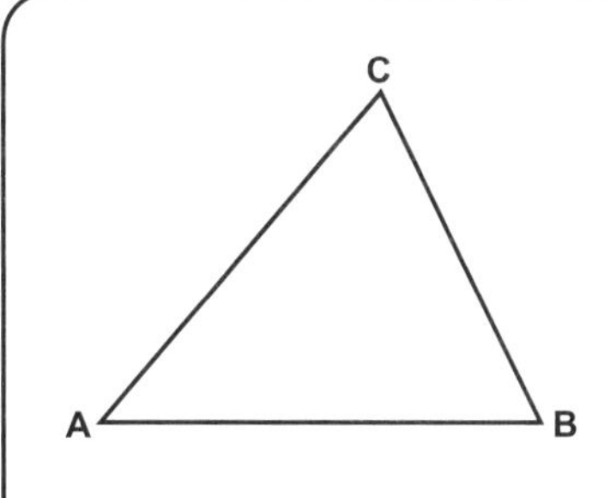

- If one side of a triangle is longer than another side, then the angle opposite the first side is larger than the angle opposite the shorter side.
- If one angle of a triangle is larger than another angle, then the side opposite the first angle is longer than the side opposite the smaller angle.
- The sum of any two sides of a triangle is greater than the length of the third side.

Is it possible for a triangle to have sides with the following lengths?

1. 20, 9, 8
2. 6, 6, 20
3. 5, 5, 10.2
4. 3, 4, 5
5. 15, 15, .03
6. 9, 12, 15

Which angle would be the largest?

7. 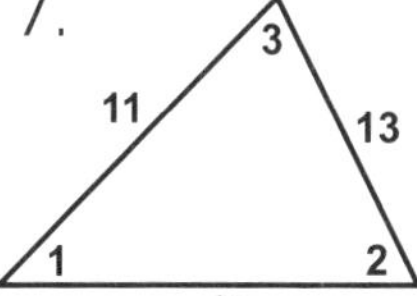

8. 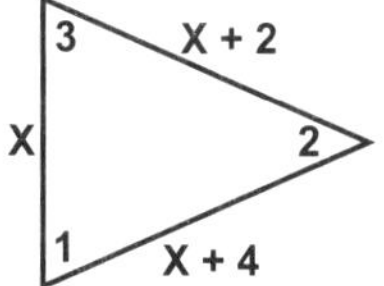

9. 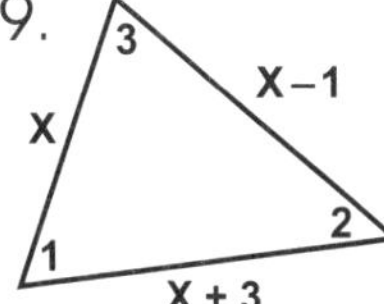

10. 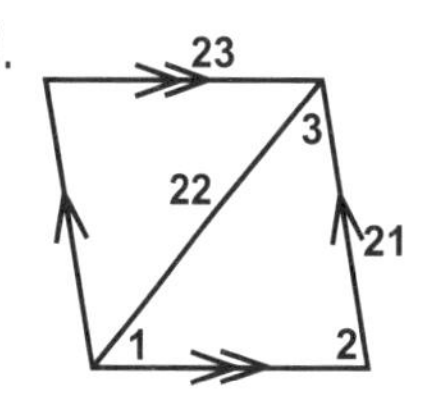

Which segment is the longest?

11. 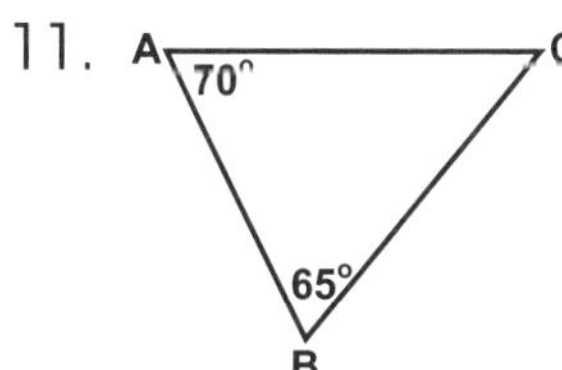

12. 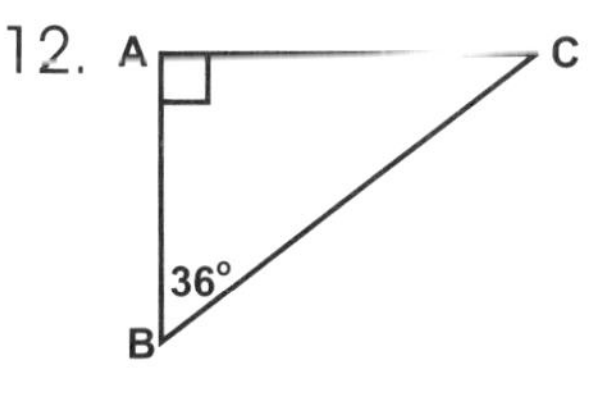

13. 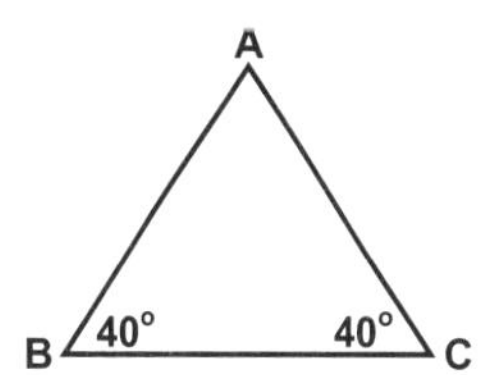

14. 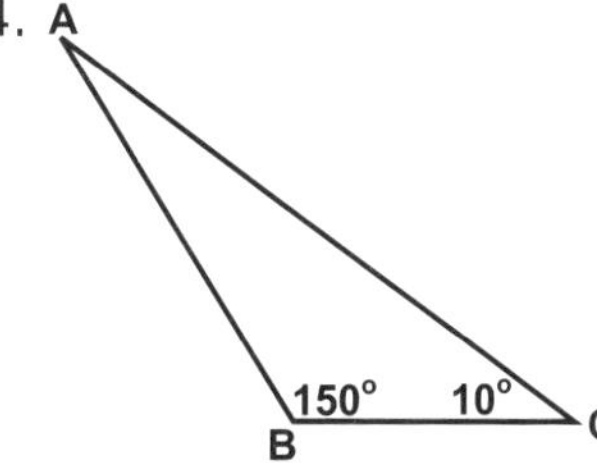

15. 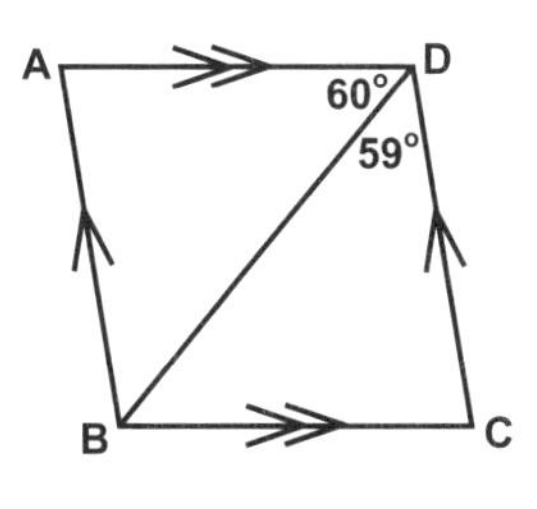

Name______________________________ HSG-CO.C.9, HSG-CO.C.10

Proofs in Column Form

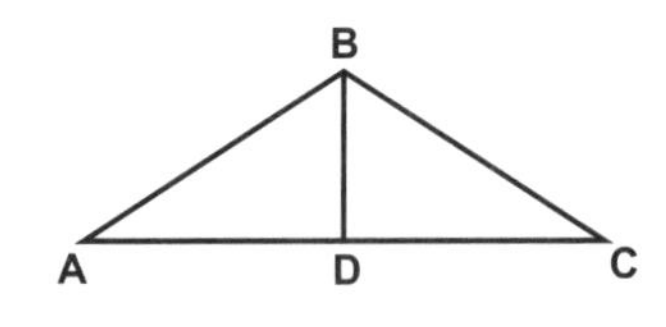

Given: D is the midpoint of $\overline{AC}$ and $\overline{AB} \cong \overline{BC}$.
Prove: $\triangle ABD \cong \triangle CBD$

Statements	Reasons
1. D is the midpoint of $\overline{AC}$	1. Given
2. $\overline{AD} \cong \overline{CD}$	2. Definition of Midpoint
3. $\overline{AB} \cong \overline{CB}$	3. Given
4. $\overline{BD} \cong \overline{BD}$	4. Reflexive Property
5. $\triangle ABD \cong \triangle CBD$	5. SSS

In each proof, the Statements are in order but the Reasons are scrambled. Write the Reasons in the correct order.

Given: $\overline{GH}$ and $\overline{FJ}$ bisect each other.
Prove: $\triangle FGI \cong \triangle JHI$

Statements	Scrambled Reasons	Reasons
1. $\overline{GH}$ and $\overline{FJ}$ bisect each other	1. Vertical angles are congruent.	________
2. $\overline{GH} \cong \overline{HI}$; $\overline{FI} \cong \overline{JI}$	2. Given	________
3. $\angle GIF \cong \angle HIJ$	3. SAS	________
4. $\triangle FGI \cong \triangle JHI$	4. Definition of Bisect	________

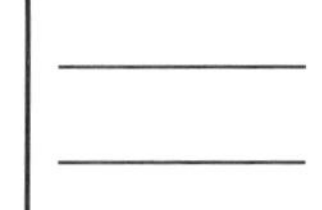

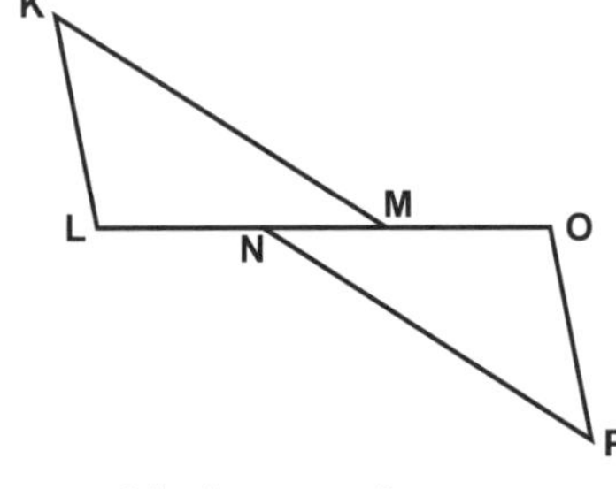

Given: KL = PO; LN = OM; KM = PN
Prove: $\triangle KLM \cong \triangle PON$

Statements	Scrambled Reasons	Reasons
1. LN = OM	1. Addition Property of Equality	________
2. LN + NM = NM + MO	2. Given	________
3. LN + NM = LM; NM + MO = NO	3. SSS	________
4. LM = NO	4. Definition of Between	________
5. KL = PO; KM = PN	5. Given	________
6. $\triangle KLM \cong \triangle PON$	6. Substitution Property	________

Name________________________________ HSG-CO.C.9, HSG-CO.C.10, HSG-CO.C.11

More Practice with Proofs

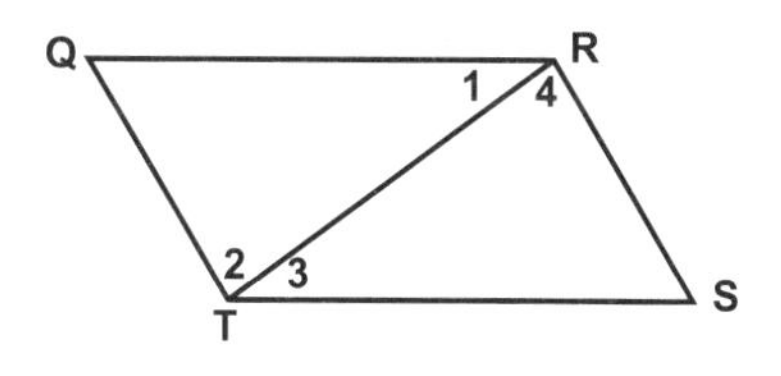

Complete the following proofs.
Given: $m\angle 1 = 40°$; $m\angle 3 = 40°$, $\angle 2 \cong \angle 4$
Prove: $\triangle RTQ \cong \triangle TRS$

Statements	Reasons
1. $m\angle 1 = 40°$; $m\angle 3 = 40°$, $\angle 2 \cong \angle 4$	1. ______________
2. $\angle 1 \cong \angle 3$	2. ______________
3. $\overline{RT} \cong \overline{TR}$	3. ______________
4. $\triangle RTQ \cong \triangle TRS$	4. ______________

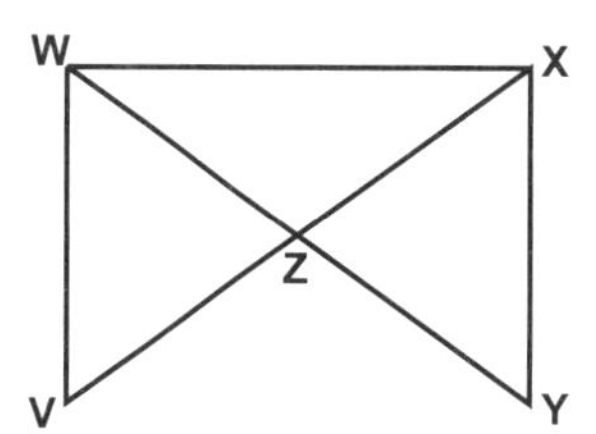

Given: $\overline{WY} \cong \overline{XV}$; $\overline{VW} \perp \overline{WX}$; $\overline{YX} \perp \overline{WX}$
Prove: $\triangle XWV \cong \triangle WXY$

Statements	Reasons
1. $\overline{VW} \perp \overline{WX}$ and $\overline{YX} \perp \overline{WX}$	1. ______________
2. ______________	2. Definition Perpendicular Lines
3. $\triangle XWV$, $\triangle WXY$ are right $\triangle$s	3. ______________
4. ______________	4. Given
5. $\overline{WX} \cong \overline{WX}$	5. ______________
6. $\triangle XWV \cong \triangle WXY$	6. ______________

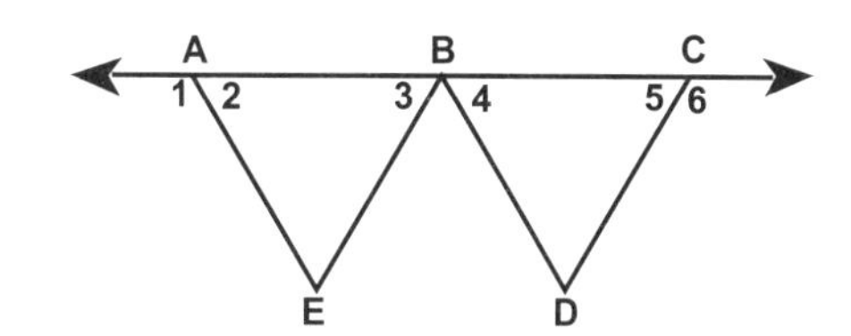

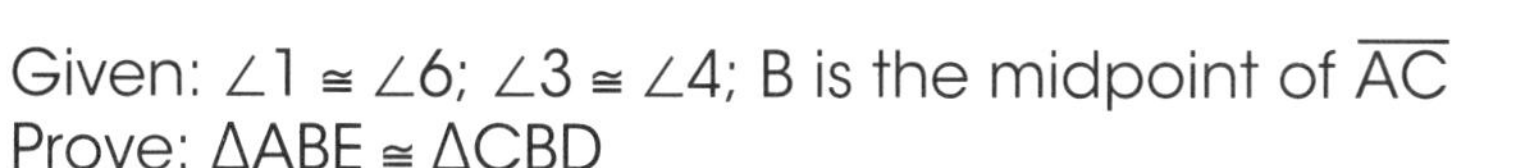
Given: $\angle 1 \cong \angle 6$; $\angle 3 \cong \angle 4$; B is the midpoint of $\overline{AC}$
Prove: $\triangle ABE \cong \triangle CBD$

Statements	Reasons
1. ______________	1. Given
2. $\overline{AB} \cong \overline{BC}$	2. ______________
3. ______________	3. Definition of Supplementary
4. $\angle 5$ is supplementary to $\angle 6$	4. ______________
5. $\angle 2 \cong \angle 5$	5. ______________
6. $\triangle ABE \cong \triangle CBD$	6. ______________

Name____________________________________ HSG-CO.D.12

Fractal: Koch Curve

In 1975, Benoit Mandelbrot used the term **fractal** to describe natural phenomena that appear to be chaotic, fragmented, and irregular but self-similar. Fractal designs can be created by iteration. An **iteration** is a repeated operation in which the output of one step becomes the input of the next. The starting object is called the **seed**.

Example: Draw a rectangle. Perform the iteration of connecting the midpoints of the adjacent sides. Every interior rectangle looks like the original—self-similar.

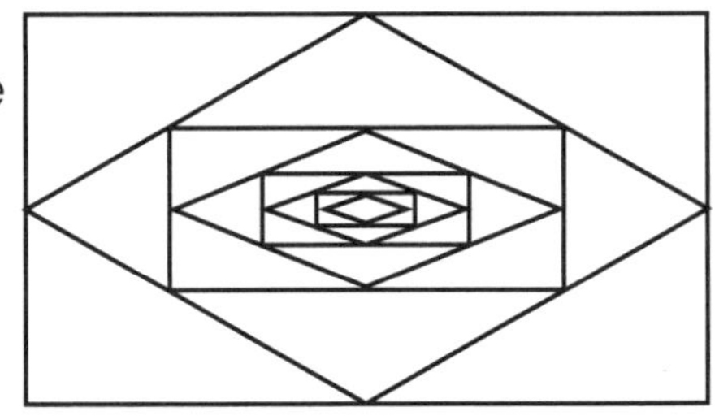

On another sheet of paper, complete Steps 0-3 to begin the Koch Curve.

Step 0 Draw a line segment 6 inches long. Consider its length to be one unit.

Step 1 Draw an equilateral triangle whose base is the middle third of the line segment. Do not draw the base.

Step 2 Draw an equilateral triangle on each segment so the base (not drawn) of each triangle is the middle third of the corresponding segment.

Step 3 Repeat Step 2.

Ex. Step 0

Step 1

Complete the table.

Step	Number of Segments	Length of 1 Segment	Total Length
0	1	1	
1	4	$\frac{1}{3}$	
2			
3			

1. Describe the pattern in each column.

A. Number of segments ______________________________

B. Length of 1 segment ______________________________

C. Total length ______________________________

2. What would the values be for Step 5? ______________________________

Historical Comment: The curve is the basis for the Koch Snowflake designed by Helge von Koch in 1904. Step 0 starts with an equilateral triangle. Steps 1, 2, 3, etc., are the same.

Name________________________________

8.EE.C.7b, HSG-GPE.B.6

Length, Links, and Midpoint Magic

Find the length of segment $\overline{FM}$

$\overline{FM} = |^-5 - 2| = |^-7| = 7$

Find the coordinate of the midpoint of segment $\overline{IO}$.

$$\text{Midpoint} = \frac{(^-2 + 4)}{2} = 1$$

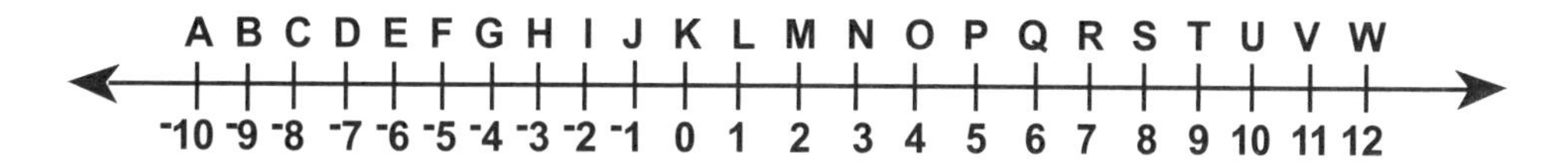

Find the length of each segment and link the segments in Columns A and B that have equal lengths.

Column A

1. Segment $\overline{GW}$
2. Segment $\overline{BN}$
3. Segment $\overline{PV}$
4. Segment $\overline{KR}$
5. Segment with endpoints $\frac{3}{4}$ and $5\frac{3}{4}$
6. Segment with endpoints $^-3$ and $^-7\frac{1}{2}$
7. Segment with endpoints $-\frac{1}{4}$ and $3\frac{1}{4}$
8. Segment with endpoints $\frac{1}{4}$ and $3\frac{1}{4}$

Column B

A. Segment $\overline{EL}$
B. Segment $\overline{AF}$
C. Segment $\overline{CS}$
D. Segment $\overline{MS}$
E. Segment with endpoints $^-2$ and 10
F. Segment with endpoints $^-1\frac{1}{2}$ and 3
G. Segment with endpoints $^-1\frac{3}{4}$ and $1\frac{3}{4}$
H. Segment with endpoints $^-5$ and $^-2$

In a Magic Square, each row, column, and diagonal has the same sum: the Magic Sum. Find the length of the segments and determine the Magic Sum: _______.

Endpoints $1\frac{1}{2}$ and 2	Endpoints $^-3$ and $4\frac{1}{2}$	Segment $\overline{DK}$	Segment $\overline{HJ}$
Segment $\overline{HN}$	Segment $\overline{QT}$	Endpoints $6\frac{1}{2}$ and 3	Endpoints $5\frac{3}{4}$ and $1\frac{1}{4}$
Segment $\overline{DH}$	Segment $\overline{JO}$	Endpoints $1\frac{3}{4}$ and $^-3\frac{3}{4}$	Endpoints $1\frac{1}{2}$ and $^-1$
Endpoints $^-2\frac{1}{4}$ and $4\frac{1}{4}$	Endpoints $^-5\frac{1}{2}$ and $^-4$	Segment $\overline{JK}$	Segment $\overline{AI}$

Name____ 8.G.A.2, HSG-CO.A.1, HSG-CO.C.9

Congruence of Segments and Addition Properties

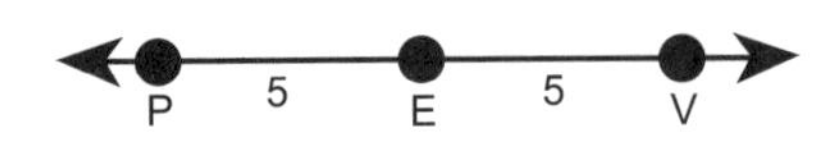

PV = PE + EV
PV = 5 + 5

Congruent Segments
PE = 5 and EV = 5
Segment addition: $\overline{PE} \cong \overline{EV}$
PV = 10

True or False?

1. $\overline{TV} \cong \overline{ML}$
2. $\overline{KJ} \cong \overline{TV}$
3. $\overline{LB} \cong \overline{JV}$
4. $\overline{TV} \cong \overline{BV}$
5. $\overline{VB} \cong \overline{LB}$
6. $\overline{KJ} \cong \overline{VB}$

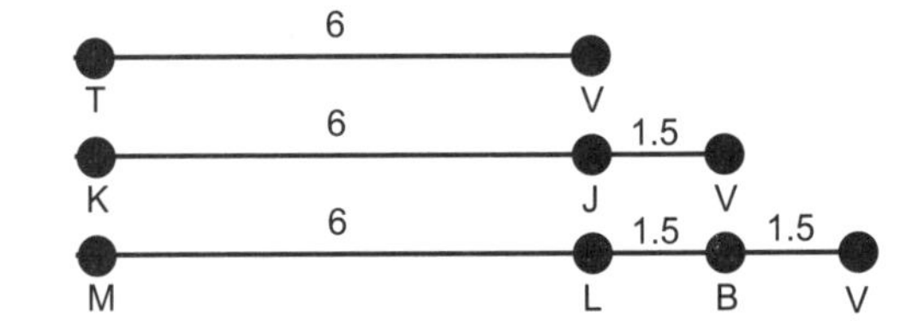

Complete.

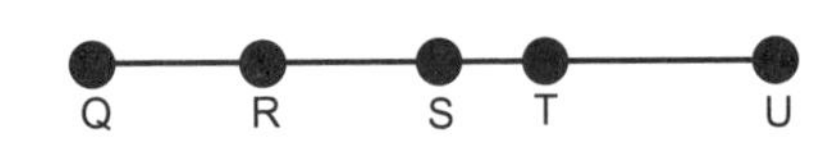

7. QR + RS =
8. RU – SU =
9. RS + SU =
10. QS + SU =
11. QU – QR – TU =
12. QR + RS + ST =

Find the length of the indicated segments.

13. J G D B (1.4, 7, 4) JD = ______ GB = ______ JB = ______

14. M S K D T (6, 10; 23, 16.5) SK = ______ DT = ______ MT = ______

15. A B C D E (71; 28, 28; 52) BC = ______ DE = ______ AE = ______

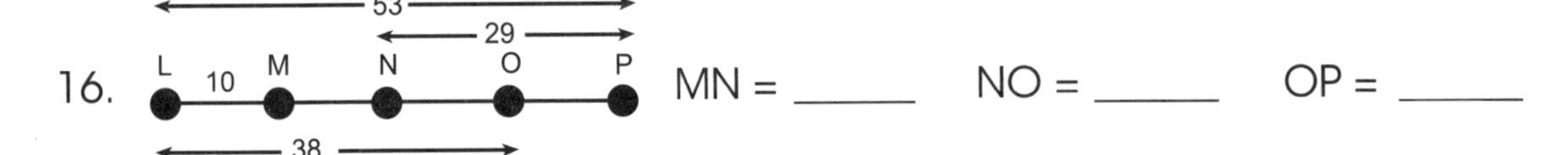

16. MN = ______ NO = ______ OP = ______

17. Which segments are congruent in #15?

Name____________________________________ 8.G.A.5, HSG-CO.A.1

Angles

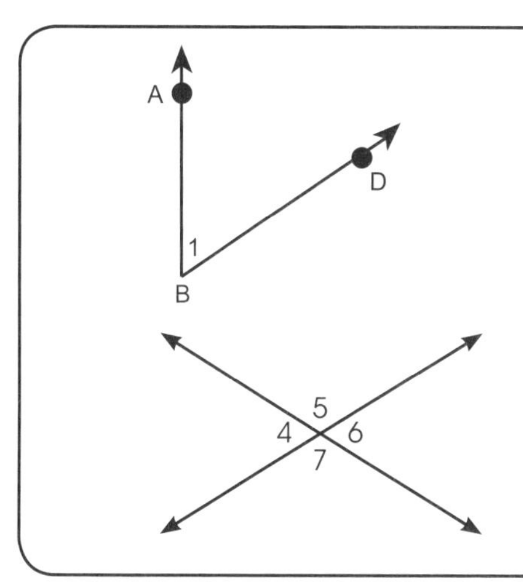

B is the vertex.
$\overrightarrow{BA}$ & $\overrightarrow{BD}$ are the sides.
4 names for the angle:
∠1, ∠B, ∠ABD, ∠DBA

Vertical angle pairs:
∠4 & ∠6; ∠5 & ∠7
Adjacent angle pairs:
∠4 & ∠5; ∠6 & ∠7

Names for each angle:
1) ∠2 or ∠FEG or ∠GEF
2) ∠3 or ∠GEH or ∠HEG
3) ∠FEH or ∠HEB

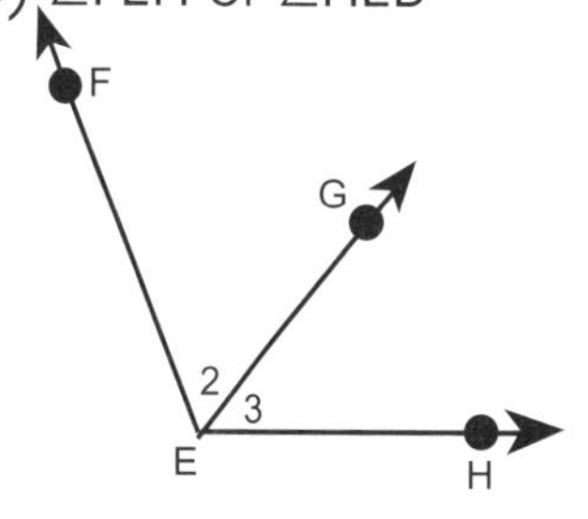

Name the indicated angle.

1. 2. 3. 4. 5.

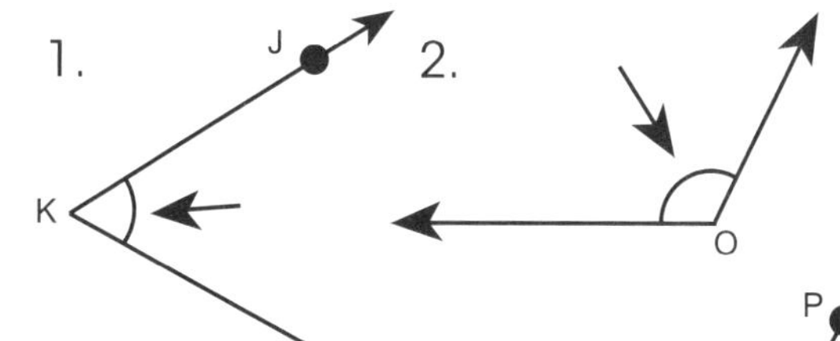

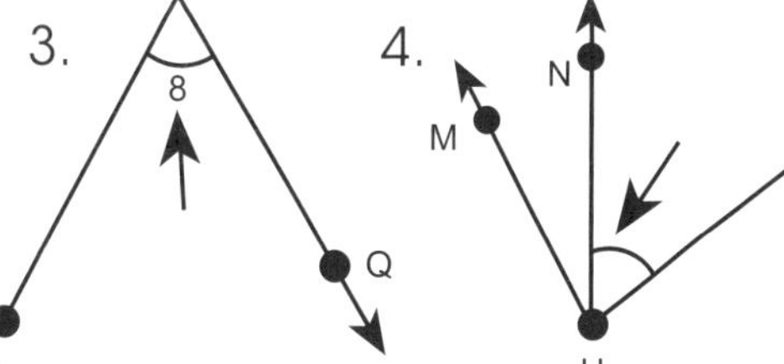

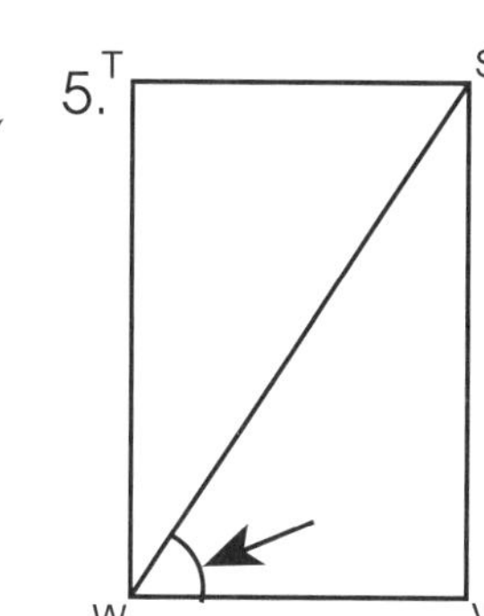

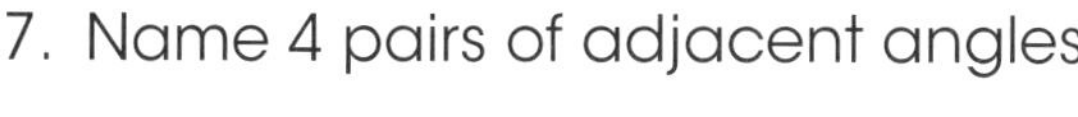

6. Name 2 pairs of vertical angles.

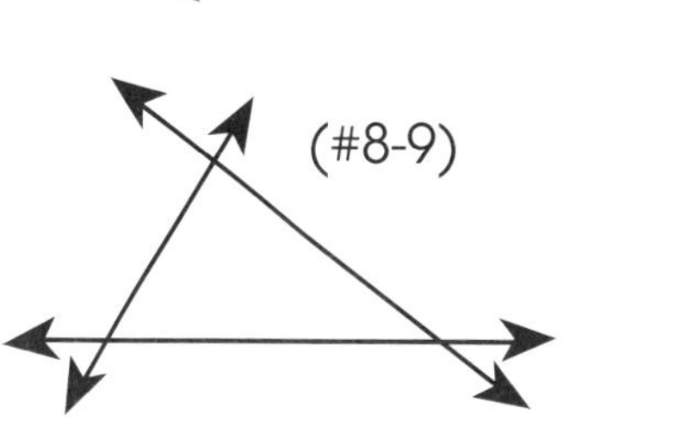

7. Name 4 pairs of adjacent angles.

8. How many pairs of vertical angles are pictured?

(#8-9)

9. How many pairs of adjacent angles are pictured?

10. Name 2 angles adjacent to ∠RES.

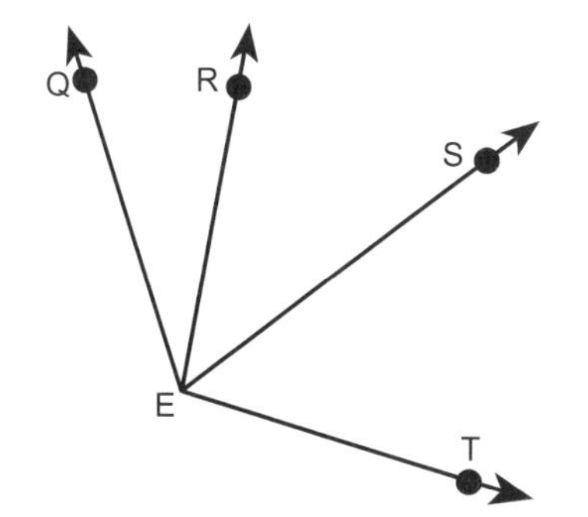

Name______________________________

8.G.A.5, HSG-CO.A.1, HSG-CO.C.9

Congruence of Angles and Addition Properties

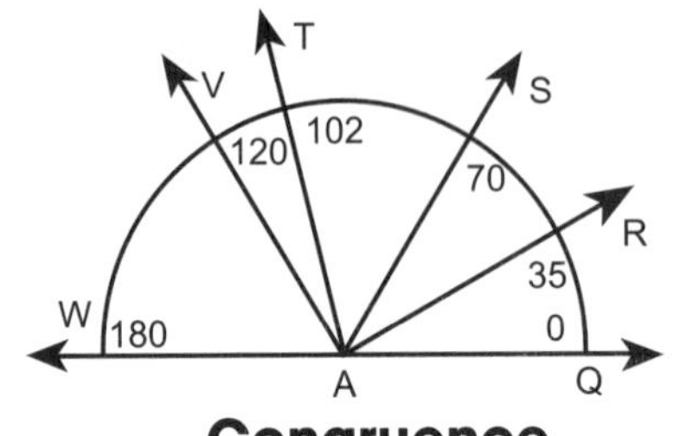

Angle Measures

m∠QAT = 102 – 0 = 102

m∠TAR = 102 – 35 = 67

m∠WAV = 180 – 120 = 60

Congruence

m∠SAR = 35, m∠RAQ = 35

∠SAR ≅ ∠RAQ

Angle Addition

m∠VAT + m∠TAS = m∠VAS

18 + 32 = 50

Find the values of each of the following.

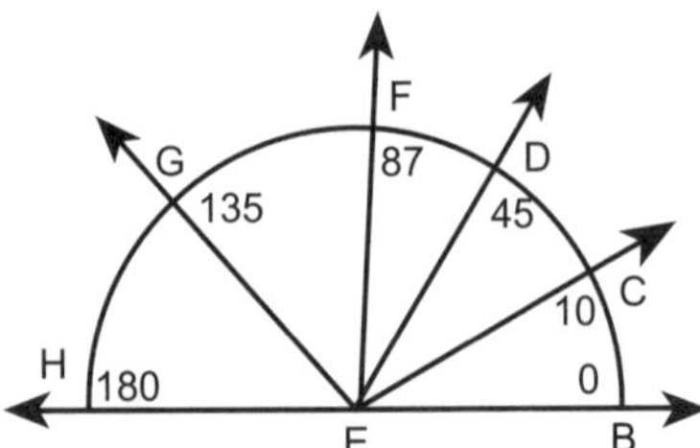

1. m∠CEB = __________
2. m∠FED = __________
3. m∠BEG = __________
4. m∠HEF = __________
5. m∠BEC + m∠CEF = __________
6. m∠DEF + m∠GEF = __________
7. m∠HEG + m∠CED = __________
8. m∠GEB – m∠DEB = __________
9. m∠GED + m∠DEC = __________
10. m∠HEG + m∠FEC = __________
11. m∠HEF – m∠HEG = __________
12. m∠GED + m∠DEC – m∠FED = __________
13. m∠HEG + m∠CEF – m∠BEC = __________
14. m∠BEG – m∠FED – m∠BEC = __________
15. Name a pair of congruent angles. __________

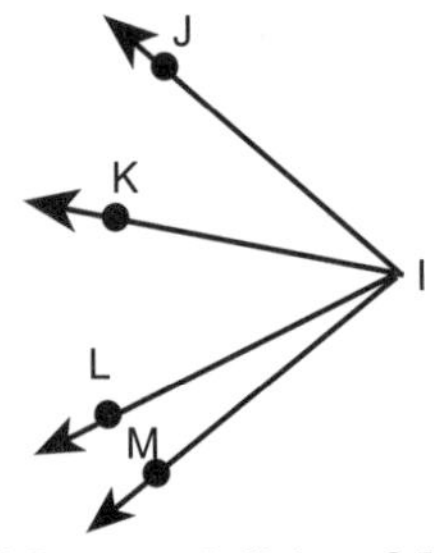

16. Name the angle with the greatest measure. ___
17. m∠JIK + m∠KIL = __________
18. m∠MIL + m∠LIJ = __________
19. m∠KIJ = 28 & m∠LIK = 39; m∠LIJ = __________
20. m∠MIJ = 81 & m∠MIL = 12; m∠LIJ = __________

Find x.

21. m∠KIL = 2x; m∠LIM = x; m∠KIM = 4x – 17 x = __________
22. m∠JIK = x; m∠KIL = 3x + 5; m∠JIL = 5x – 15 x = __________

Name______________________________

8.G.A.5, HSG-CO.A.1, HSG-CO.C.9

Classifying Angles

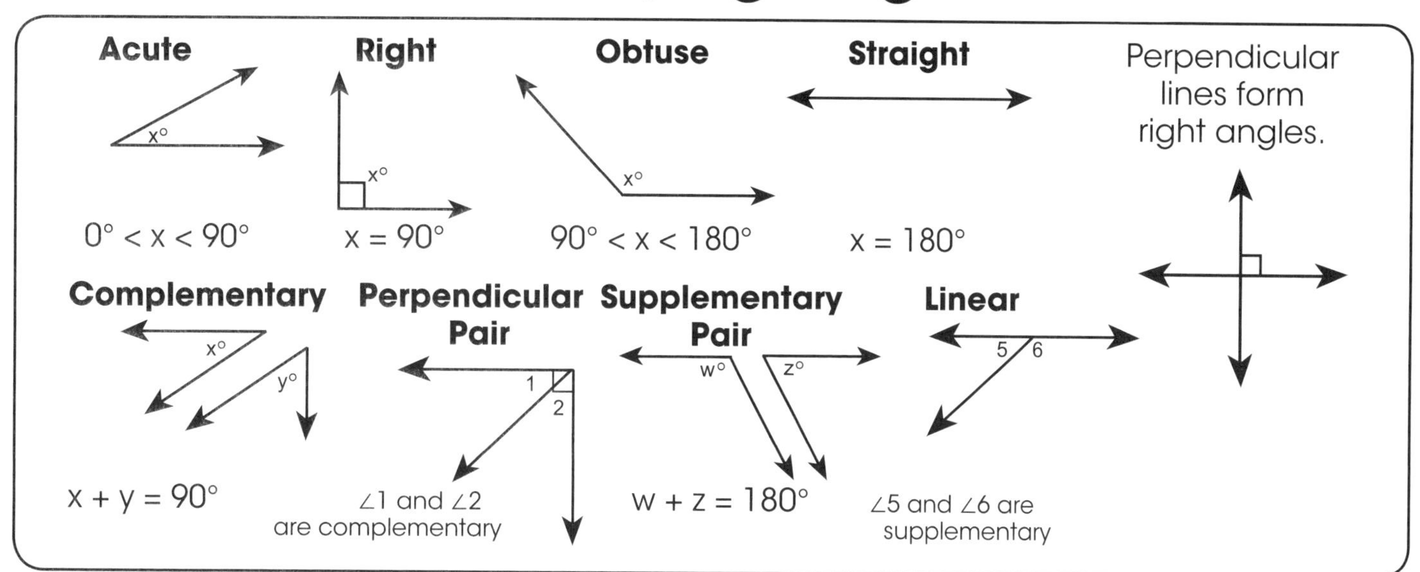

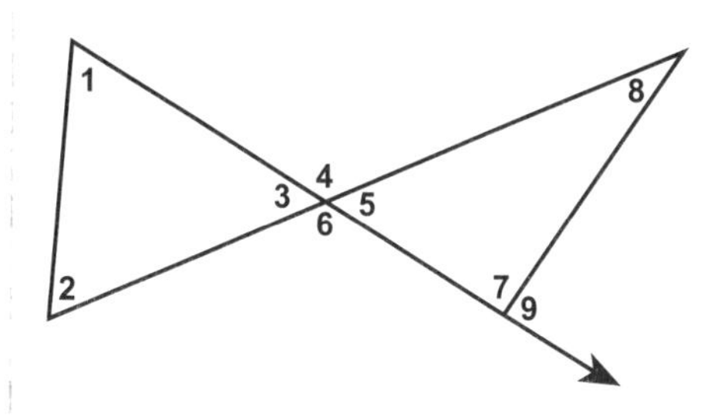

In the figure, $m\angle 7 = 90°$

1. Name the angles which appear to be:
   a. acute b. obtuse c. right

2. Name five pairs of supplementary angles. ______________________
3. ∠7 and ∠9 form a ______________________.

∠10 and ∠11 are complementary angles.

4. $m\angle 10 = 32°$; $m\angle 11 =$ ______
5. $m\angle 11 = 72°$; $m\angle 10 =$ ______
6. $m\angle 10 = 4x$; $m\angle 11 = 2x$; $x =$ ______
7. $m\angle 10 = x$; $m\angle 11 = x + 20$; $x =$ ______

∠12 and ∠13 are supplementary angles.

8. $m\angle 12 = 2y$; $m\angle 13 = 3y - 15$; $y =$ ______
9. $m\angle 12 = y + 10$; $m\angle 13 = 3y + 10$; $y =$ ______
10. The measure of ∠12 is five times the measure of ∠13. Find the measure of each angle.

∠13 and ∠14 are complementary angles, and
∠14 and ∠15 are supplementary angles.

11. $m\angle 13 = 47°$; $m\angle 14 =$ ______; $m\angle 15 =$ ______
12. $m\angle 14 = 78°$; $m\angle 13 =$ ______; $m\angle 15 =$ ______
13. $m\angle 15 = 135°$; $m\angle 13 =$ ______; $m\angle 14 =$ ______

Name________________________________ 8.G.A.5, HSG-CO.C.9

Mixed Practice with Angles

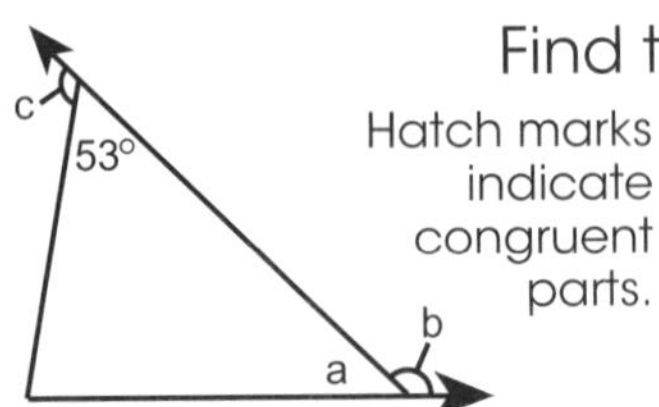

Hatch marks indicate congruent parts.

Find the measure of the lettered angles.

c = 180 – 53 = 127°
b = c = 127°
a = 180 – b = 180 – 127 = 53°

1.
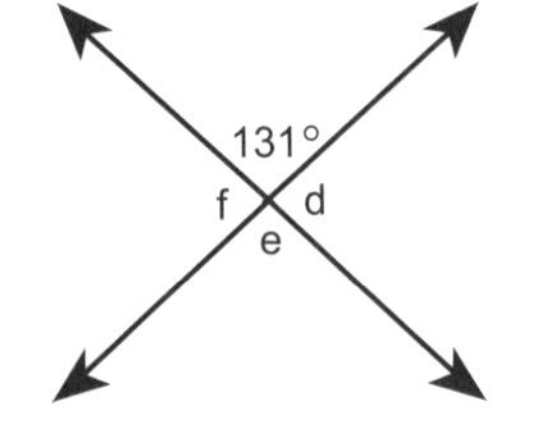

d = __________
e = __________
f = __________

2.
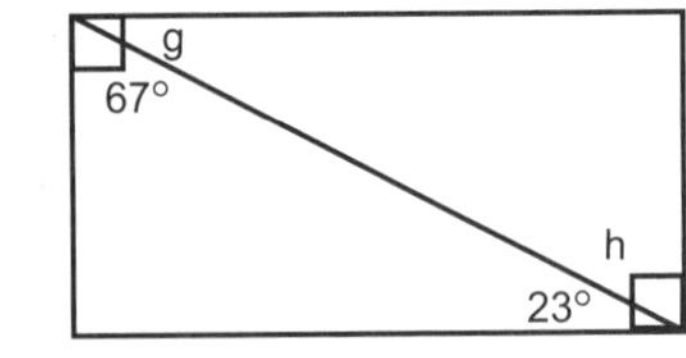

g = __________
h = __________

3.
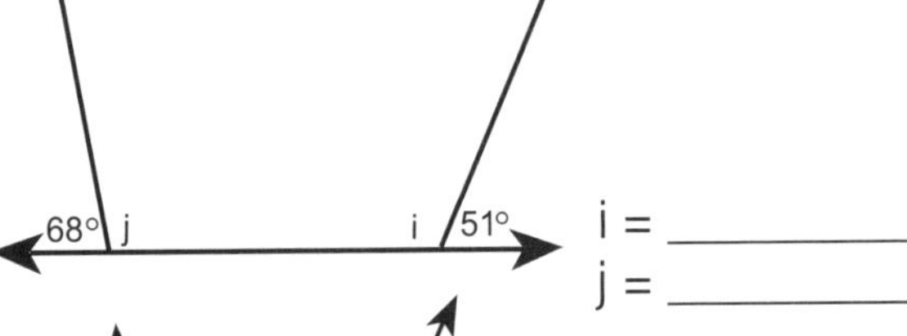

i = __________
j = __________

4.
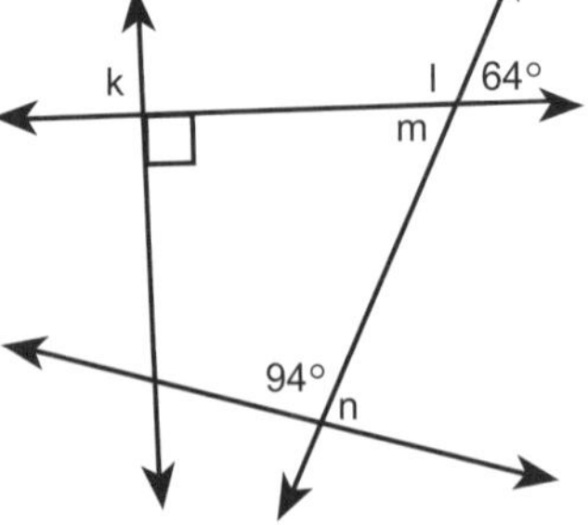

k = __________
l = __________
m = __________
n = __________

5.
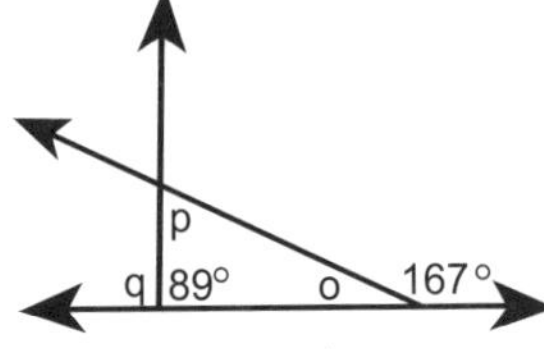

o = __________
p = __________
q = __________

6.
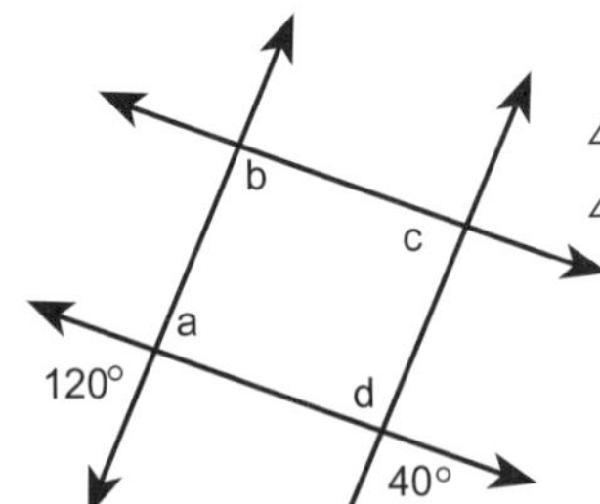

∠a is supplementary to ∠b
∠c is supplementary to ∠d

a = __________
b = __________
c = __________

7.
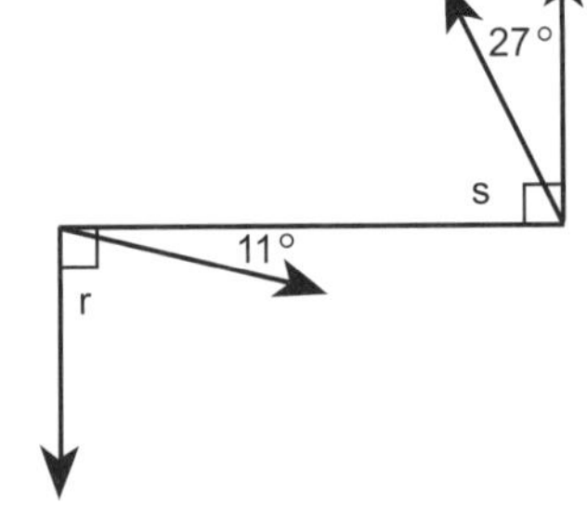

r = __________
s = __________

8.
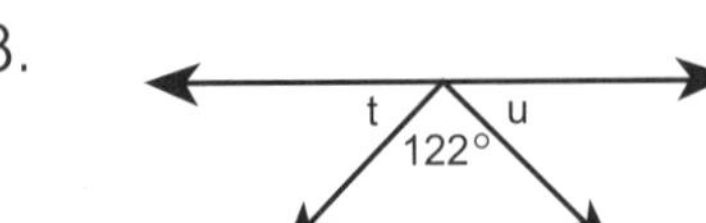

∠t ≅ ∠u

t = __________
u = __________

9.
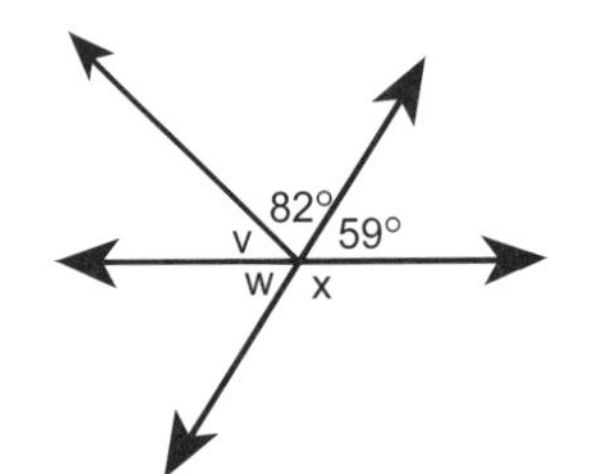

v = __________
w = __________
x = __________

10.
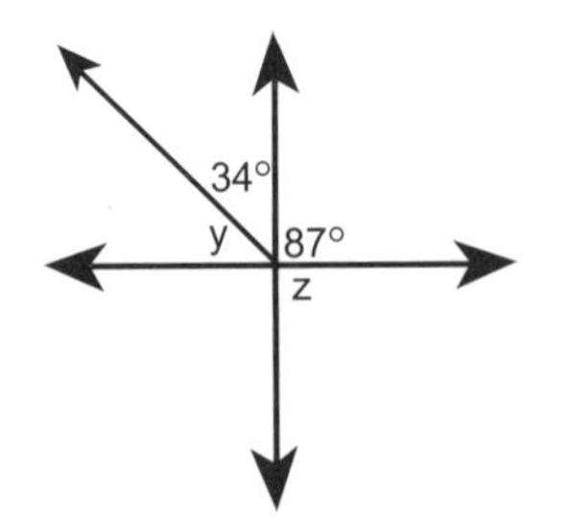

y = __________
z = __________

11.
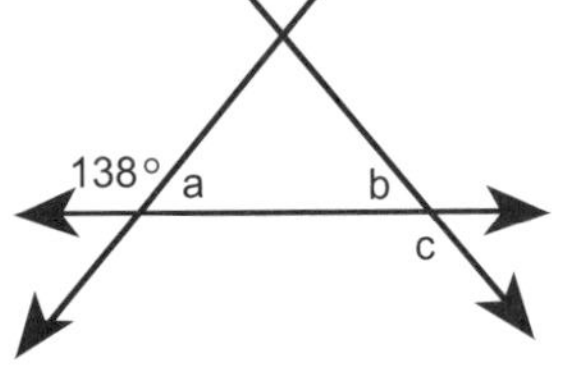
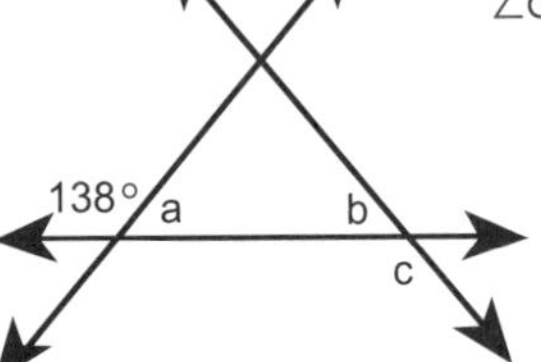

∠a is complementary to ∠b

a = __________
b = __________
c = __________

12.

a = __________
b = __________
c = __________
d = __________

Name______________________________

8.G.A.5, HSG-CO.C.9

Angles and Parallel Lines

ℓ1 and ℓ2 are **parallel lines**
(they do not intersect).
ℓ3 is a **transversal.**
Eight angles are formed:

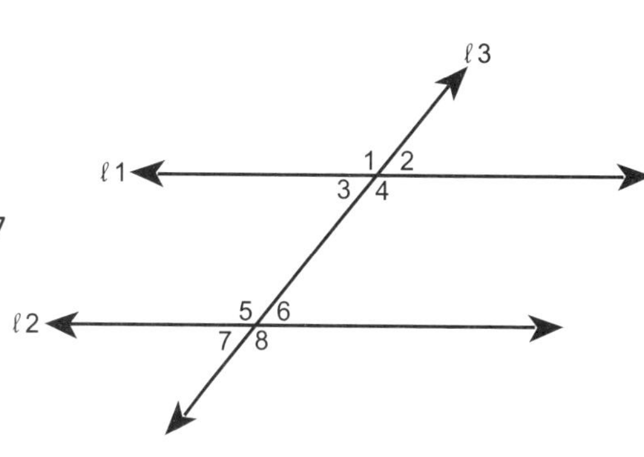

vertical angles = ∠1 & ∠4, ∠2 & ∠3, ∠5 & ∠8, ∠6 & ∠7
interior angles = ∠3, ∠4, ∠5 and ∠6
same side interior angles = ∠3 & ∠4, ∠5 & ∠6
alternate side interior angles = ∠3 & ∠6, ∠4 & ∠5
exterior angles = ∠1, ∠2, ∠7 and ∠8
same side exterior angles = ∠1 & ∠7, ∠2 & ∠8
alternate side exterior angles = ∠1 & ∠8, ∠2 & ∠7
corresponding angles = ∠1 & ∠5, ∠2 & ∠6, ∠3 & ∠7, ∠4 & ∠8

Given two parallel lines, same side interior angles are supplementary. Fill in the blanks with the correct answers.

1. Vertical angles are ______________.
2. Same side exterior angles are ______________.
3. Alternate side interior angles are ______________.
4. Alternate side exterior angles are ______________.
5. Corresponding angles are ______________.

Find the measures of the designated angles.
ℓ_1 is parallel to ℓ_2.

m∠a = __________ m∠e = __________
m∠b = __________ m∠f = __________
m∠c = __________ m∠g = __________
m∠d = __________

a ll b, c ll d

m∠1 = __________ m∠9 = __________
m∠2 = __________ m∠10 = __________
m∠3 = __________ m∠11 = __________
m∠4 = __________ m∠12 = __________
m∠5 = __________ m∠13 = __________
m∠6 = __________ m∠14 = __________
m∠7 = __________ m∠15 = __________
m∠8 = __________

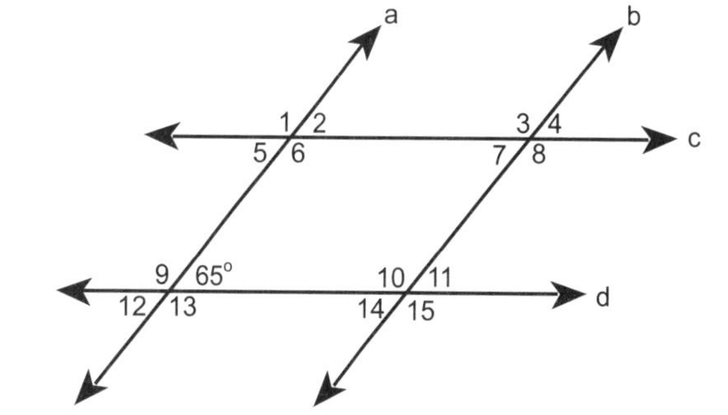

Name_______________________________

More Angles and Parallel Lines

Find the missing values.

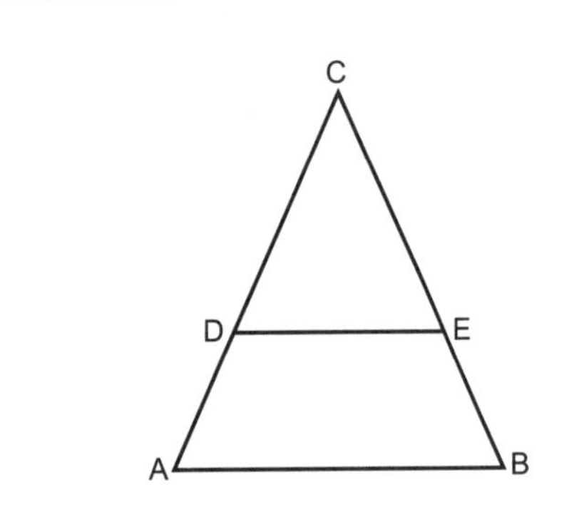

$\overline{AB} \parallel \overline{DE}$, $\angle A = 60°$, $m\angle B = 50°$

$m\angle C = \underline{70}°$

$60 + 50 + m\angle C = 180$

$m\angle C = 70$

$m\angle CDE = \underline{60}°$

$\angle CDE \cong \angle CAB$

$m\angle EDA = \underline{120}°$

$60 + m\angle EDA = 180$

$m\angle EDA = 120$

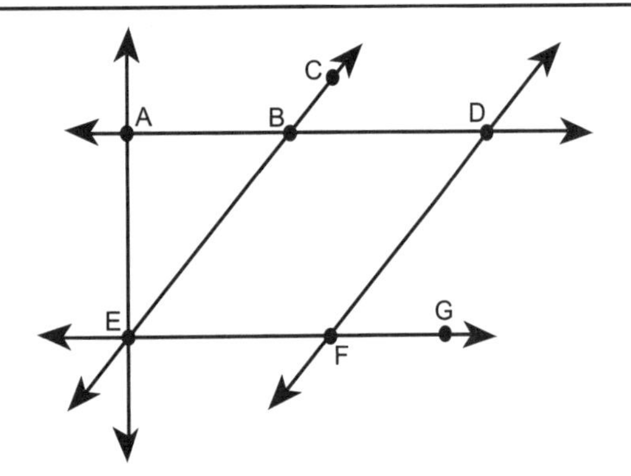

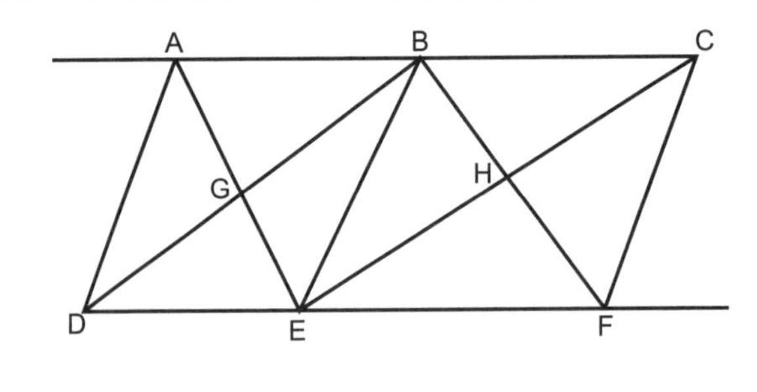

$\overline{AD} \parallel \overline{EG}$, $\overline{EC} \parallel \overline{FD}$,
$\overline{EG} \perp \overline{AE}$, $m\angle CBA = 140°$

$m\angle ABE =$ __________

$m\angle EAB =$ __________

$m\angle AEB =$ __________

$m\angle BEF =$ __________

$m\angle EFD =$ __________

$m\angle BDF =$ __________

$m\angle DFG =$ __________

$\overline{AC} \parallel \overline{DF}$, $\overline{AD} \parallel \overline{BE} \parallel \overline{CF}$, $\overline{BD} \parallel \overline{CE}$,
$\overline{AE} \parallel \overline{BF}$, $m\angle BDE = 30°$, $m\angle ADB = 20°$
$m\angle BEG = 70°$

$m\angle BAD =$ __________

$m\angle BAE =$ __________

$m\angle ABD =$ __________

$m\angle AGB =$ __________

$m\angle BED =$ __________

$m\angle BEF =$ __________

$m\angle BFE =$ __________

$\overline{MN} \parallel \overline{RS}$, $\overline{MO} \parallel \overline{NR} \parallel \overline{QS}$, $\overline{OR} \parallel \overline{MS} \parallel \overline{NQ}$,
$m\angle NMP = 75°$, $m\angle RPS = 40°$

$m\angle MNP =$ __________

$m\angle MPN =$ __________

$m\angle PMO =$ __________

$m\angle MOP =$ __________

$m\angle OPM =$ __________

$m\angle OPR =$ __________

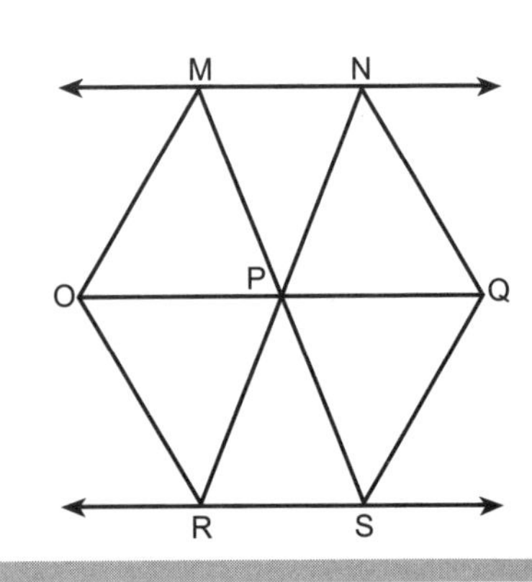

Name________________________________

8.G.A.5, HSG-CO.C.9

Proofs Using Parallel Lines

Complete the following proofs.

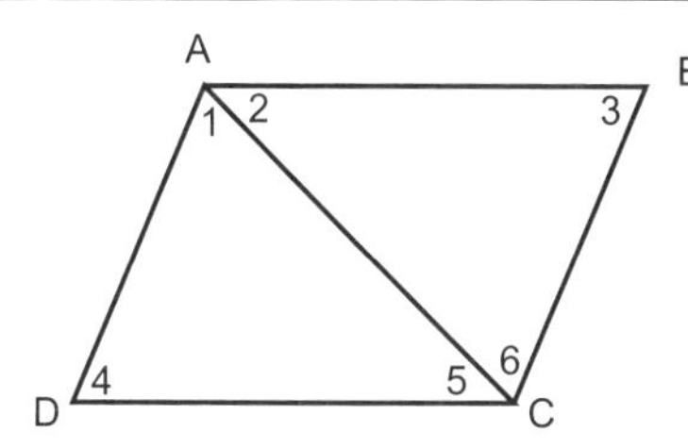

Given: $\overline{AB} \parallel \overline{DC}$, $\overline{AD} \parallel \overline{BC}$
Prove: $\triangle ABC \cong \triangle CDA$

Statements	Reasons
1. $\overline{AB} \parallel \overline{DC}$, $\overline{AD} \parallel \overline{BC}$	1. Given
2. $\angle 2 \cong \angle 5$, $\angle 1 \cong \angle 6$	2. If $\parallel$ lines, then alternate interior $\angle$s are $\cong$.
3. $\overline{AC} \cong \overline{AC}$	3. Reflexive
4. $\triangle ABC \cong \triangle CDA$	4. ASA

Given: $\overline{AB} \cong \overline{CB}$, $\overline{DB} \cong \overline{EB}$, $\angle 1 \cong \angle 4$
Prove: $\overline{DE} \parallel \overline{AC}$

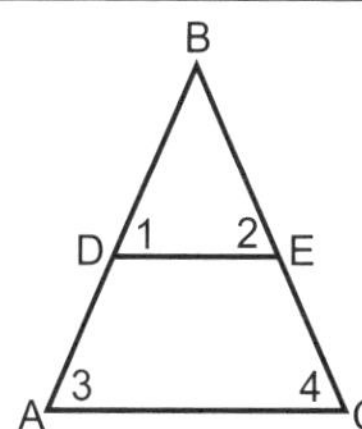

Statements	Reasons
1. ______________________	1. Given
2. ______________________	2. In a $\triangle$, $\angle$s opposite $\cong$ sides are $\cong$.
3. $\angle 1 \cong \angle 3$	3. ______________________
4. $\overline{DE} \parallel \overline{AC}$	4. ______________________

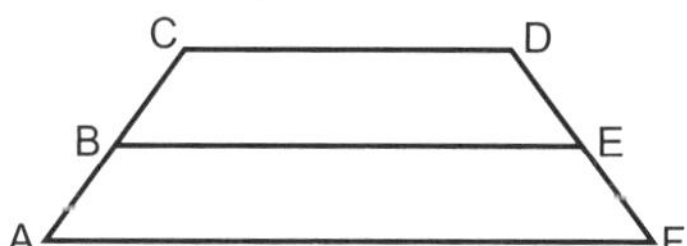

Given: $\overline{AF} \parallel \overline{CD}$, $\angle DCA \cong \angle EBA$
Prove: $\overline{AF} \parallel \overline{BE}$

Statements	Reasons
1. ______________________	1. Given
2. $\angle DCA$, $\angle EBA$ are corresponding angles	2. ______________________
3. $\overline{CD} \parallel \overline{BE}$	3. ______________________
4. $\overline{AF} \parallel \overline{BE}$	4. ______________________

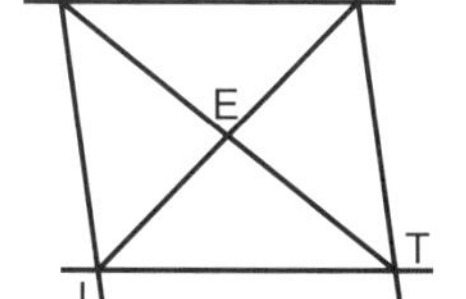

Given: $\overline{KI} \parallel \overline{AT}$, $\overline{KA} \parallel \overline{IT}$, $\overline{KI} \cong \overline{TA}$
Prove: $\triangle KEI \cong \triangle TEA$

Statements	Reasons
1. ______________________	1. Given
2. $\angle EKI \cong \angle ETA$; $\angle EIK \cong \angle EAT$	2. ______________________
3. $\triangle KEI \cong \triangle TEA$	3. ______________________

Name________________________________ 8.G.A.5, HSG-CO.C.10

More Proofs

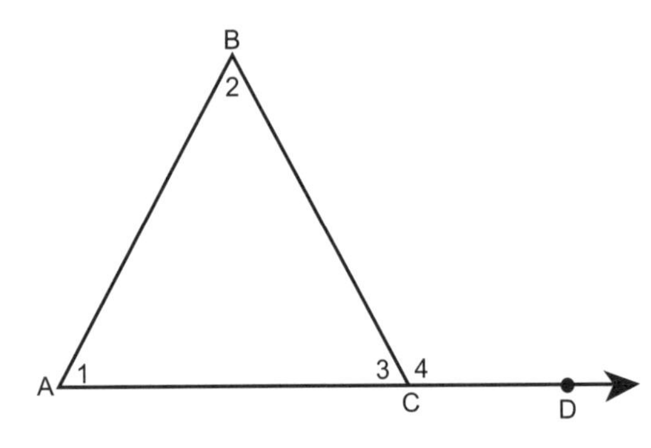

Given: ΔABC with exterior $\angle 4$
Prove: $m\angle 4 = m\angle 1 + m\angle 2$

Statements	Reasons
1. $m\angle 1 + m\angle 2 + m\angle 3 = 180°$	1. ______________________
2. ______________________	2. Two angles that form a linear pair are supplementary.
3. $m\angle 3 + m\angle 4 = 180°$	3. ______________________
4. ______________________	4. Substitution
5. $m\angle 4 = m\angle 1 + m\angle 2$	5. ______________________

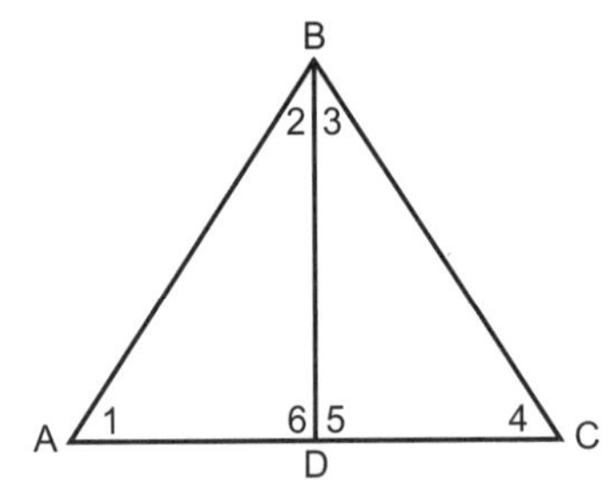

Given: $\angle ABC$ is a right angle, $\overline{AC} \perp \overline{BD}$
Prove: $\angle 1 \cong \angle 3$

Statements	Reasons
1. ______________________	1. Given
2. ______________________	2. $\perp$ lines form right $\angle$s.
3. ΔABD is a right triangle	3. ______________________
4. $\angle 1$ is complementary to $\angle 2$	4. ______________________
5. $\angle 2$ is complementary to $\angle 3$	5. ______________________
6. ______________________	6. Reflexive Property
7. $\angle 1 \cong \angle 3$	7. ______________________

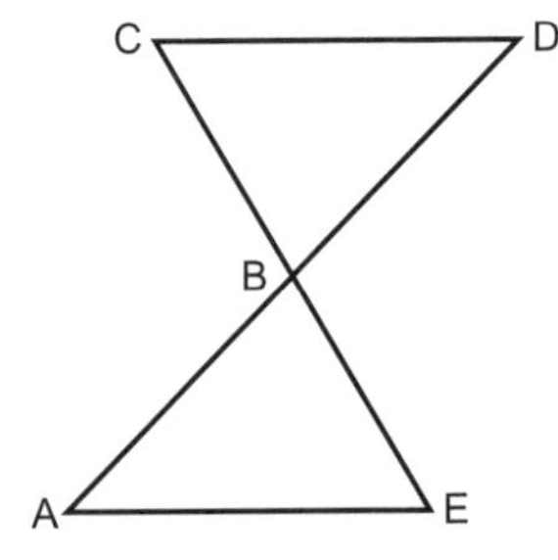

Given: B is the midpoint of $\overline{AD}$; B is the midpoint of $\overline{CE}$
Prove: $\overline{CD} \parallel \overline{AE}$

Name______________________________

Constructing Congruent Segments

Given: $\overline{AB}$

Construct a segment congruent to $\overline{AB}$.

1. Use a straight edge to draw a working line, ℓ.
2. Choose a point on ℓ and label it A′.
3. Set your compass for radius $\overline{AB}$ by placing one end at point A and another at point B. Draw an arc.
4. Using $\overline{AB}$ as radius, place one end of compass on A′ and draw an arc. Label the point of intersection B′.

$\overline{AB} \cong \overline{A'B'}$

1. Construct a segment congruent to $\overline{CD}$.

2. Construct a segment congruent to $\overline{EF}$.

3. Construct a segment congruent to $\overline{XY}$.

4. Construct a segment whose length is $\overline{CD} + \overline{EF}$.

5. Construct a segment whose length is $\overline{EF} + \overline{XY}$.

6. Construct a segment whose length is $\overline{EF} - \overline{CD}$.

Name________________________________ 8.G.A.5, HSG-CO.D.12

Constructing Perpendicular Bisectors

Given: $\overline{AB}$

Construct the perpendicular bisector of $\overline{AB}$.

1. Copy segment $\overline{AB}$.
2. Choose a radius greater than $\frac{1}{2}$ $\overline{AB}$ and less than AB. Using A as center, draw 2 arcs, one above $\overline{AB}$ and one below $\overline{AB}$. Repeat using B as center.
3. Draw $\overleftrightarrow{CD}$.

$\overleftrightarrow{CD}$ is the perpendicular bisector of $\overline{AB}$.

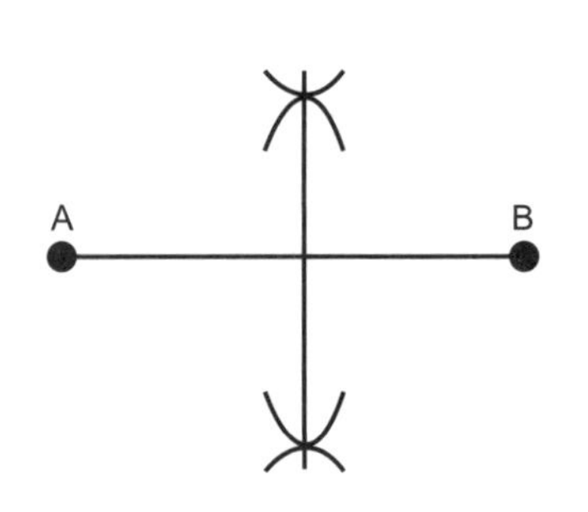

I. Construct the perpendicular bisector of the following.

1.

2.

3.

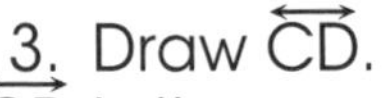

II. Bisect side $\overline{YZ}$ of ΔXYZ.

4.

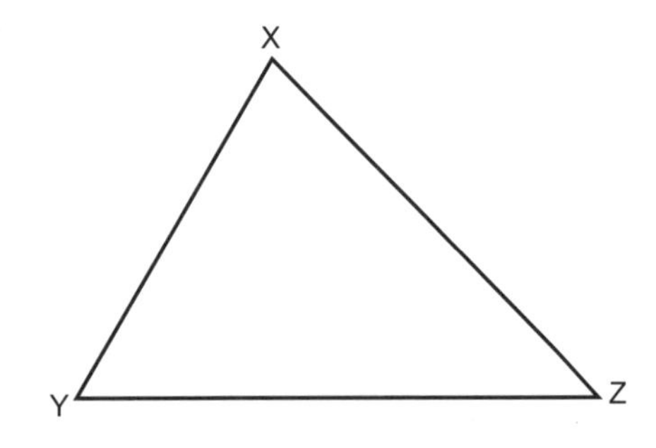

5. Construct a segment whose length equals $\overline{XY} + \overline{YZ} + \overline{XZ}$.

Name____________________________

8.G.A.5, HSG-CO.D.12

Constructing Perpendiculars from Point on Line

Given: Point A on line ℓ

Construct the perpendicular to point A.

1. Copy ℓ.
2. Using A as center, choose any radius less than ℓ. Draw arcs intersecting ℓ. Label them C and D.
3. Using C as center, choose a radius greater than CA. Draw an arc above ℓ. Repeat using D as center with same radius.
4. Draw $\overleftrightarrow{XA}$.

$\overleftrightarrow{XA}$ is perpendicular to ℓ at point A.

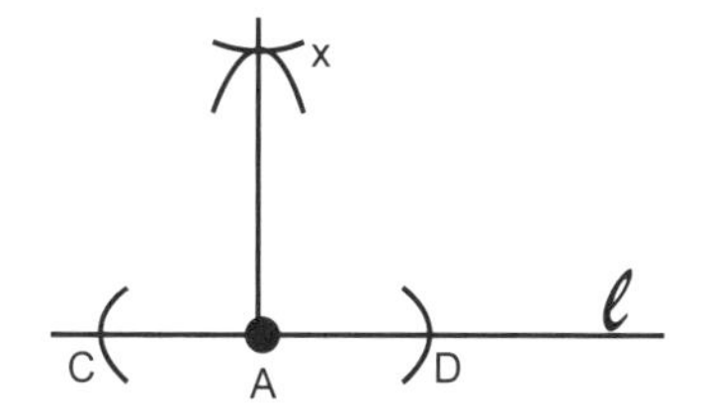

I. Construct perpendicular lines to the given points.

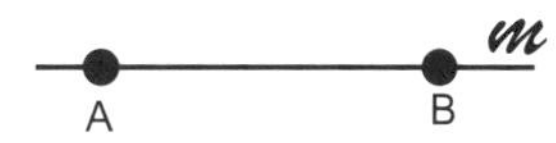

1. point A

2. point B

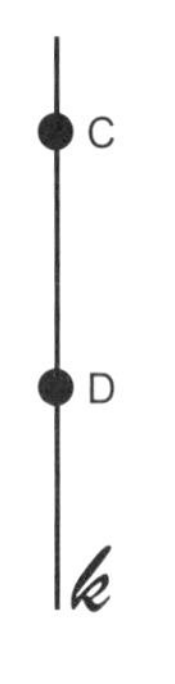

3. point C

4. point D

Name________________________________ 8.G.A.5, HSG-CO.D.12

Constructing Perpendiculars from Point Not on Line

Given: Point P outside line ℓ

Construct a line perpendicular from P to ℓ.

1. Copy ℓ.
2. Using P as center, draw two arcs intersecting ℓ. Label them A and B.
3. Choose a radius greater than $\frac{1}{2}$ AB. Using A as center, draw arc below ℓ. Repeat, using B as center with same radius. Label X.
4. Draw $\overleftrightarrow{PX}$.

$\overleftrightarrow{PX}$ is perpendcular to ℓ.

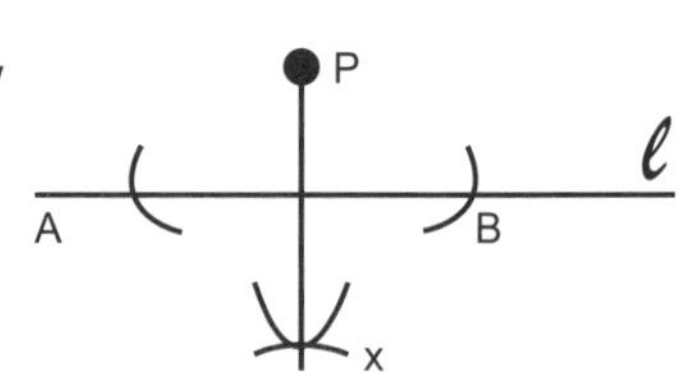

I. Construct perpendicular lines to ℓ from:

1. point A

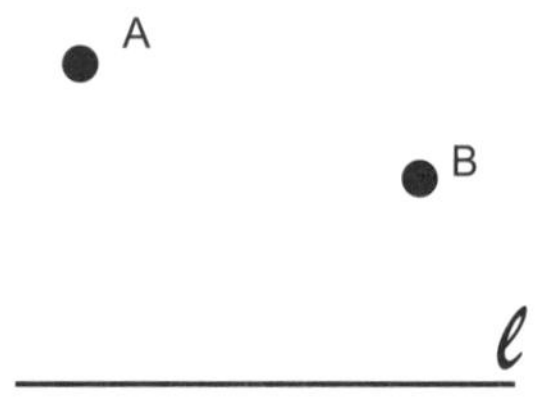

2. point B

II. Construct the perpendicular lines from each vertex to the opposite side in ΔABC.

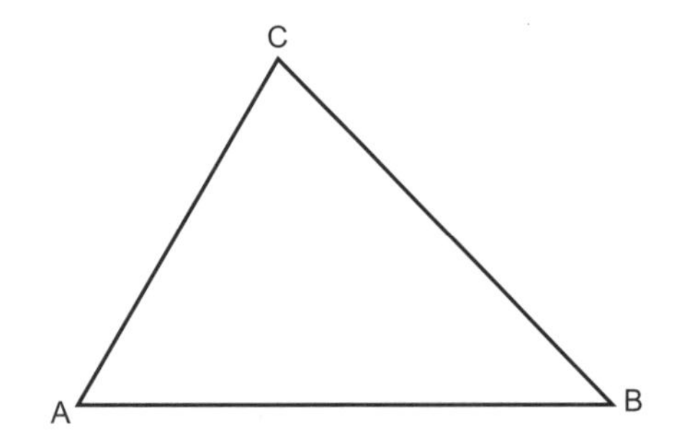

Name______________________________

8.G.A.5, HSG-CO.D.12

Constructing Congruent Angles

Given: ∠XYZ

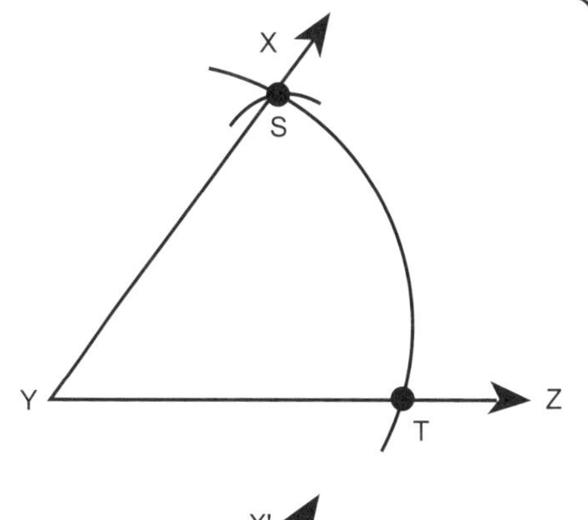

Construct an angle congruent to ∠XYZ.

1. Draw a ray. Label it Y′Z′.
2. Using Y as center, choose any radius and draw an arc that intersects $\overrightarrow{YX}$ and $\overrightarrow{YZ}$. Label points S and T.
3. Using Y′ as center and the same radius, draw an arc intersecting $\overrightarrow{Y'Z'}$. Label the point of intersection Q.
4. Using T as center, find radius equal to TS. Draw arc through point S.
5. Using Q as center, draw arc using radius equal to TS. Label point of intersection P.
6. Draw $\overrightarrow{Y'P}$.

∠XYZ ≅ ∠PY′Z′.

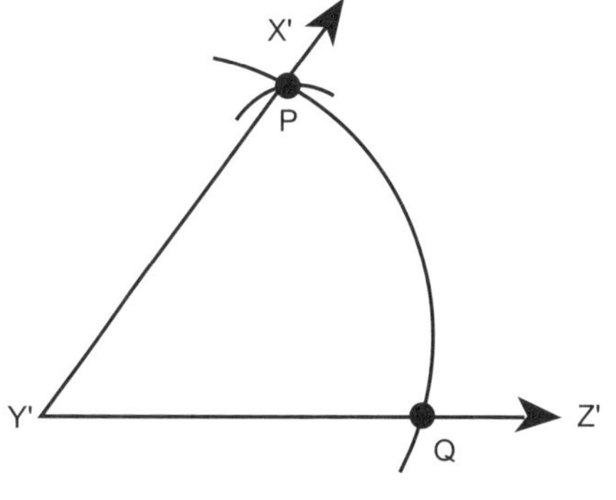

1. Construct a congruent angle to ∠ABC.

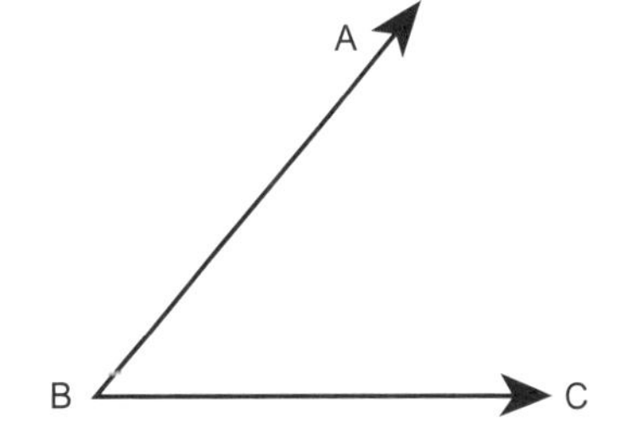

2. Construct a congruent angle to ∠XYZ.

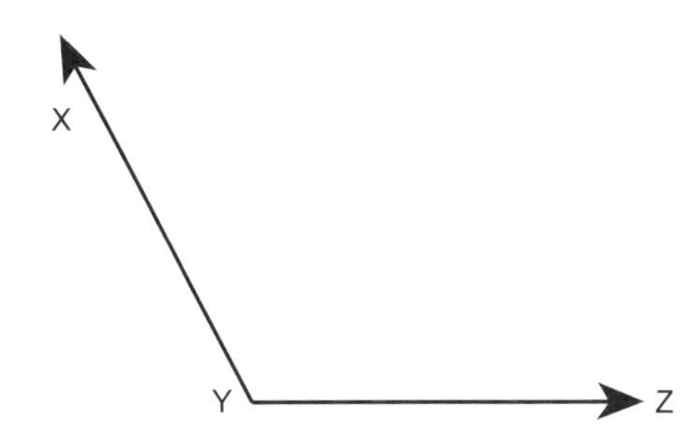

3. Construct ΔABC using ∠A and ∠B.

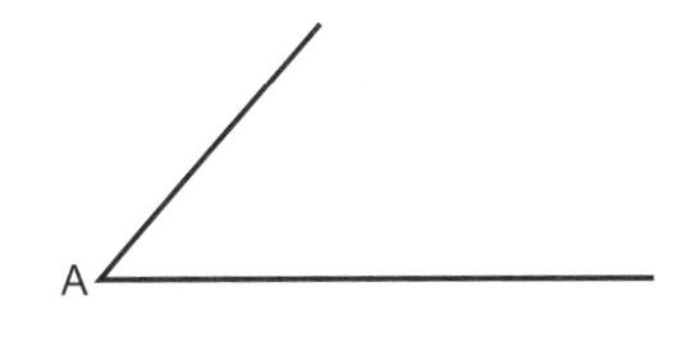

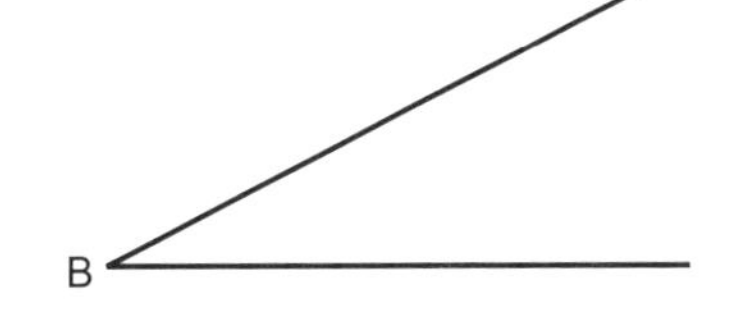

Name____________________

8.G.A.5, HSG-CO.D.12

Constructing Angle Bisectors

Given: ∠ABC

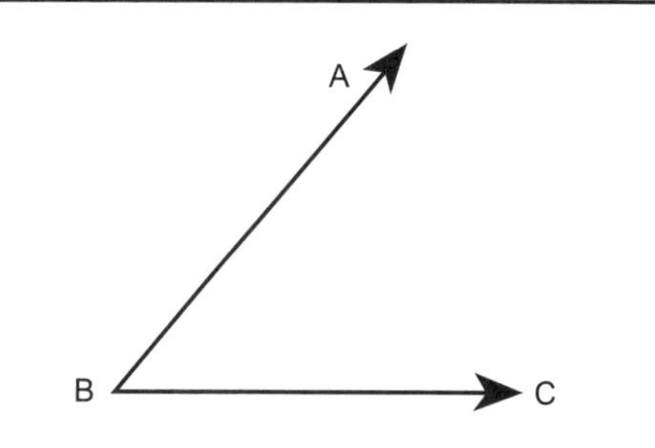

Construct an angle bisector.

1. Copy ∠ABC.
2. Using B′ as center, choose any radius, and draw an arc intersecting $\overrightarrow{B'A'}$ and $\overrightarrow{B'C'}$.
3. Using X as center, choose a radius greater than $\frac{1}{2}$ XY. Draw an arc in the interior of ∠A′B′C′. Repeat using Y as center and same radius. Label point of intersection Z.
4. Draw $\overrightarrow{B'Z'}$.

$\overrightarrow{B'Z}$ bisects ∠A′B′C′.

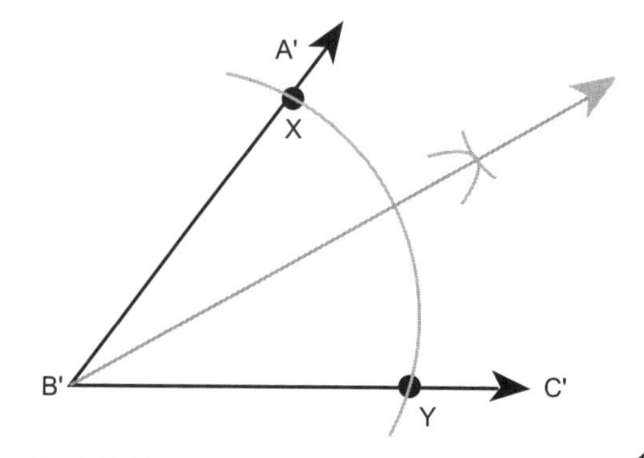

1. Bisect ∠XYZ.

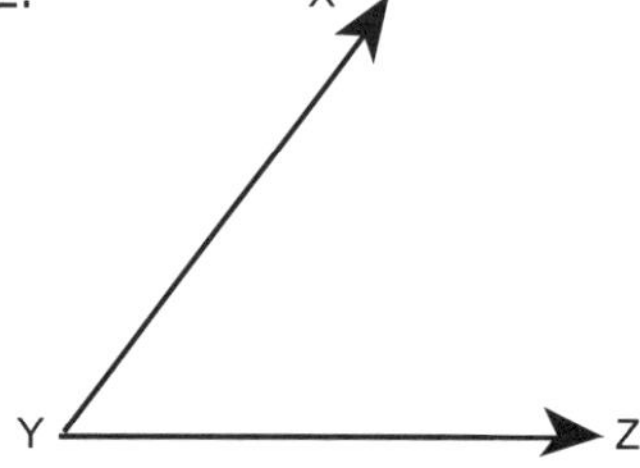

2. Bisect ∠ABC.

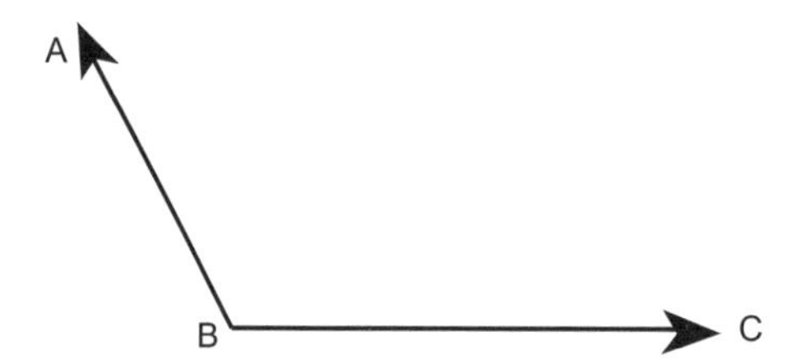

3. Construct a 45° angle.
   Hint: Construct perpendicular lines first.

4. Construct an equilateral Δ. Use AB as the length of each side.

5. What is the measurement of each angle in #4? __________

6. Construct a 30° angle.
   Hint: Use your equilateral Δ.

Name________________________________ 8.G.A.5, HSG-CO.D.12

Constructing Parallel Lines

Given: Point X and line ℓ

Construct a line parallel to ℓ containing x.

1. Copy line ℓ and point X.
2. Place points A and B anywhere on line ℓ. Draw $\overrightarrow{XA}$.
3. At point X, construct $\angle 1$ so that $\angle 1$ is congruent to $\angle XAB$. Let m be the line drawn for $\angle 1$.

$m \parallel \ell$

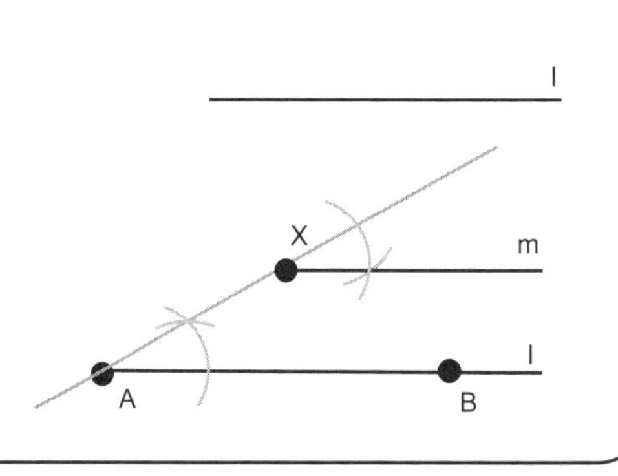

1. Construct a line parallel to $\overline{AB}$ at point K.

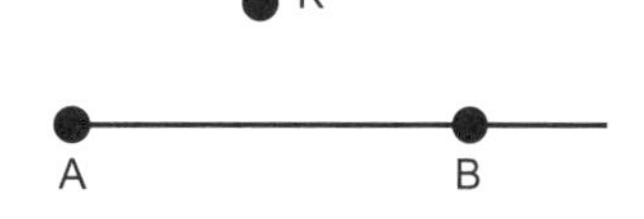

2. Construct a line parallel to $\overline{XY}$ at point A.

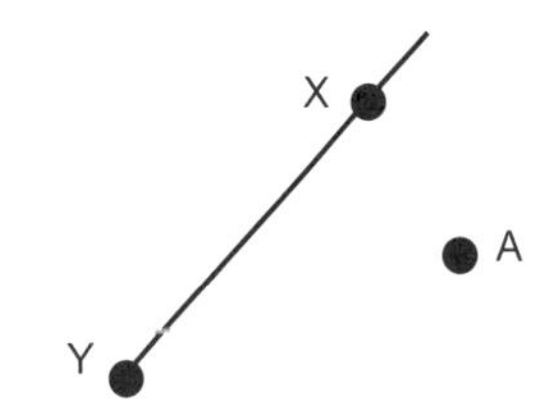

3. Construct a rectangle with length equal to $\overline{RT}$ and width equal to $\overline{RS}$.

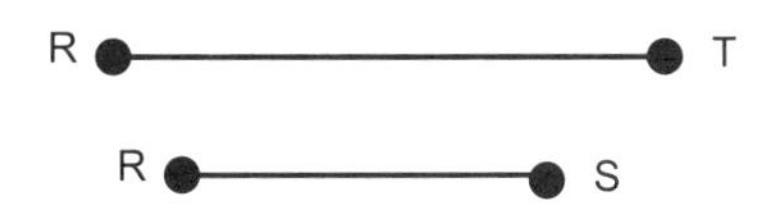

4. Construct a parallelogram with a 45° angle, length equal to AB, and any width.

Name______________________________ 8.G.A.5, HSG-CO.A.3

Properties of Parallelograms

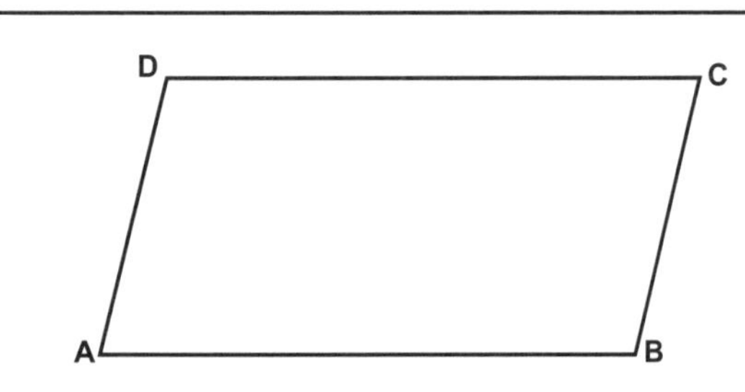

- Four sides.
- Both pairs of opposite sides are parallel.
- Both pairs of opposite sides are congruent.
- Both pairs of opposite angles are congruent.
- Diagonals bisect each other.

Complete the following ▱ABCD.

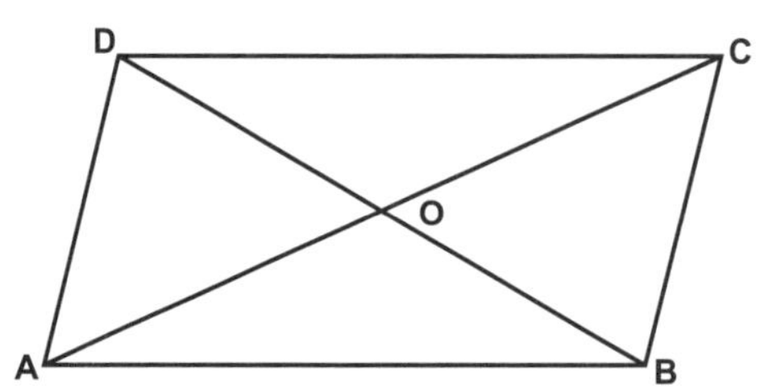

1. $\overline{AB} \parallel$ ______
2. $\overline{AB} \cong$ ______
3. $\angle A \cong$ ______
4. $\overline{OA} \cong$ ______
5. $\overline{OB} \cong$ ______
6. $\overline{AD} \cong$ ______

Find the missing values for each parallelogram.

7.

8.

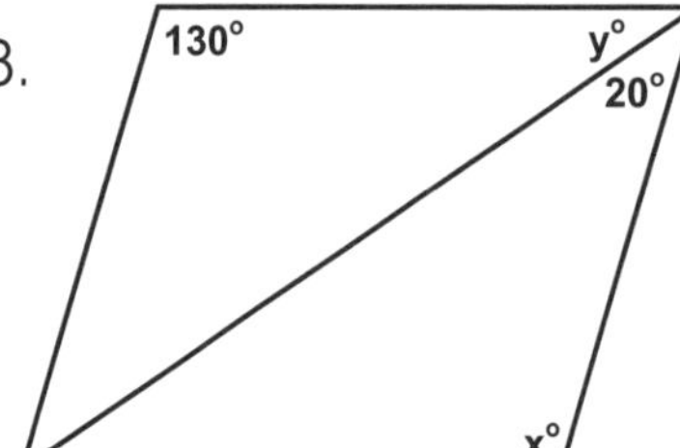

9.

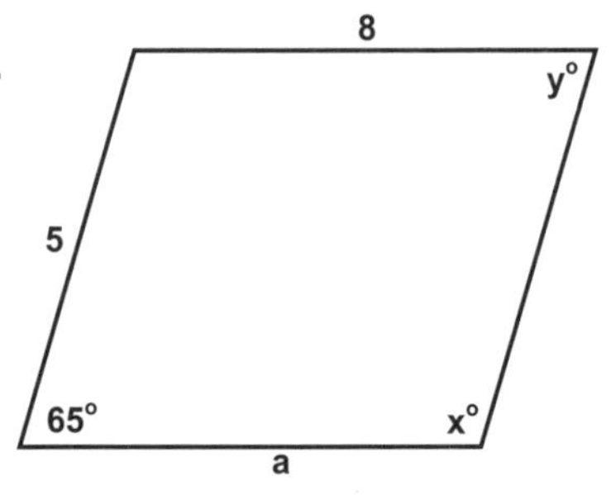

10.

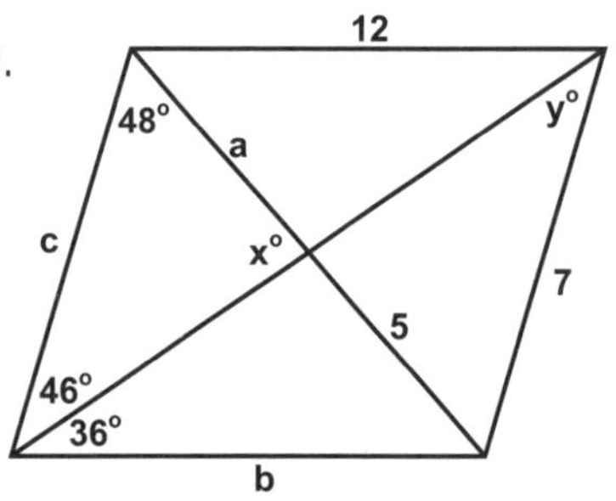

11.

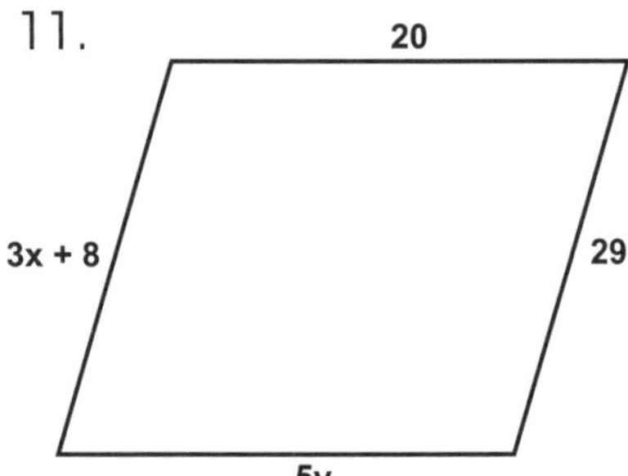

12.

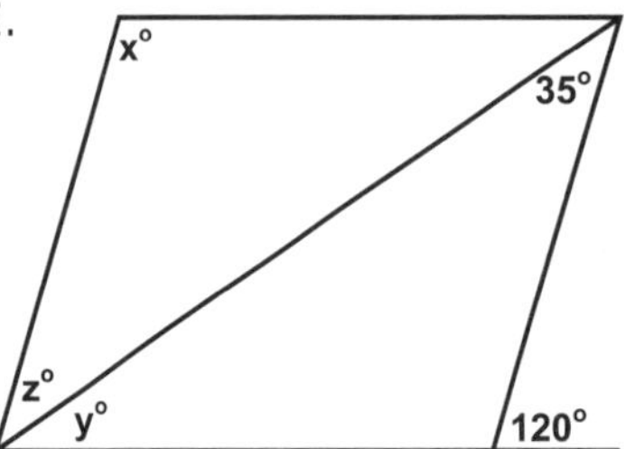

13.

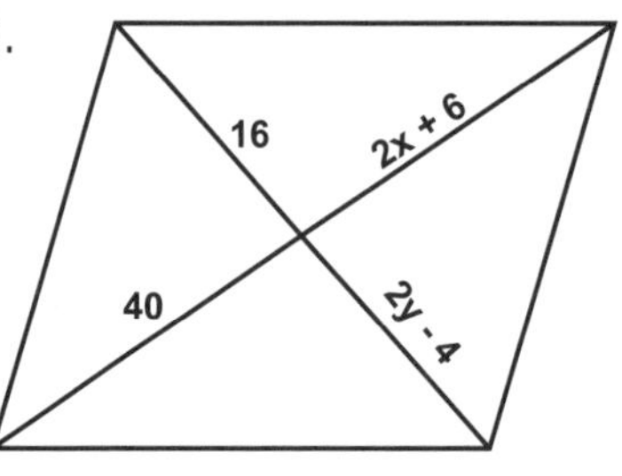

Name________________________________ 8.G.A.5, HSG-CO.C.11

Two-Column Proofs: Parallelograms

Five Ways to Prove that a Quadrilateral is a Parallelogram:

1. Show both pairs of opposite sides are parallel.
2. Show both pairs of opposite sides are congruent.
3. Show one pair of opposite sides are both congruent and parallel.
4. Show both pairs of opposite angles are congruent.
5. Show that diagonals bisect each other.

Given: ABCD is a ▱.
$\angle 1 \cong \angle 2$, $\overline{DF} \cong \overline{EB}$
Prove: EBFD is a ▱.

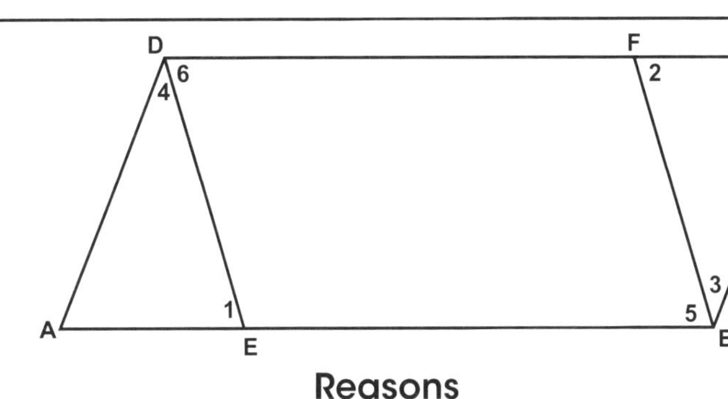

Statements	Reasons
1. ____________	1. Given
2. AD ≅ CB	2. ____________
3. $\angle A \cong \angle C$	3. ____________
4. ____________	4. AAS
5. $\overline{DE} \cong \overline{FB}$	5. ____________
6. EBFD is a ▱.	6. ____________

Given: $\overline{AD} \parallel \overline{CB}$
$\angle DCA \cong \angle BAC$
Prove: ABCD is a ▱.

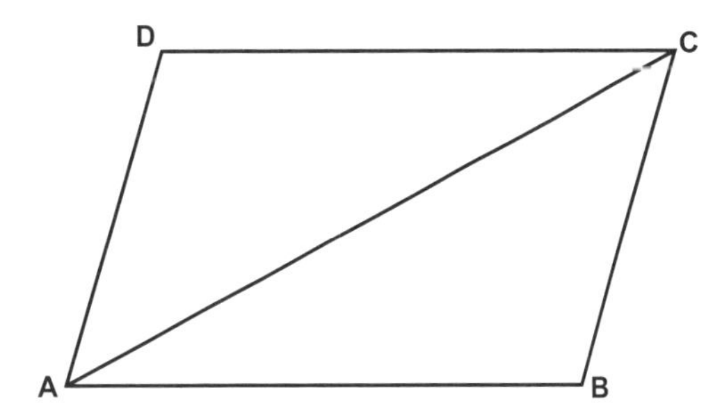

Statements	Reasons
1. ____________	1. Given
2. $\overline{AC} \cong \overline{AC}$	2. ____________
3. $\angle DAC \cong \angle BCA$	3. ____________
4. $\Delta DAC \cong \Delta BCA$	4. ____________
5. $\overline{DA} \cong \overline{BC}$	5. ____________
6. ABCD is a ▱.	6. ____________

Name________________________________

8.G.A.5, HSG-CO.C.11

More Two-Column Proofs: Parallelograms

Given: ABCE is a ▱.
FB ⊥ AD; DC ⊥ BC
Prove: FBCD is a ▱.

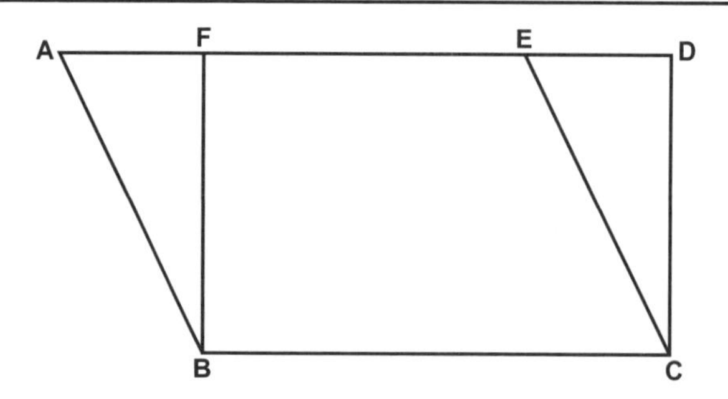

Statements	Reasons
1. ____________________	1. Given
2. $\overline{AD} \parallel \overline{BC}$	2. ____________________
3. $\overline{FB} \parallel \overline{DC}$	3. ____________________
4. FBCD is a ▱.	4. ____________________

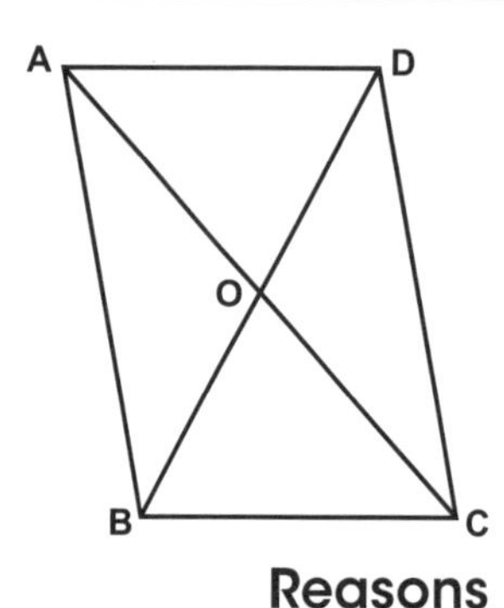

Given: $\overline{AB} \parallel \overline{DC}$
$\overline{OB} \cong \overline{OD}$
Prove: ABCD is a ▱.

Statements	Reasons
1. ____________________	1. Given
2. ∠AOB ≅ ∠COD	2. ____________________
3. ∠ABD ≅ ∠CDB	3. ____________________
4. ΔABO ≅ ΔCDO	4. ____________________
5. $\overline{AB} \cong \overline{CD}$	5. ____________________
6. ABCD is a ▱.	6. ____________________

On your own.

Given: $\overline{AD} \cong \overline{BC}$; ∠EBC ≅ ∠ECB
E is the midpoint of AD
∠1 ≅ ∠2
Prove: ABCD is a ▱.

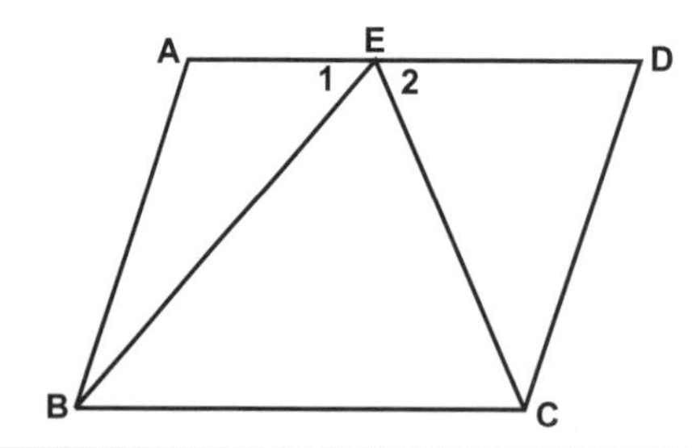

Name______________________________

8.G.A.5, HSG-CO.A.3

Special Parallelograms

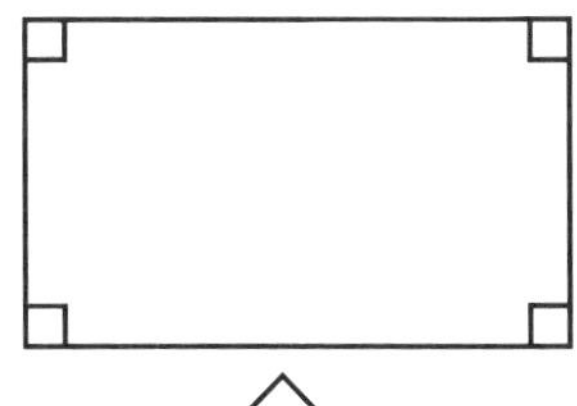

Rectangle
- Parallelogram with four right $\angle$s.
- Diagonals are congruent.

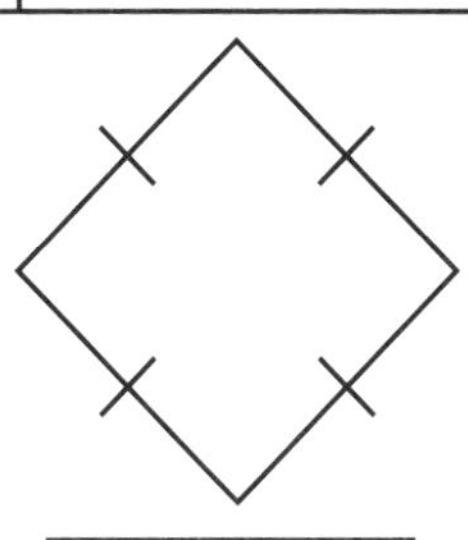

Rhombus
- Parallelogram with four congruent sides.
- Diagonals are perpendicular.
- Diagonals bisect the $\angle$s of the rhombus.

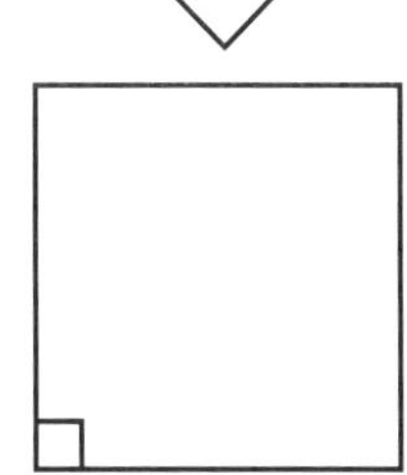

Square
- Parallelogram with four right $\angle$s and four congruent sides.
- Diagonals are perpendicular.
- Diagonals bisect the $\angle$s of the square.

In problems 1-8, list the letters of the quadrilaterals that the property holds true for:
a) Parallelogram b) Rectangle c) Rhombus d) Square

1. Diagonals bisect each other.
2. All $\angle$s are right $\angle$s.
3. All sides are congruent.
4. Opposite sides are congruent.
5. Opposite angles are congruent.
6. Diagonals are congruent.
7. Diagonals are perpendicular.
8. Opposite sides are parallel.

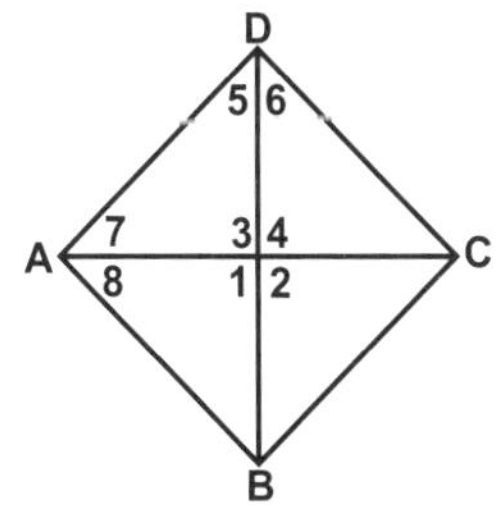

9. ABCD is a rhombus. If $m\angle 8 = 35$, find the measures of $\angle 1$, $\angle 2$, $\angle 3$, $\angle 4$, $\angle 5$, $\angle 6$, $\angle 7$.

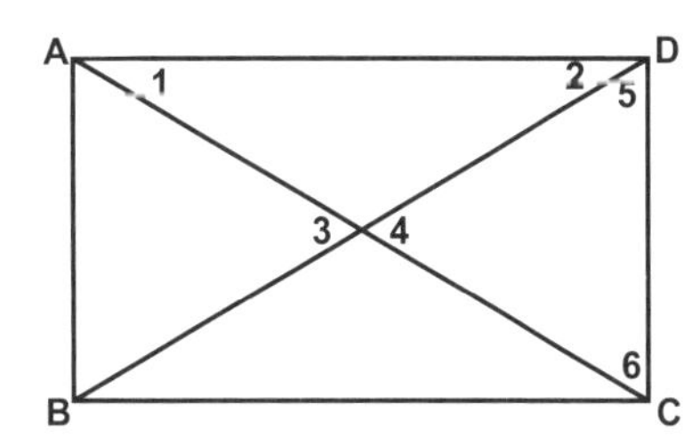

10. ABCD is a rectangle. If $m\angle 1 = 20$, find the measures of $\angle 2$, $\angle 3$, $\angle 4$, $\angle 5$, $\angle 6$.

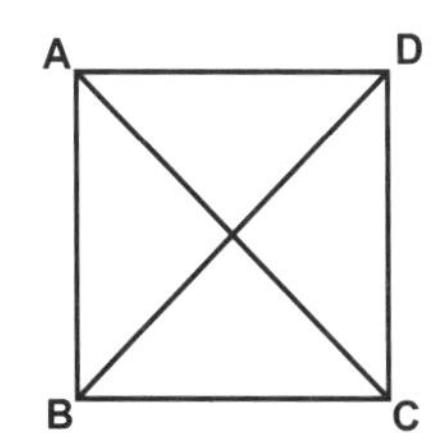

11. ABCD is a square. If $\overline{AC} = 16$ and $\overline{BD} = 2x + 4$, find x.

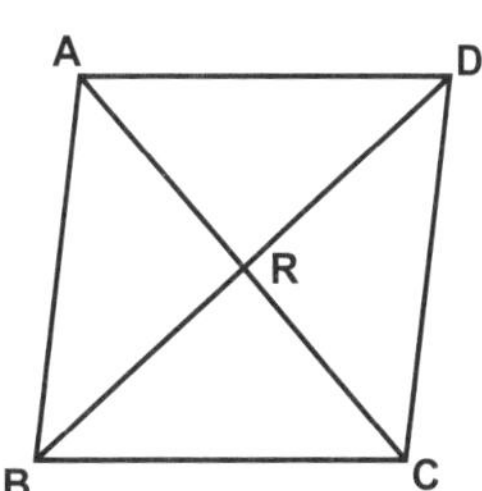

12. ABCD is a parallelogram. $\overline{AR} = 2x + 3$, $\overline{RC} = 35$, $\overline{BR} = 4y - 10$ $\overline{DR} = 90$. Find x and y.

Name________________________________ 8.G.A.5, HSG-CO.A.3

Trapezoids

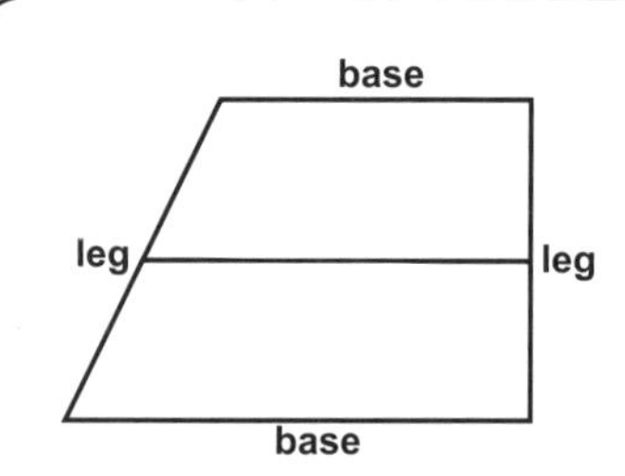

- Four sides
- Exactly one pair of parallel sides
- The median is parallel to the bases.
- The median has a length equal to the average of the bases.

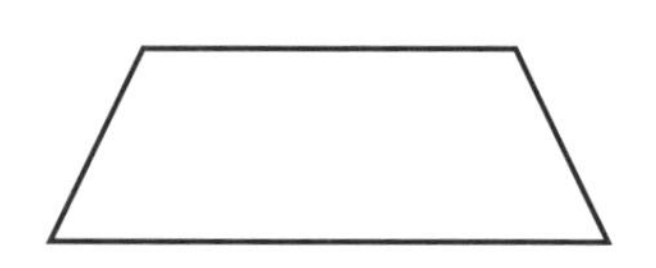

Isosceles Trapezoid

- A trapezoid with congruent legs
- Opposite ∠s are supplementary.

Find the missing values.

1. 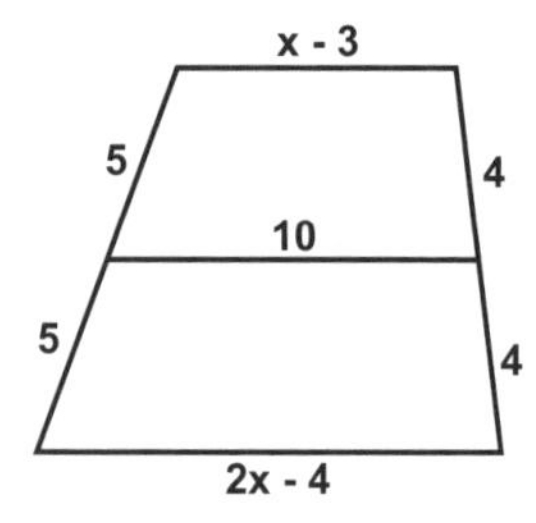

2. 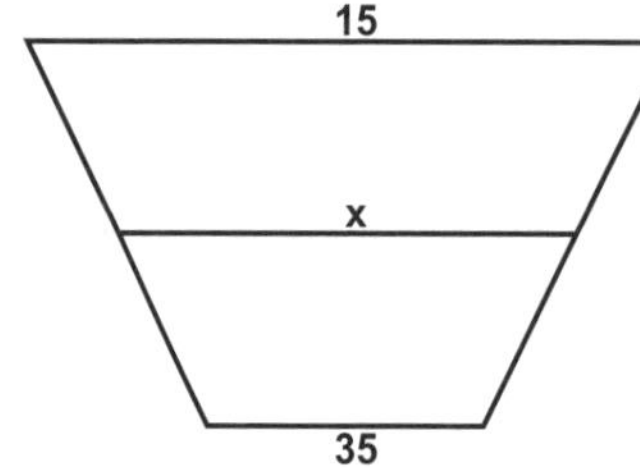

3. 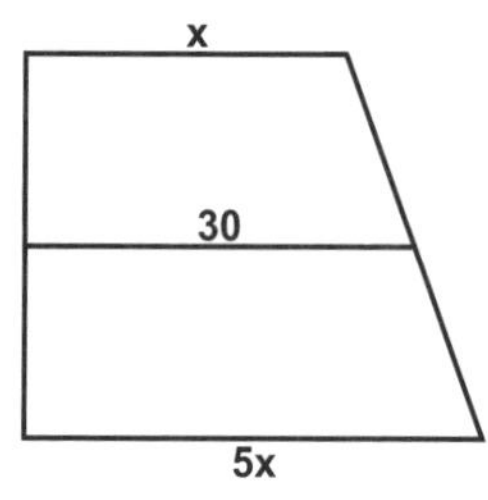

4. 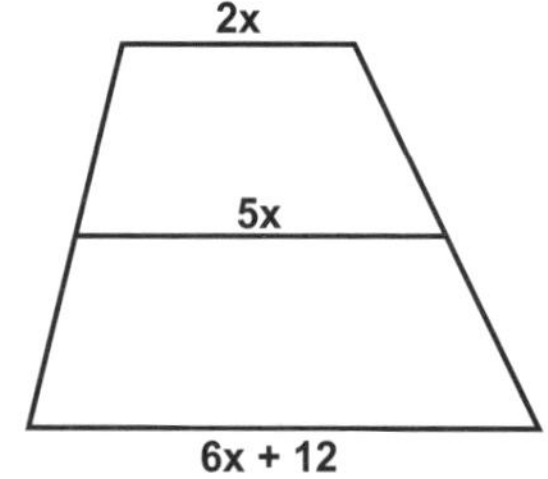

5. 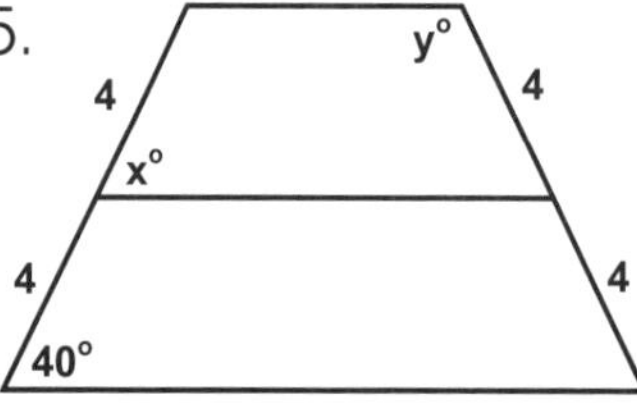

6. 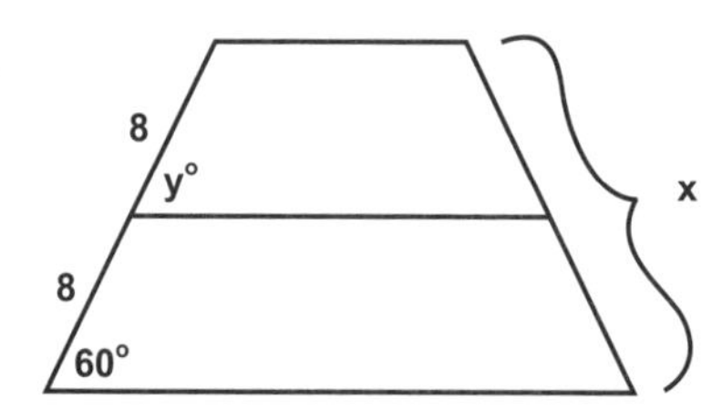

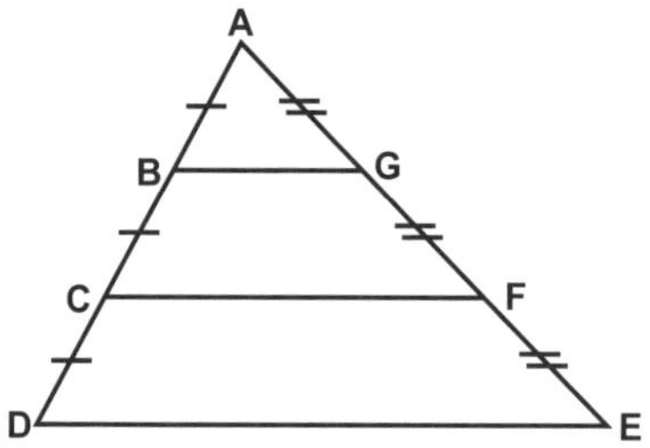

A, B, G, C, F, D, E

7. If BG = 8, then CF = ______ and DE = ______.
8. If CF = 10, then BG = ______ and DE = ______.
9. If DE = 15 and BG = 7, then CF = ______.
10. If CF = 2x + 4, BG = 2x + 1, and DE = 3x + 2, then x = ______.

Name________________________ 8.G.A.5, HSG-CO.C.9, HSG-CO.C.10

More Two-Column Proofs

Complete the following proofs.

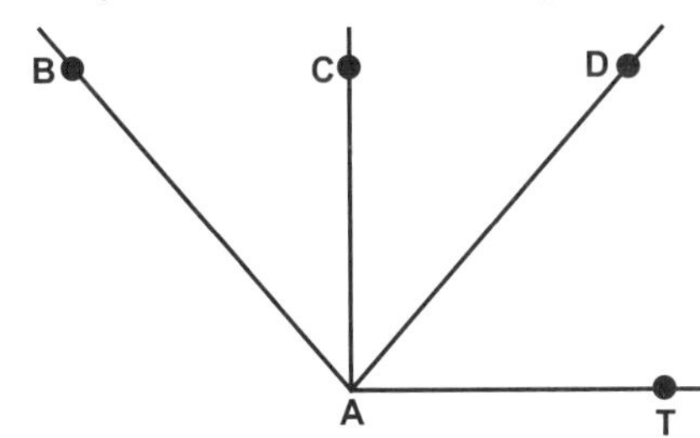

Given: $\overline{CA} \perp \overline{AT}$, m∠BAC + m∠DAT = 90°
Prove: ∠BAC ≅ ∠CAD

Statements	Reasons
1. ________	1. Given
2. ________	2. If ⊥ lines, then right ∠.
3. ________	3. AAP
4. m∠CAT = 90°	4. ________
5. ________	5. Substitution
6. ________	6. Substitution
7. m∠CAD = m∠BAC	7. ________
8. ∠BAC ≅ ∠CAD	8. ________

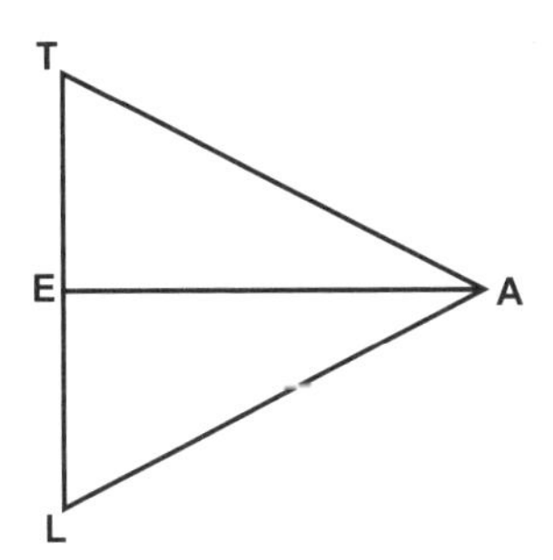

Given: ΔTAL is isosceles, $\overline{TE} = \overline{LE}$
Prove: ΔTEA ≅ ΔLEA

Statements	Reasons
1. ________	1. Given
2. $\overline{AT} \cong \overline{AL}$	2. ________
3. ________	3. If =, then ≅
4. $\overline{AE} \cong \overline{EA}$	4. ________
5. ΔTEA ≅ ΔLEA	5. ________

Name______________________________ (8.G.A.5, HSG-CO.C.9, HSG-CO.C.10)

More Practice with Proofs

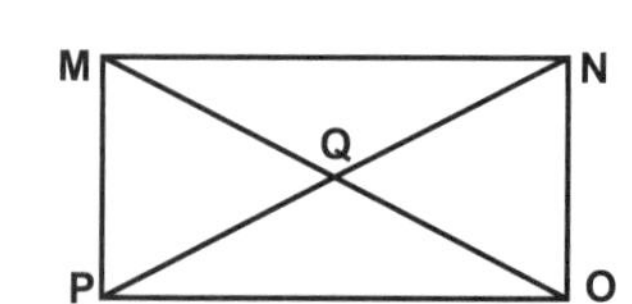

Given: MNOP is a rectangle; $\overline{MO}$ and $\overline{PN}$ are diagonals
Prove: $\triangle MQN \cong \triangle PQO$

Statements	Reasons
1. MNOP is a rectangle; $\overline{MO}$ and $\overline{PN}$ are diagonals	1. ____________________
2. $\overline{MN} \parallel \overline{PO}$, $\overline{MP} \parallel \overline{NO}$	2. ____________________
3. $\angle NMO \cong \angle POM$, $\angle MNP \cong \angle OPN$	3. ____________________
4. ____________________	4. In a rectangle, opposite sides are $\cong$.
5. $\triangle MQN \cong \triangle PQO$	5. ____________________

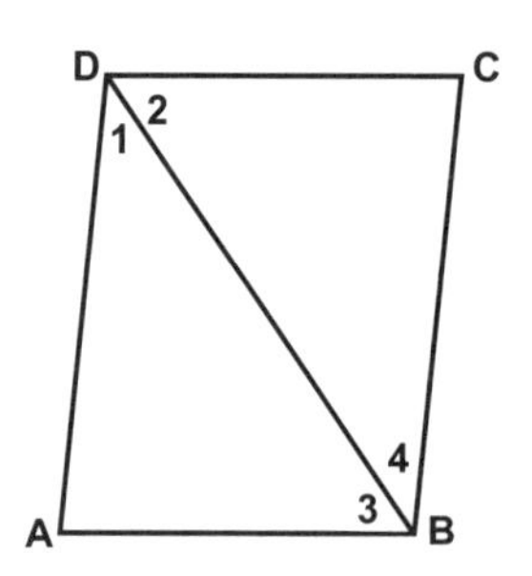

Given: $\overline{AB} \parallel \overline{DC}$, $\angle 1 \cong \angle 4$
Prove: $\triangle ABD \cong \triangle CDB$

Statements	Reasons
1. $\overline{AB} \parallel \overline{DC}$, $\angle 1 \cong \angle 4$	1. ____________________
2. $\angle 3 \cong \angle 2$	2. ____________________
3. $\overline{BD} \cong \overline{BD}$	3. ____________________
4. $\triangle ABD \cong \triangle CDB$	4. ____________________

Name________________________________ 8.G.A.5, HSG-CO.C.10

More Two-Column Proofs

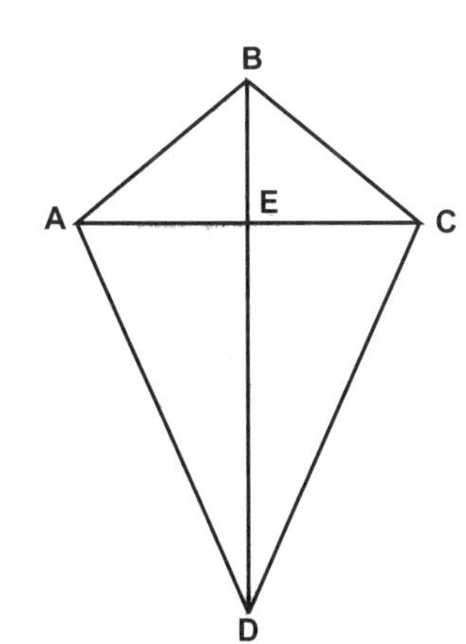

Complete the following proofs.

Given: $\overline{AB} \cong \overline{CB}$, BD bisects $\angle ABC$
Prove: $\overline{BD} \perp \overline{AC}$

Statements	Reasons
1. $\overline{AB} \cong \overline{CB}$, $\overline{BD}$ bisects $\angle ABC$	1. ______________________
2. $\angle ABE \cong \angle CBE$	2. ______________________
3. $\overline{BE} \cong \overline{BE}$	3. ______________________
4. $\Delta ABE \cong \Delta CBE$	4. ______________________
5. ______________________	5. CPCTC
6. ______________________	6. If 2 $\angle$s form a linear pair, they are supplementary.
7. $\angle BEA$ and $\angle BEC$ are right $\angle$s.	7. ______________________
8. $\overline{BD} \perp \overline{AC}$	8. ______________________

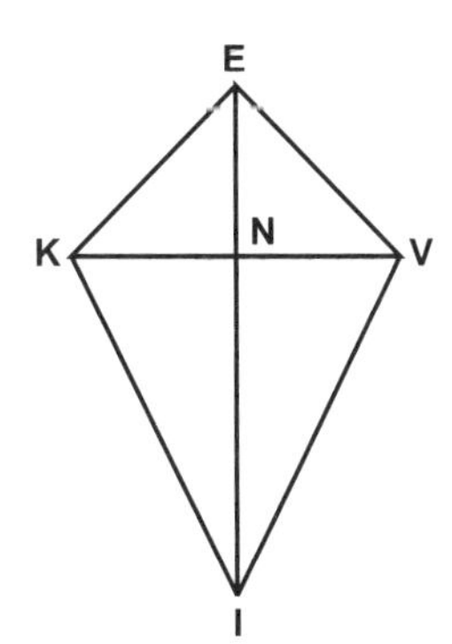

Given: $\overline{EI}$ bisects $\angle KEV$, $\overline{EI}$ bisects $\angle KIV$
Prove: $\overline{KE} \cong \overline{VE}$, $\overline{KI} \cong \overline{VI}$

Statements	Reasons
1. ______________________	1. Given
2. ______________________	2. If bisected, then two $\cong$ angles.
3. $\overline{EI} \cong \overline{EI}$	3. ______________________
4. $\Delta KEI \cong \Delta VEI$	4. ______________________
5. $\overline{KE} \cong \overline{EV}$, $\overline{KI} \cong \overline{VI}$	5. ______________________

Name____ 8.G.A.5, HSG-CO.C.10, HSG-CO.C.11

More Two-Column Proofs

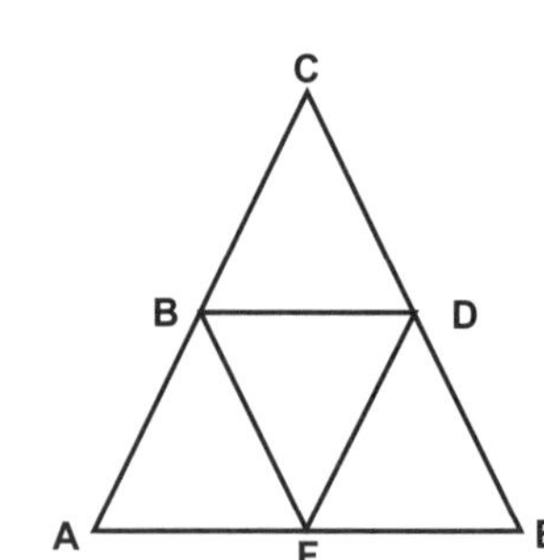

Given: B is the midpoint of $\overline{AC}$, D is the midpoint of $\overline{CE}$,
F is the midpoint of $\overline{AE}$

Prove: $\Delta CBD \cong \Delta BAF \cong \Delta DFE \cong \Delta FDB$

Statements	Reasons
1. ______________________	1. Given
2. $BF = \frac{1}{2}CE$, $BD = \frac{1}{2}AE$, $FD = \frac{1}{2}AC$	2. ______________________
3. ______________________	3. If midpoint, then two ≅ segments.
4. AB = BC, AF = FE, CD = DE	4. ______________________
5. ______________________	5. Definition of Between
6. AB +AB = AC, CD + CD = CE, AF + AF = AE	6. ______________________
7. ______________________	7. Combining Similar Terms
8. $AB = \frac{1}{2}AC$, $CD = \frac{1}{2}CE$, $AF = \frac{1}{2}AE$	8. ______________________
9. ______________________	9. Substitution
10. $\overline{AB} \cong \overline{FD}$, $\overline{CD} \cong \overline{BF}$, $\overline{AF} \cong \overline{BD}$	10. ______________________
11. ______________________	11. Substitution
12. $\Delta CBD \cong \Delta BAF \cong \Delta DFE \cong \Delta FDB$	12. ______________________

How are the four small triangles and the one large triangle related?

Name________________________________

8.G.A.5, HSG-CO.C.9

More Two-Column Proofs

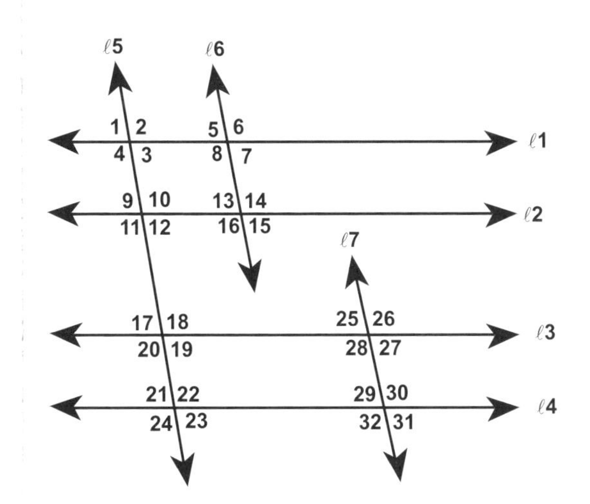

Given: $\angle 1 \cong \angle 7$, $\angle 1 \cong \angle 15$, $\angle 17 \cong \angle 27$, $\angle 17 \cong \angle 31$, $\angle 9 \cong \angle 17$

Prove: $\ell_1 \parallel \ell_4$ and $\ell_5 \parallel \ell_7$

Statements	Reasons
1. ______________________	1. Given
2. $\ell_5 \parallel \ell_6$	2. ______________________
3. $\angle 7 \cong \angle 15$	3. ______________________
4. ______________________	4. If corresponding $\angle$s are $\cong$, then $\parallel$ lines.
5. ______________________	5. If corresponding $\angle$s are $\cong$, then $\parallel$ lines.
6. ______________________	6. Substitution
7. $\ell_3 \parallel \ell_4$	7. ______________________
8. $\ell_2 \parallel \ell_3$	8. ______________________
9. $\ell_1 \parallel \ell_3$	9. ______________________
10. $\ell_1 \parallel \ell_4$ and $\ell_5 \parallel \ell_7$	10. ______________________

Assume $\ell_1 \parallel \ell_2 \parallel \ell_3 \parallel \ell_4$ and $\ell_5 \parallel \ell_6 \parallel \ell_7$.

1. If $m\angle 9 = 70°$, then $m\angle 15 =$ __________.
2. If $m\angle 25 = 73°$, then $m\angle 22 =$ __________.
3. If $m\angle 18 = 120°$, then $m\angle 2 =$ __________.
4. If $m\angle 32 = 80°$, then $m\angle 12 =$ __________.
5. If $m\angle 3 = 84°$, then $m\angle 17 =$ __________.
6. If $m\angle 11 = 75°$, then $m\angle 23 =$ __________.
7. If $m\angle 28 = 100°$, then $m\angle 13 =$ __________.
8. If $m\angle 30 = 101°$, then $m\angle 19 =$ __________.

Name______________________________

8.G.A.5, HSG-SRT.A.2, HSG-SRT.A.3

Ways to Prove Triangles Similar

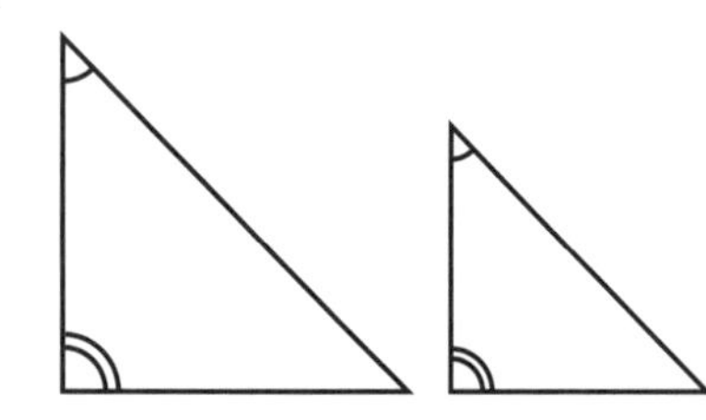

AA (angle, angle) or **AAA(angle, angle, angle)** = 2 or 3 angles of one triangle congruent to the corresponding angles of another triangle ⇒ ~ Δs (corresponding sides are proportional).

SAS (side, angle, side) = two sides of one triangle are proportional to the corresponding sides of another triangle and the included angles are congruent ⇒ ~ Δs.

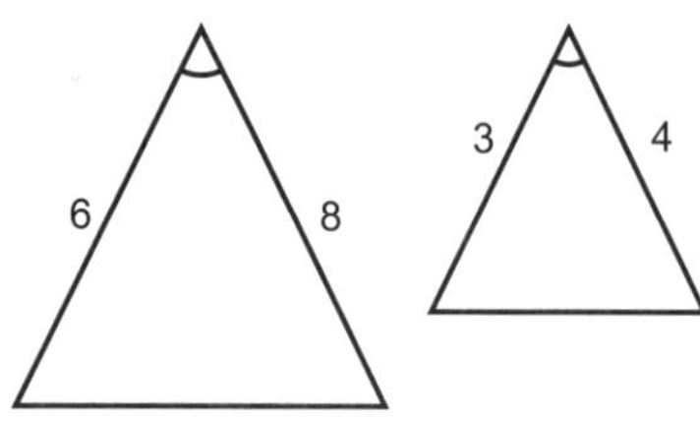

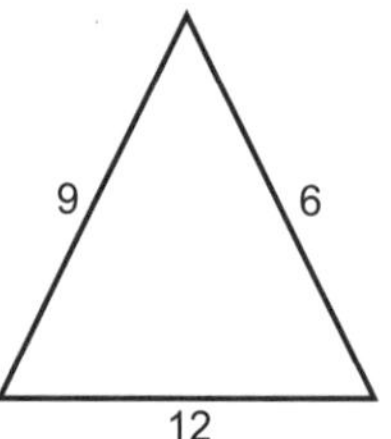

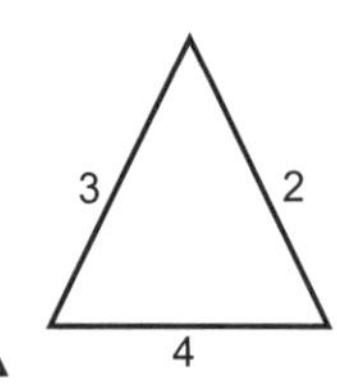

SSS (side, side, side) = three sides of one triangle are proportional to the corresponding sides of another triangle ⇒ ~ Δs.

Identify which property will prove these triangles similar.

1. 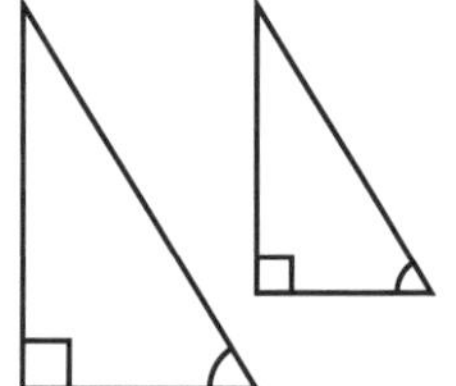

2. 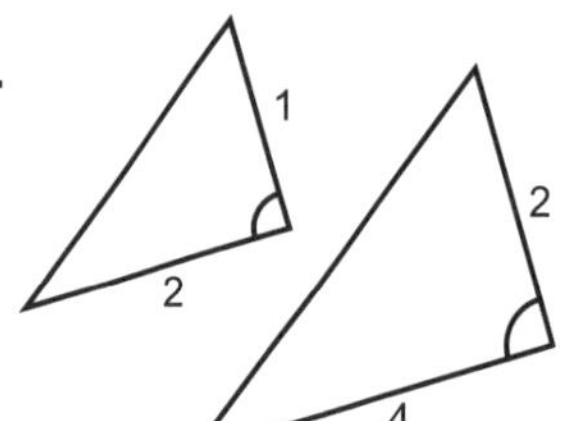

3. 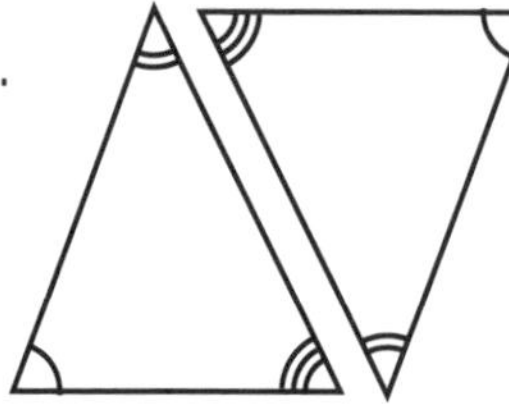

4. 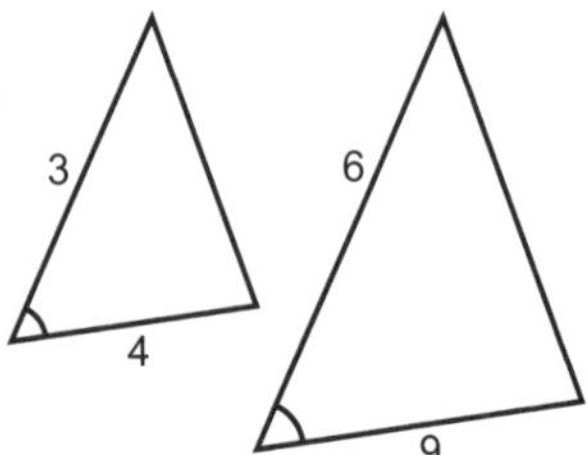

5. 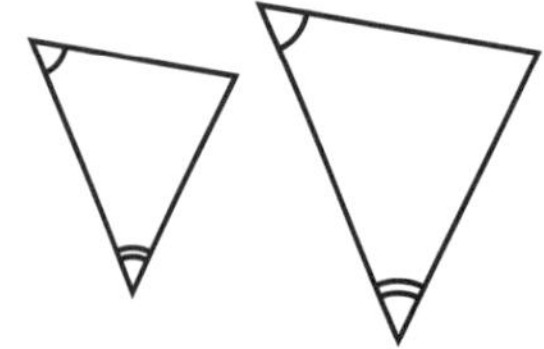

6. 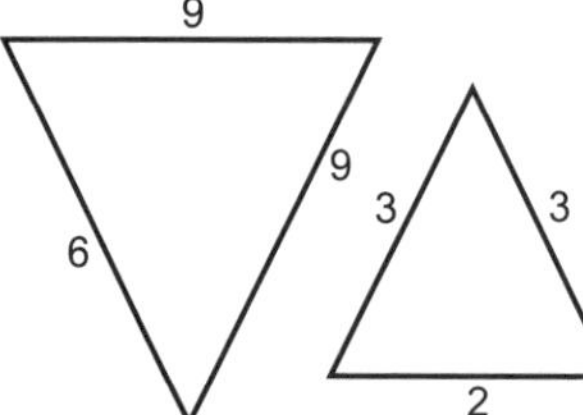

7. 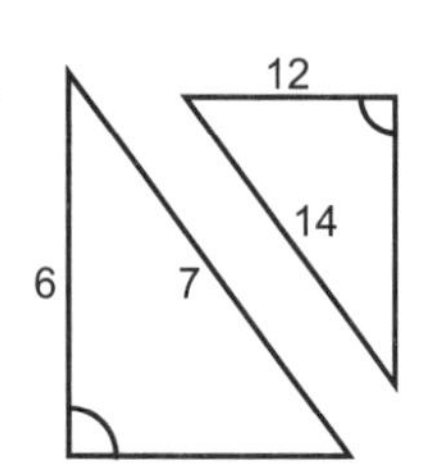

8. 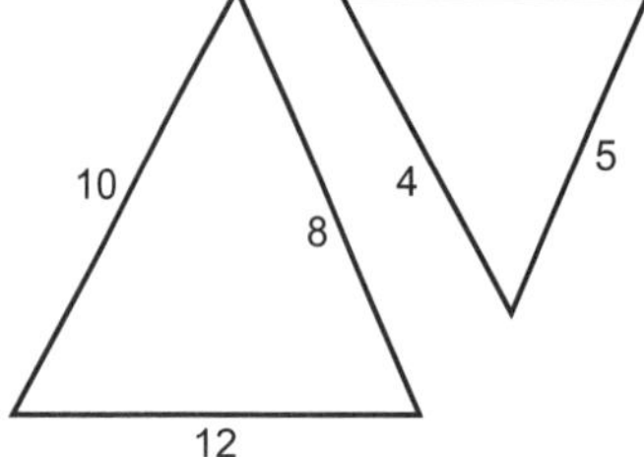

9. 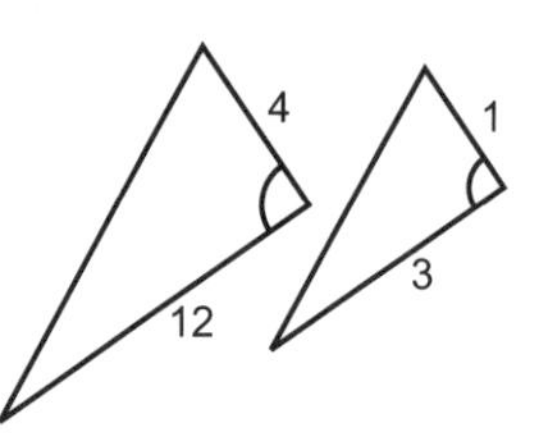

10. 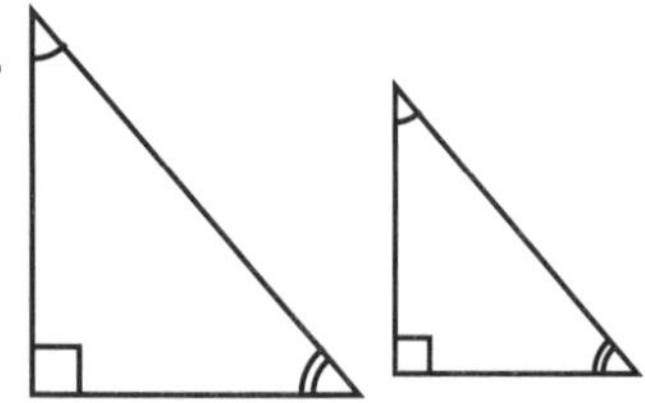

11. 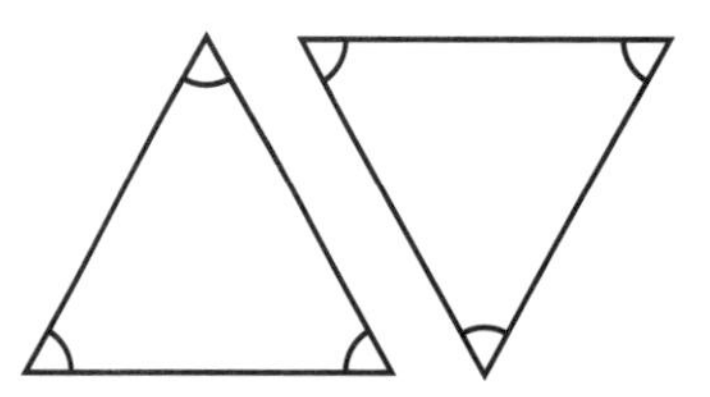

12. 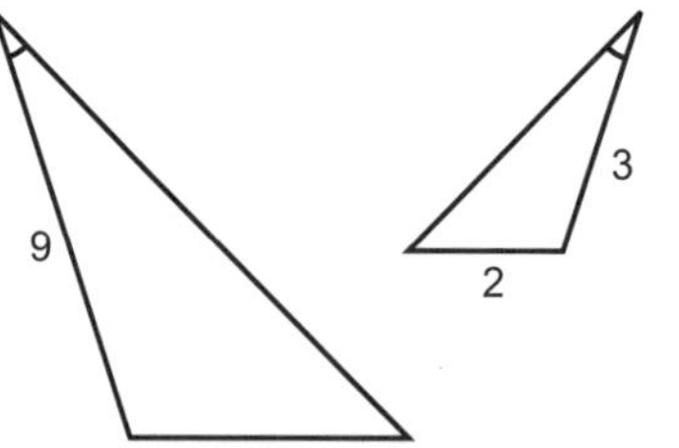

Name________________________________ 8.G.A.5, HSG-SRT.A.2, HSG-SRT.A.3

Practice with Similar Triangles

State whether each pair of triangles is similar by AA~, SSS~ or SAS~. If none of these apply, write N.

1. 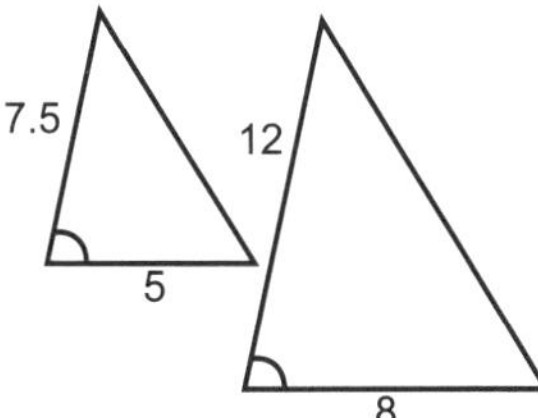

2. 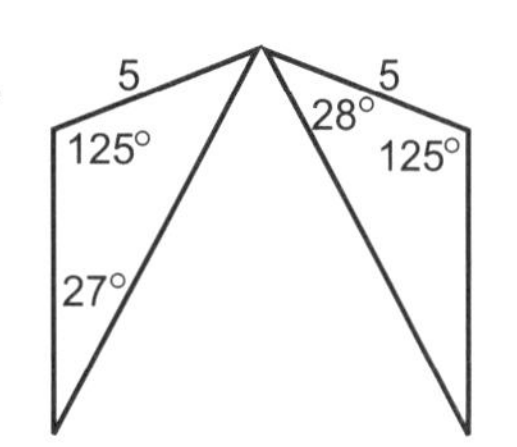

3. 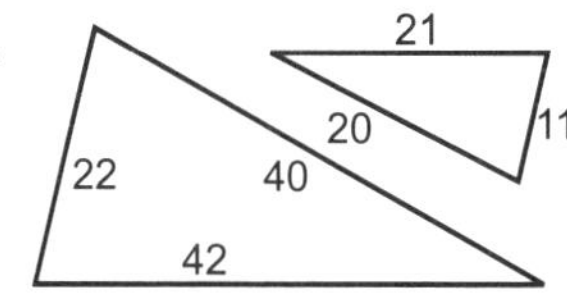

4. 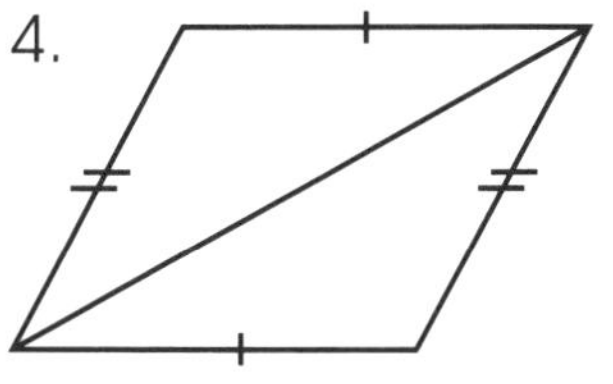

5. 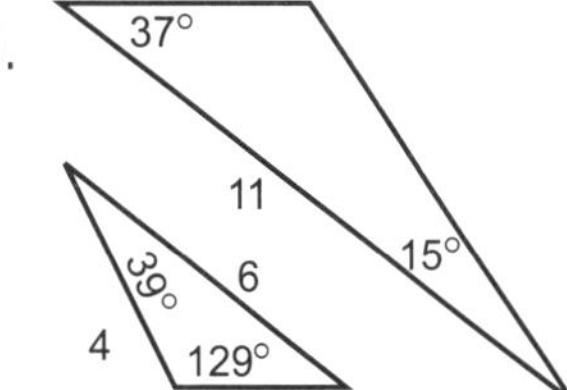

6. 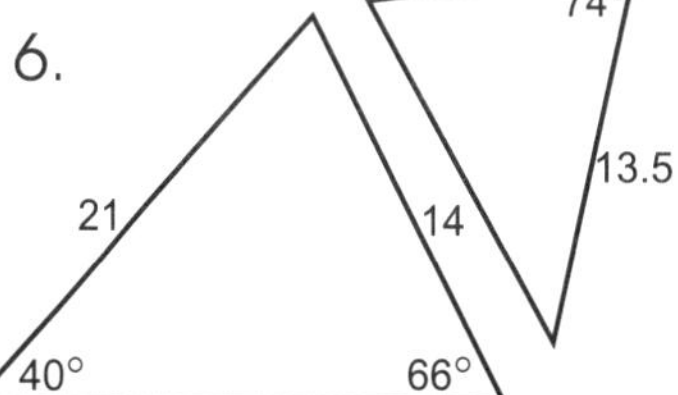

7. 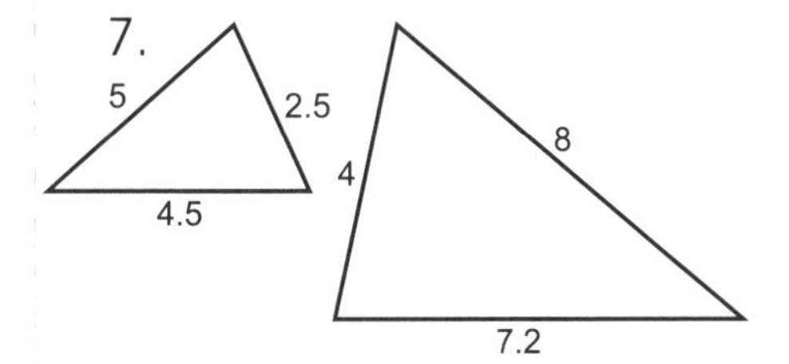

8. 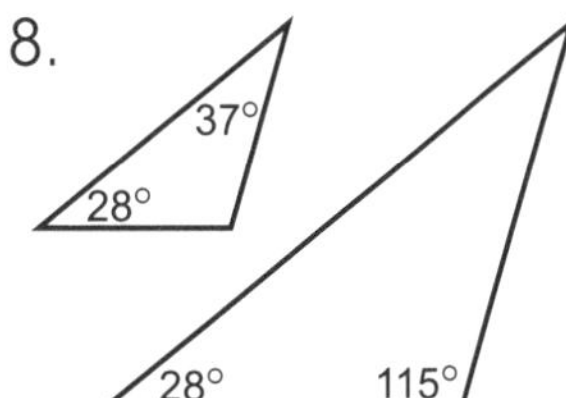

9. 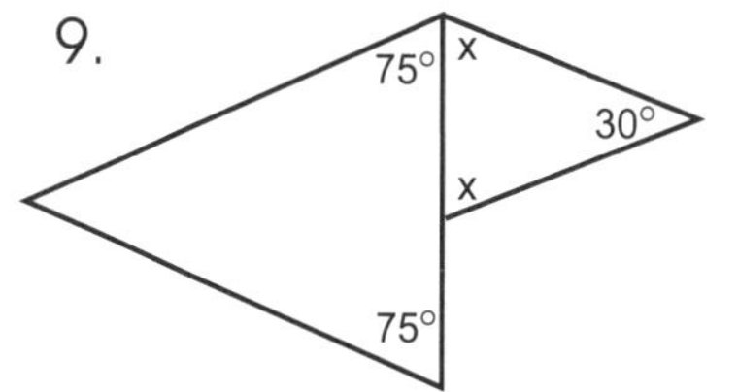

10. 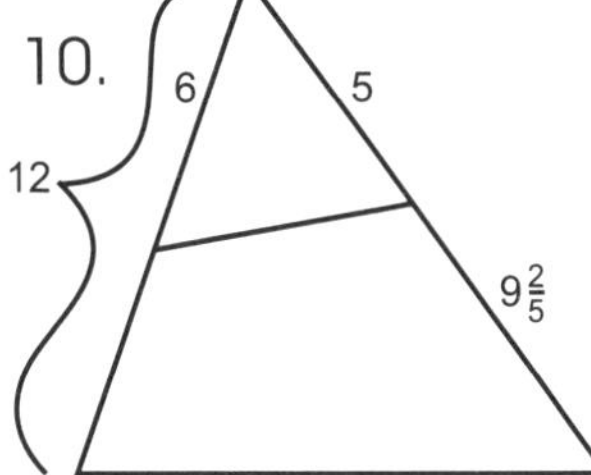

11. 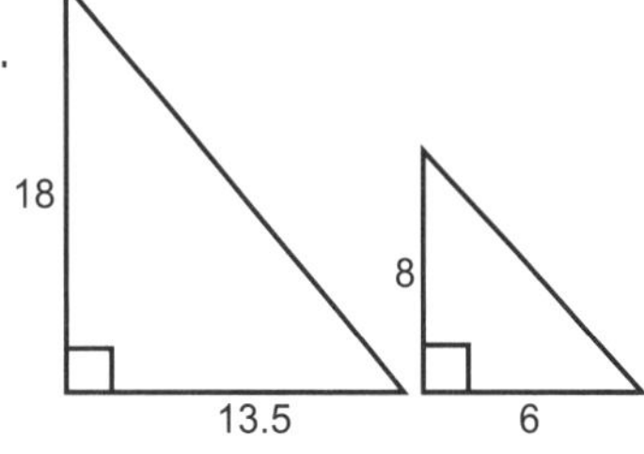

12. 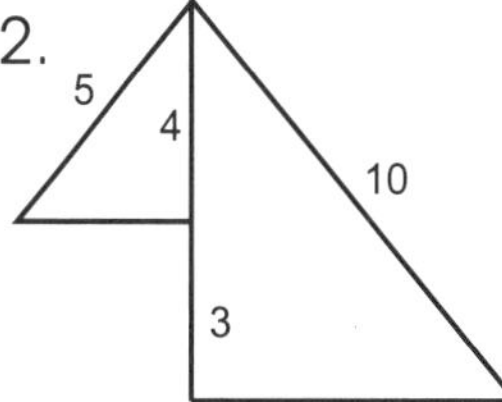

13. 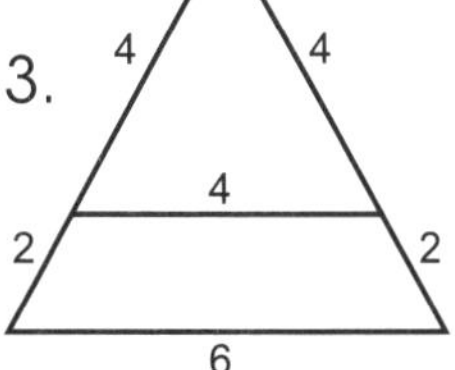

14. 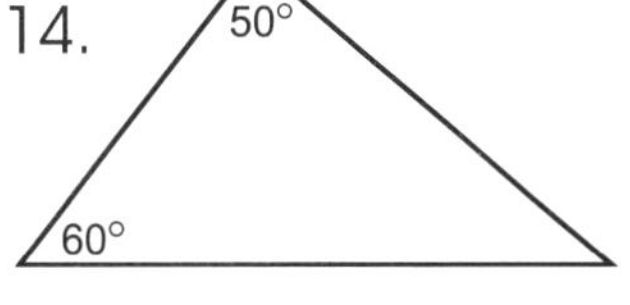

Name________________________________ HSG-SRT.A.2

Identifying Similar Triangles

SAS–Side, Angle, Side
2 pairs of corresponding sides in proportion and the included angles ≅

SSS–Side, Side, Side
3 pairs of corresponding sides in proportion

Tell whether the triangles are similar by SAS~ or by SSS~.

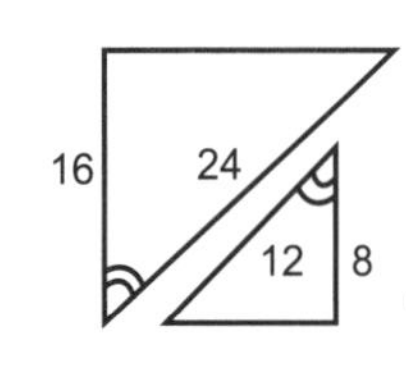

$\frac{12}{24} = \frac{8}{16}$

$192 = 192$

SAS~

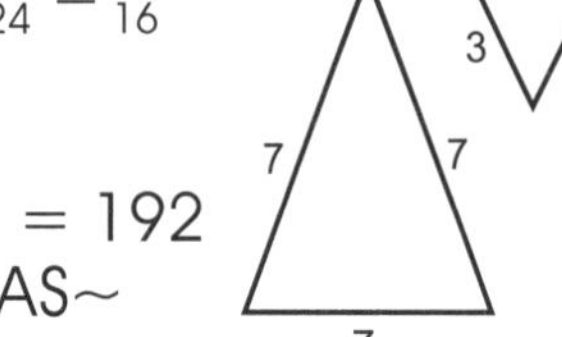

$\frac{3}{7} = \frac{3}{7} = \frac{3}{7}$

SSS~

1.

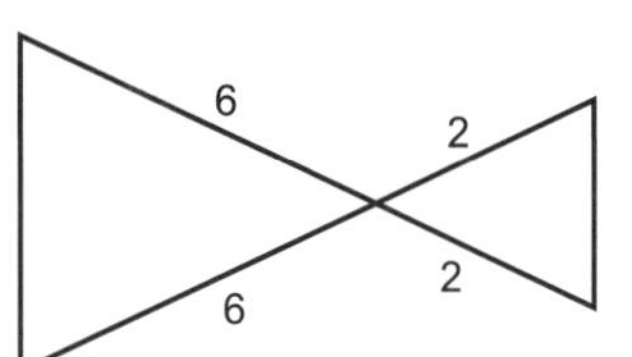

2.

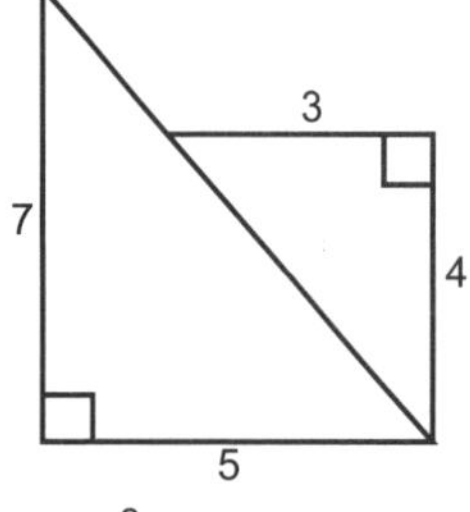

3.

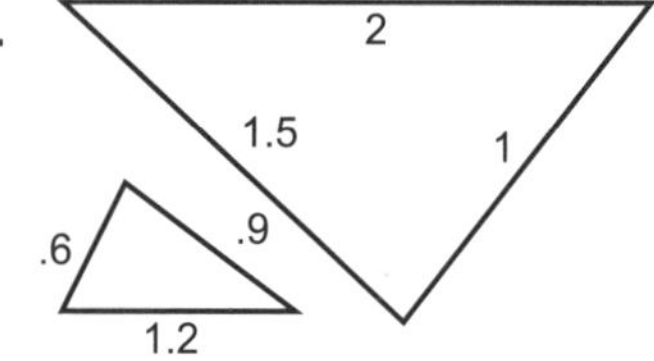

4.

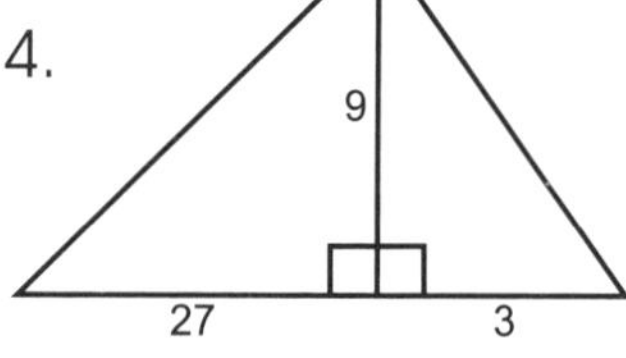

5.

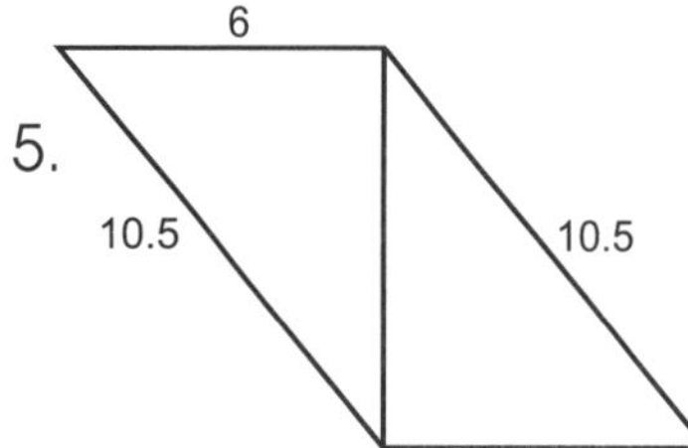

6.

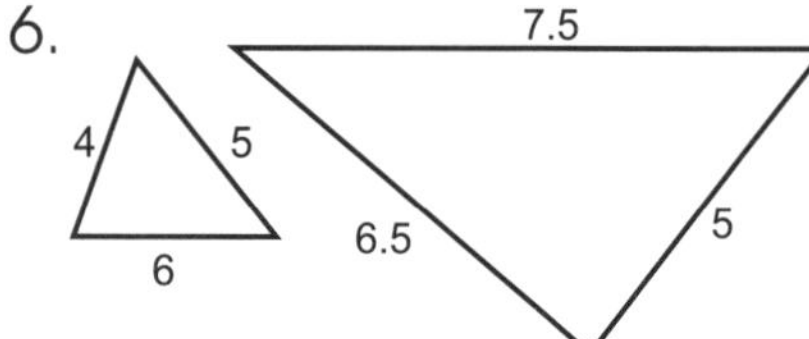

7.

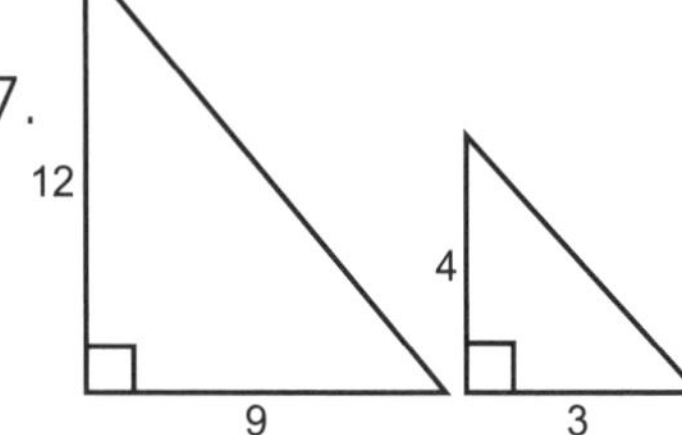

8.

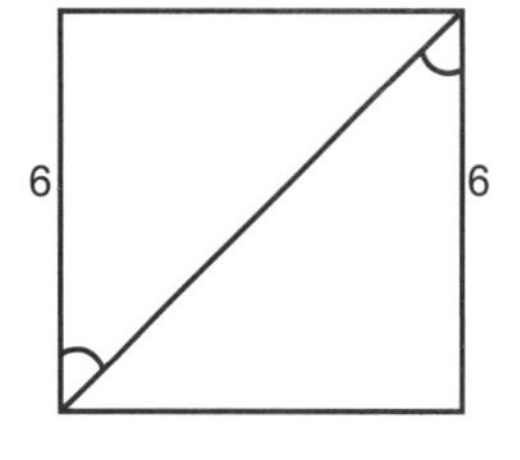

9.

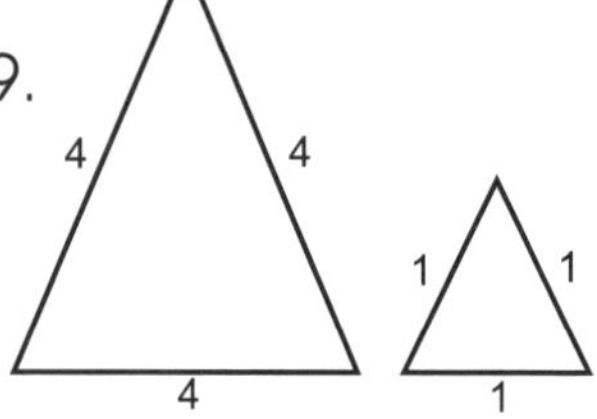

10.

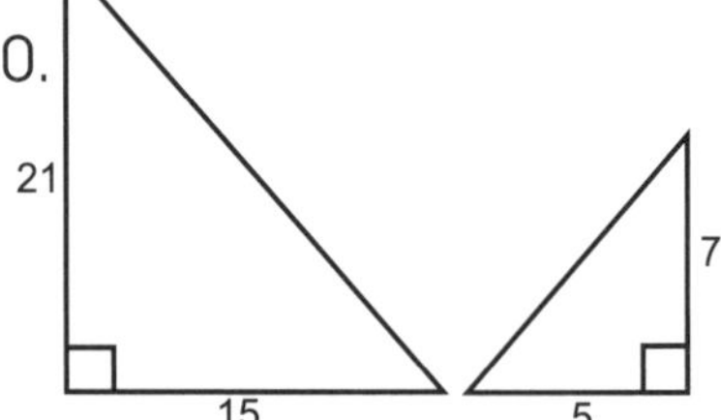

11.

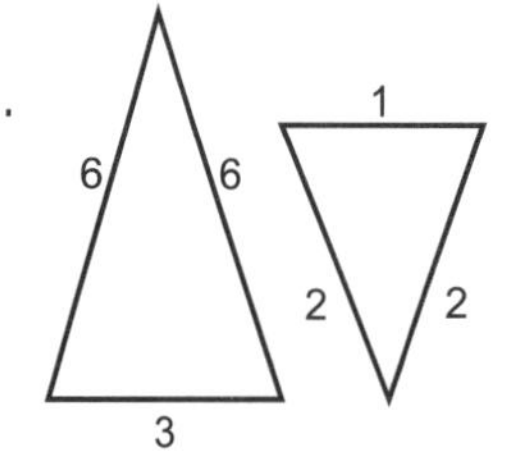

12.

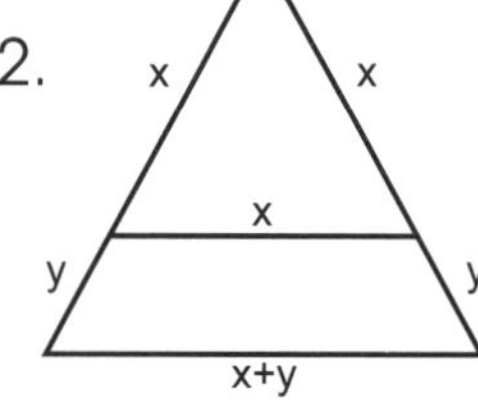

Name_______________________________ 8.G.A.5, HSG-SRT.A.2, HSG-SRT.A.3

Working with Similar Triangles

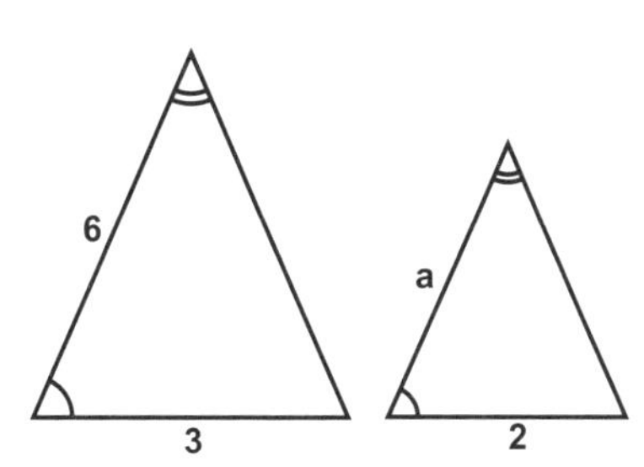

$$\frac{2}{3} = \frac{a}{6}$$

$3a = 2\,(6)$

$3a = 12$

$a = 4$

Find the labeled lengths.

1.

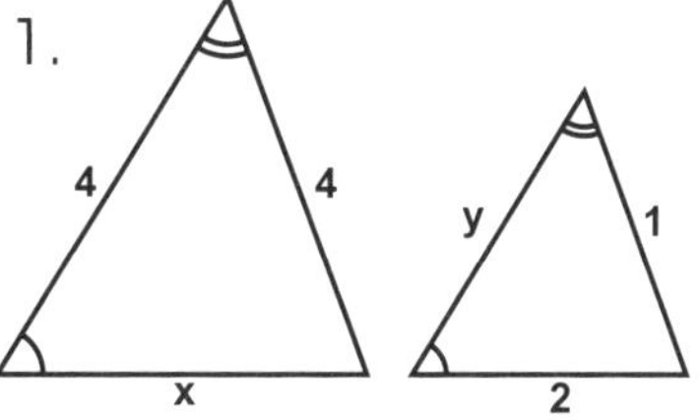

2.

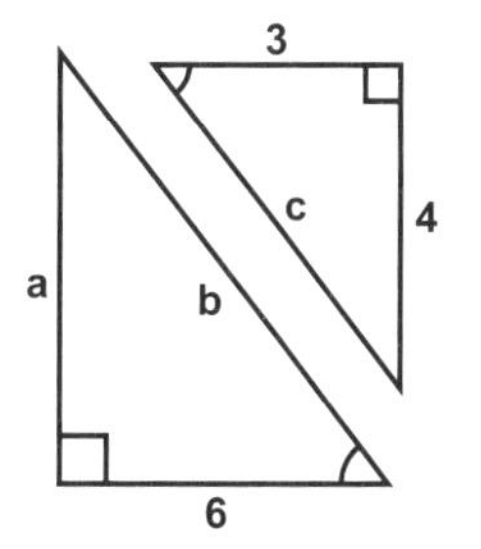

3.

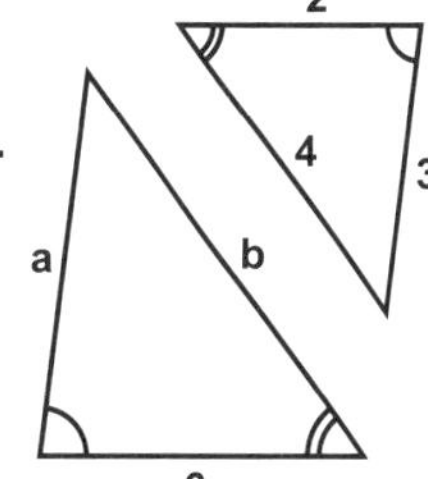

4.

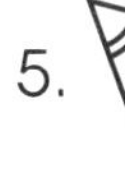

5.

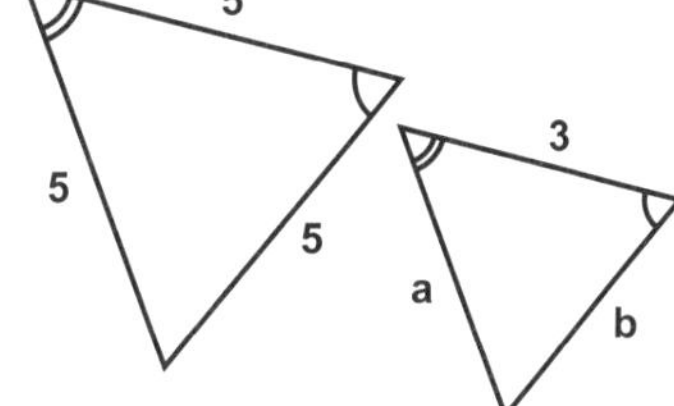

6.

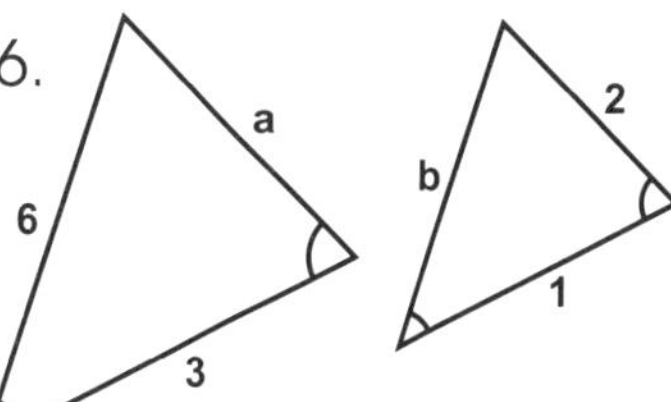

7.

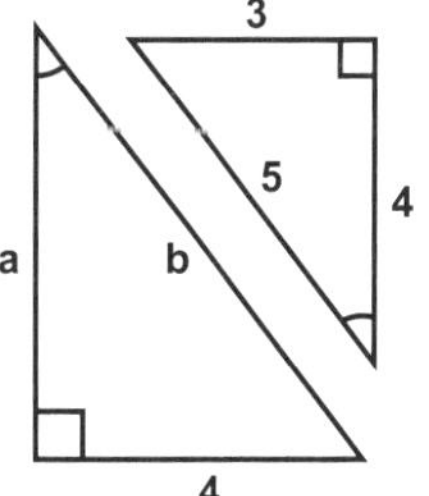

8.

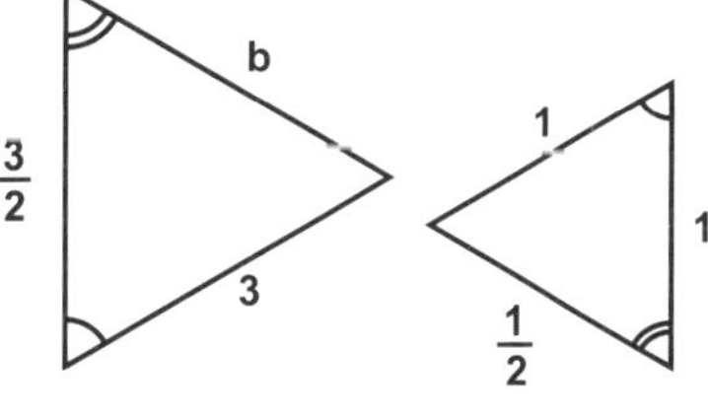

9.

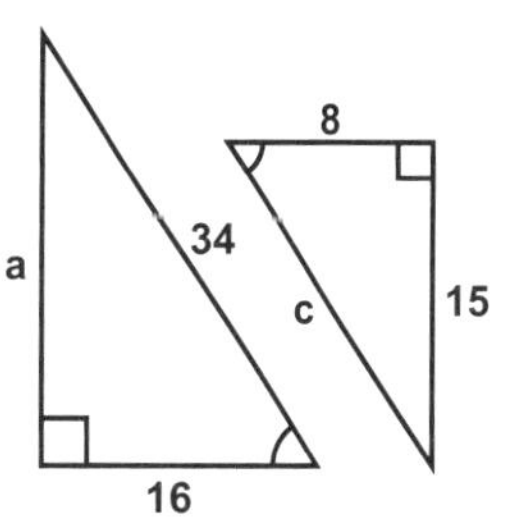

10.

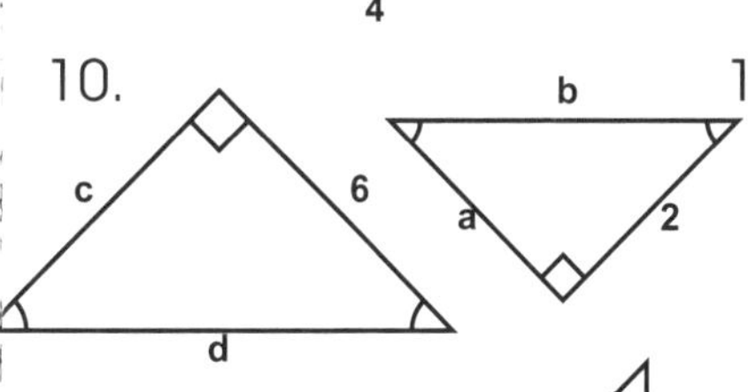

11.

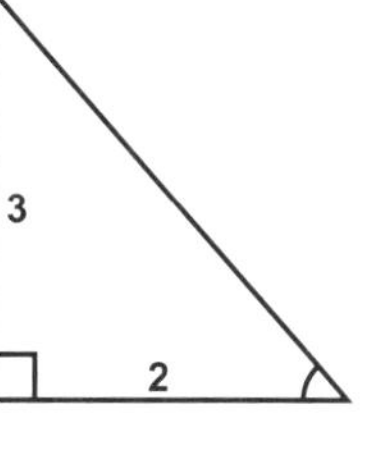

12.

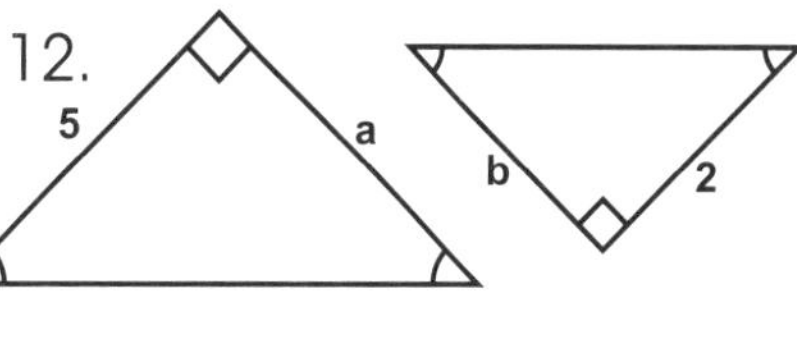

13.

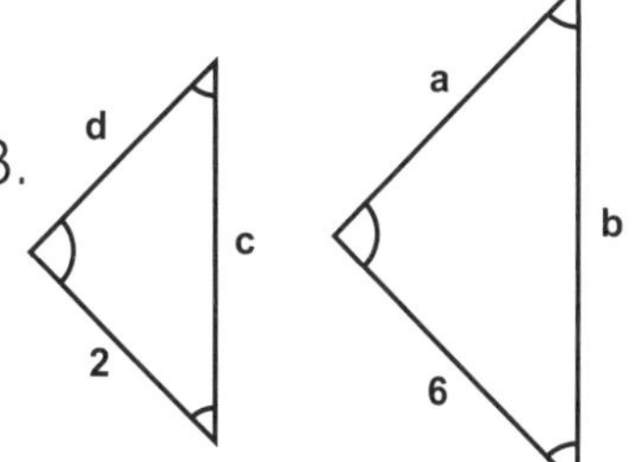

14.

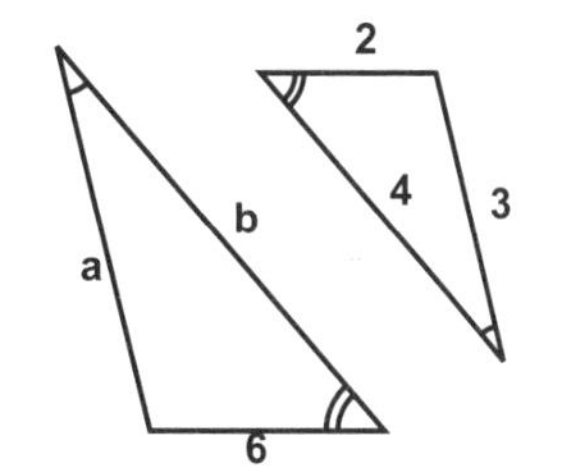

15.

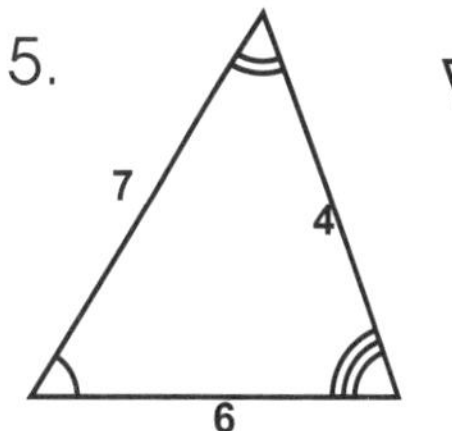

Name________________________________ 8.G.A.5, HSG-SRT.A.2, HSG-SRT.A.3

More Similar Triangles

Find the area of the following triangles. **Hint:** $A = \frac{1}{2}bh$

1. 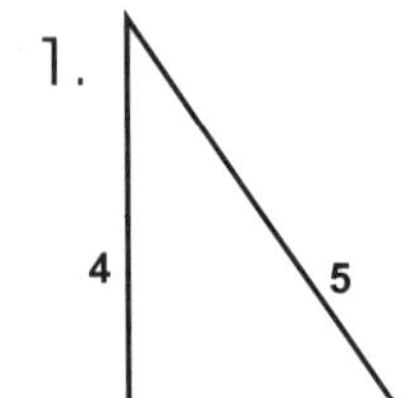

2. 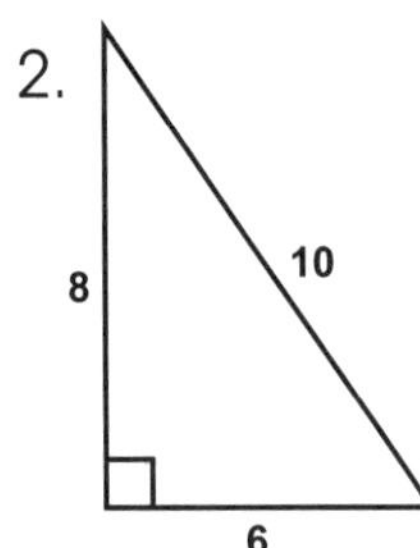

3. 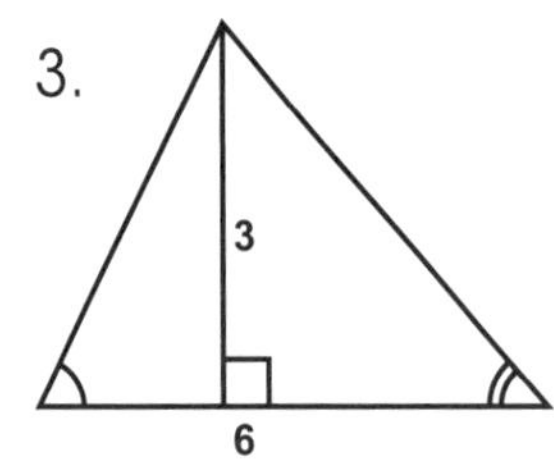

4. 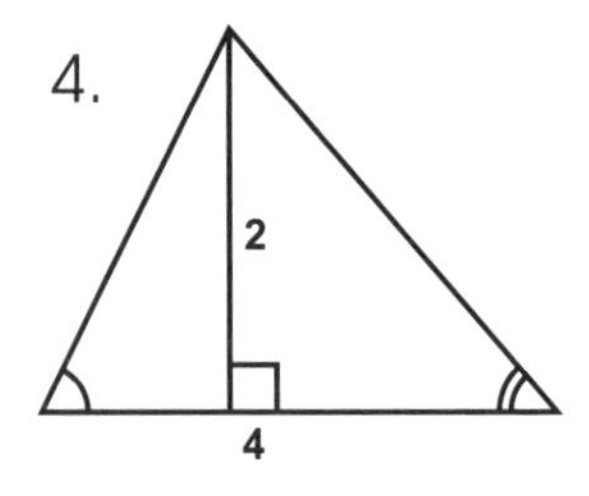

5. What is the ratio of the sides in #1 and #2?____________________
6. What is the ratio of the sides in #3 and #4?____________________
7. What is the ratio of the areas in #1 and #2? ____________________
8. What is the ratio of the areas in #3 and #4?____________________
9. What can you conclude about this?____________________

Find the ratio of the areas in the following sets of similar triangles with corresponding sides labeled.

10. 

11. 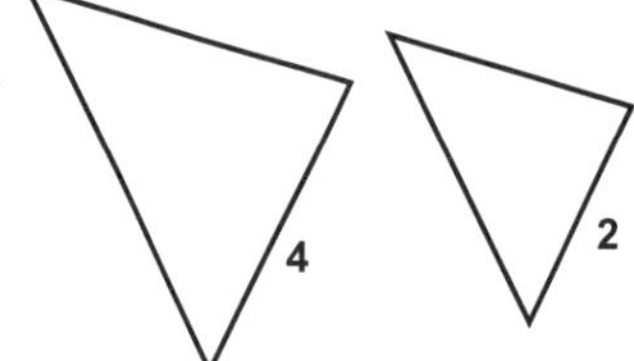

12. 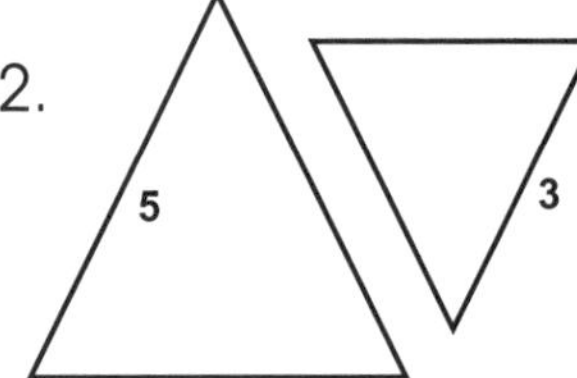

13. 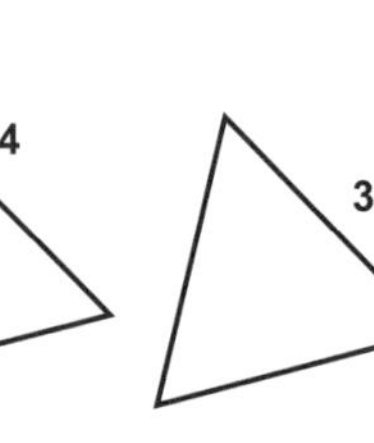

14. 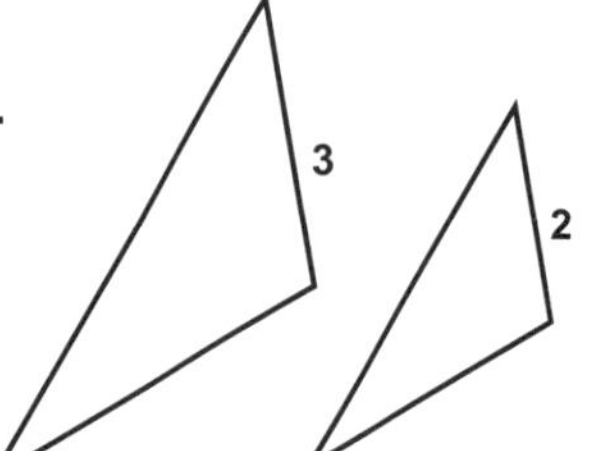

15. 

16. 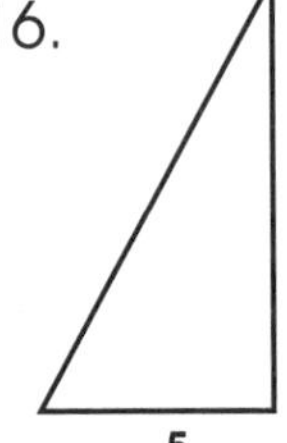

17. 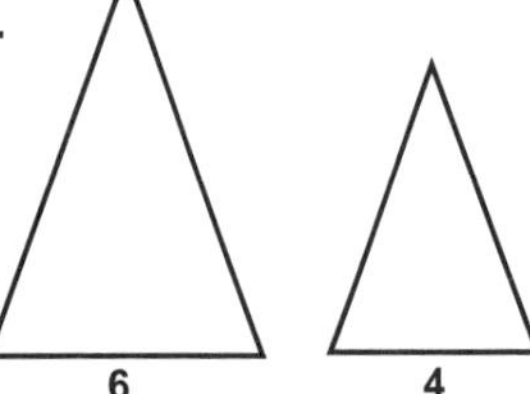

18. 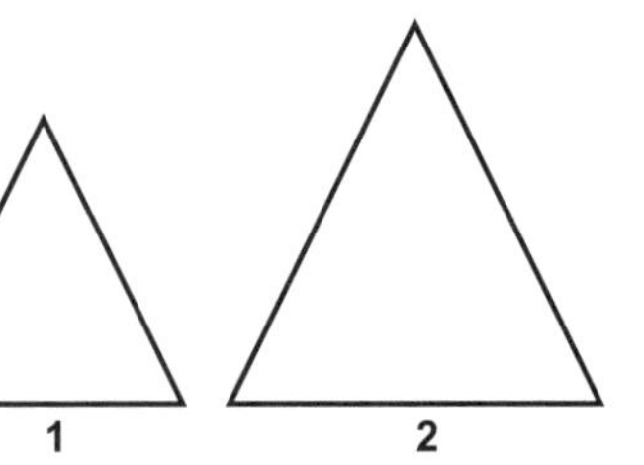

Name____________________________________

Two-Column Proofs

Given: $AB > AC$, $BD = EC$
Prove: $BE > CD$

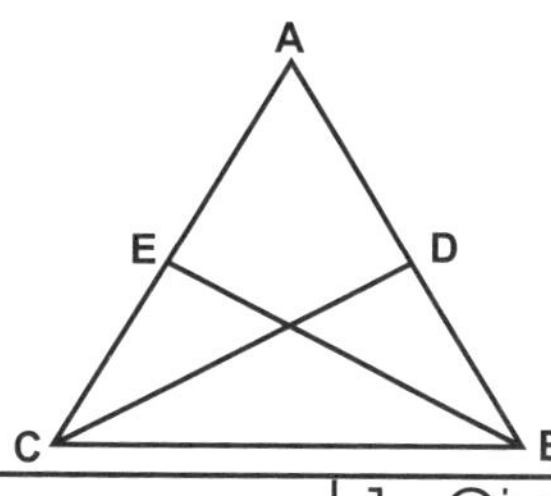

Statements	Reasons
1. $AB > AC$, $BD = EC$	1. Given
2. $m\angle ACB > m\angle ABC$	2. In a Δ, if two $\angle$s are not $\cong$, then the bigger side is opposite the bigger $\angle$.
3. $\overline{BC} \cong \overline{BC}$	3. Reflexive Property
4. $\overline{BD} \cong \overline{EC}$	4. If =, then $\cong$.
5. $BE > CD$	5. Hinge Theorem

Complete the following proofs.

Given: $m\angle ABD > m\angle DBC$
Prove: $AD > BD$

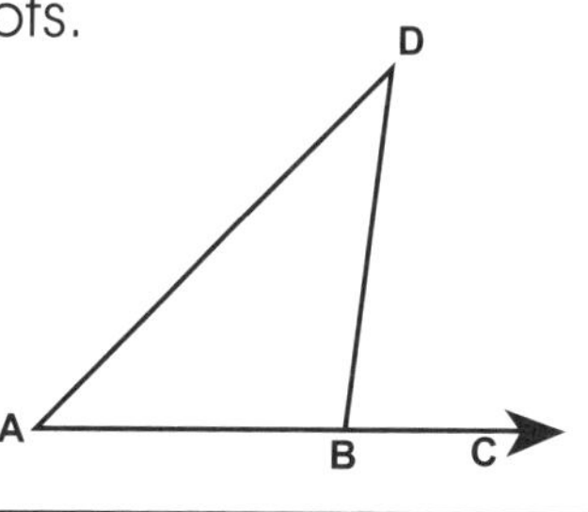

Statements	Reasons
1. ________________	1. Given
2. $m\angle DBC > m\angle DAB$	2. ________________
3. ________________	3. Substitution
4. $AD > BD$	4. ________________

Given: $\overline{BO} \cong \overline{HN}$, $\overline{OH} \cong \overline{BN}$
Prove: $\angle O \cong \angle N$

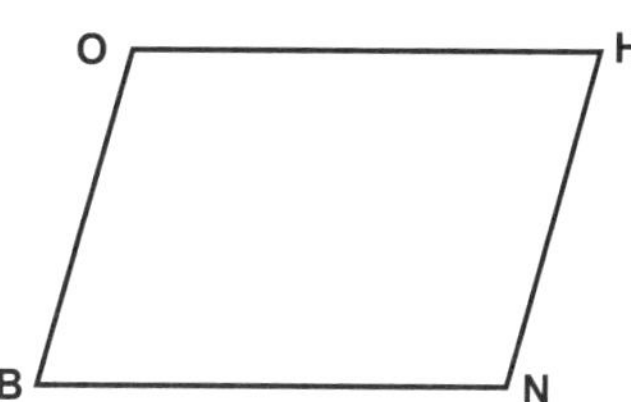

Statements	Reasons
1. Construct $\overline{BH}$	1. ________________
2. ________________	2. Given
3. ________________	3. Reflexive Property
4. ________________	4. SSS
5. $\angle O \cong \angle N$	5. ________________

Name________________________________ HSG-SRT.B.4

More Two-Column Proofs

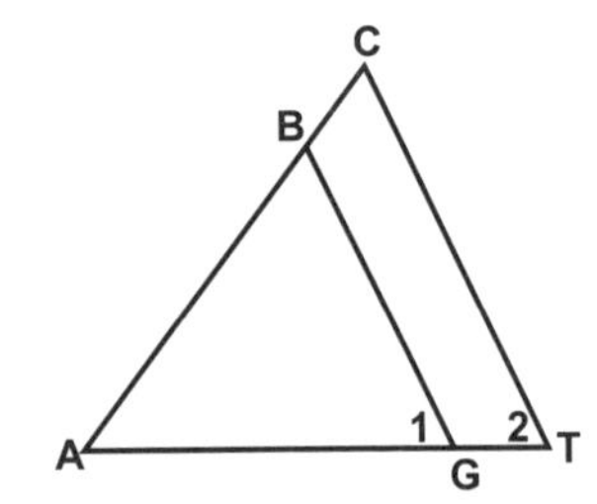

Given: $\overline{CT} \parallel \overline{BG}$
Prove: $\Delta CAT \sim \Delta BAG$

Statements	Reasons
1. ________________	1. Given
2. ________________	2. If ∥ lines, then corresponding ∠s are ≅ .
3. ________________	3. Reflexive Property
4. $\Delta CAT \sim \Delta BAG$	4. ________________

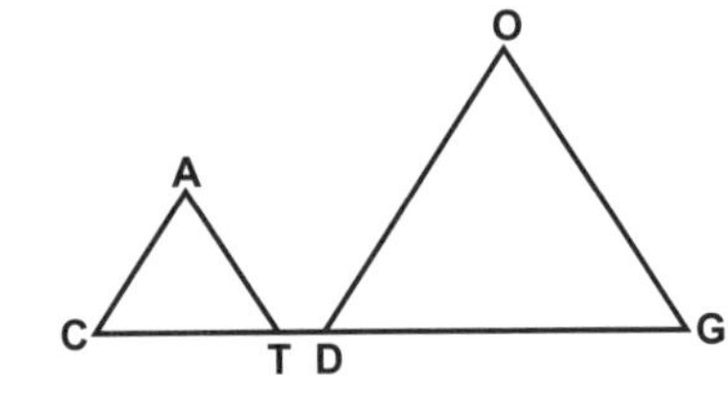

Given: $\overline{AC} \cong \overline{AT}$, $\overline{OD} \cong \overline{OG}$, $\overline{AC} \parallel \overline{OD}$
Prove: $\Delta CAT \sim \Delta DOG$

Statements	Reasons
1. ________________	1. Given
2. $\angle ACT \cong \angle ODG$	2. ________________
3. $\angle ACT \cong \angle ATC$, $\angle ODG \cong \angle OGD$	3. ________________
4. ________________	4. Substitution
5. $\Delta CAT \sim \Delta DOG$	5. ________________

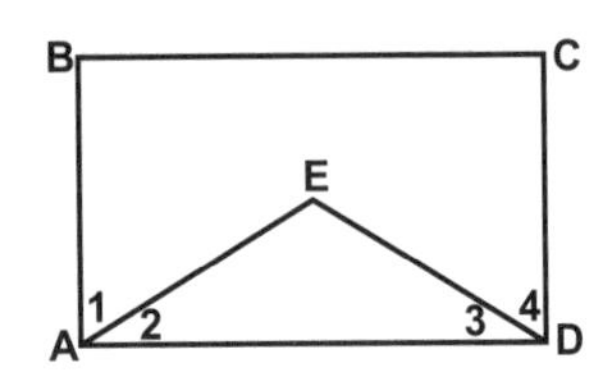

Given: $\angle BAD \cong \angle CDA$, $\angle 1 \cong \angle 4$
Prove: $\angle 2 \cong \angle 3$

Statements	Reasons
1. ________________	1. Given
2. ________________	2. AAP
3. $m\angle BAD = m\angle CDA$, $m\angle 1 = m\angle 4$	3. ________________
4. ________________	4. Substitution
5. ________________	5. APOE
6. $\angle 2 \cong \angle 3$	6. ________________

Name________________________________ HSG-SRT.C.6, HSG-SRT.C.7

Right Triangle Trigonometry

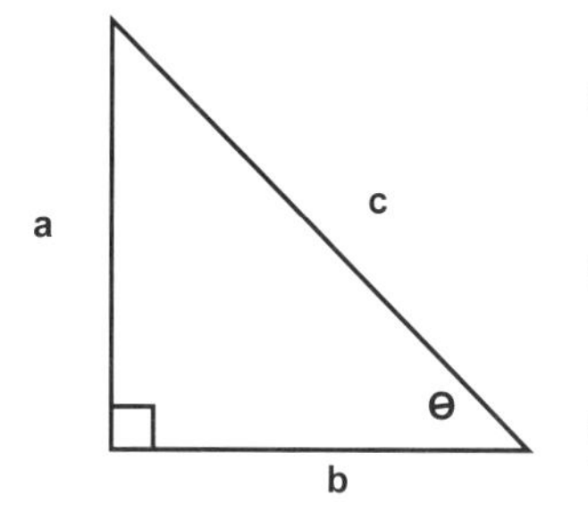

sine	$\sin\theta = \dfrac{\text{opposite}}{\text{hypotenuse}} = \dfrac{a}{c}$
cosine	$\cos\theta = \dfrac{\text{adjacent}}{\text{hypotenuse}} = \dfrac{b}{c}$
tangent	$\tan\theta = \dfrac{\text{opposite}}{\text{adjacent}} = \dfrac{a}{b}$
cosecant	$\csc\theta = \dfrac{1}{\sin\theta} = \dfrac{c}{a}$
secant	$\sec\theta = \dfrac{1}{\cos\theta} = \dfrac{c}{b}$
cotangent	$\cot\theta = \dfrac{1}{\tan\theta} = \dfrac{b}{a}$

Find the six trigonometric functions for the angles below.

1. 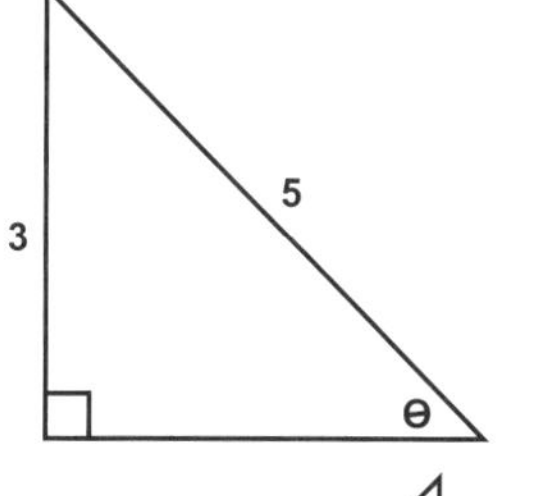

2. 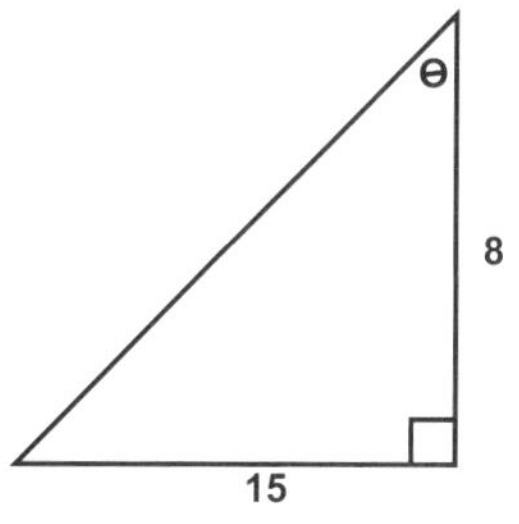

3. 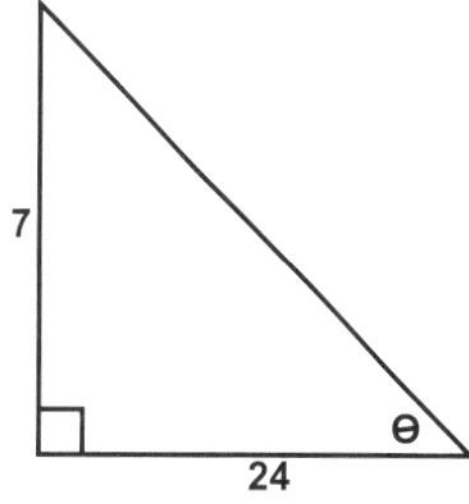

4. 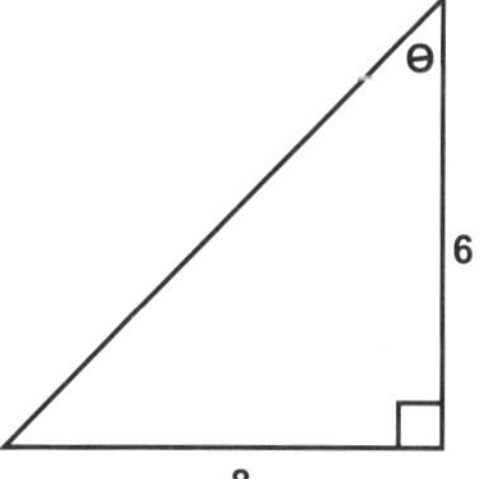

5. 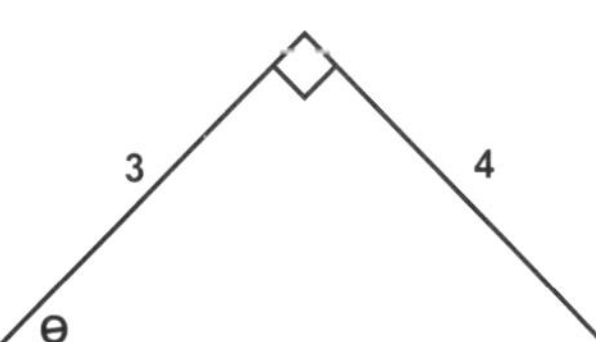

6. 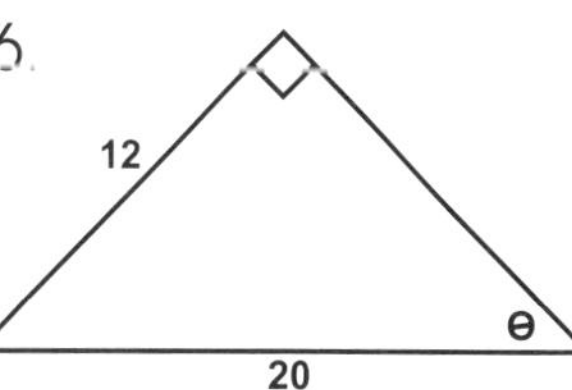

7. 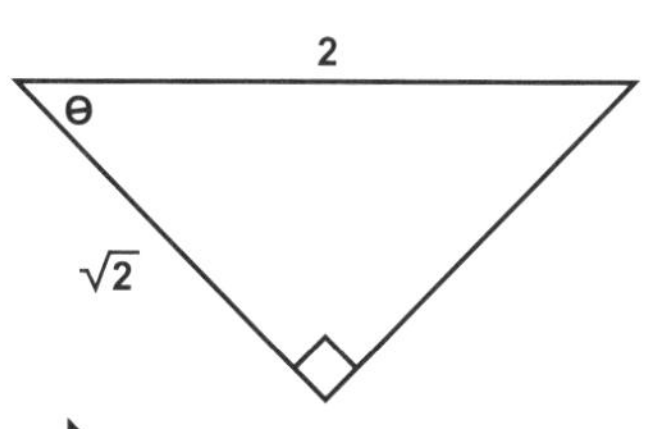

8. 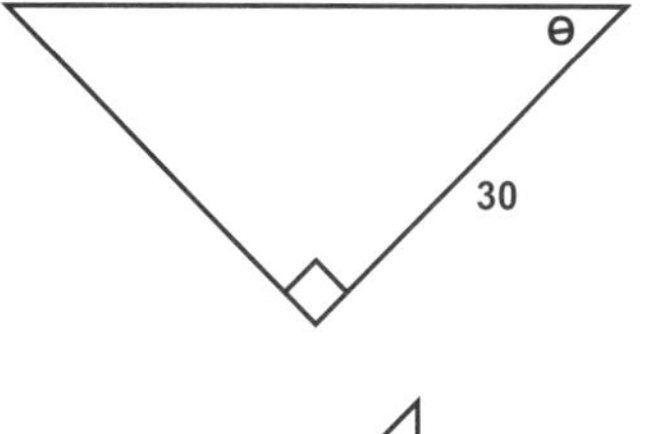

9. 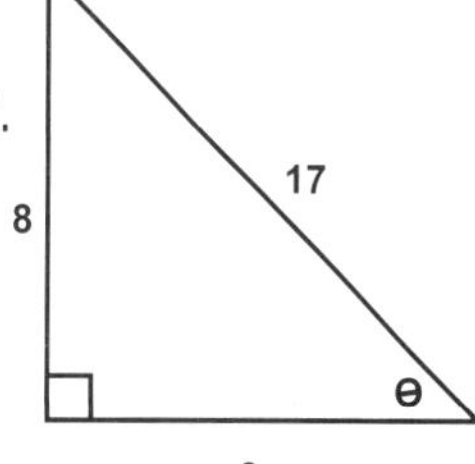

10. 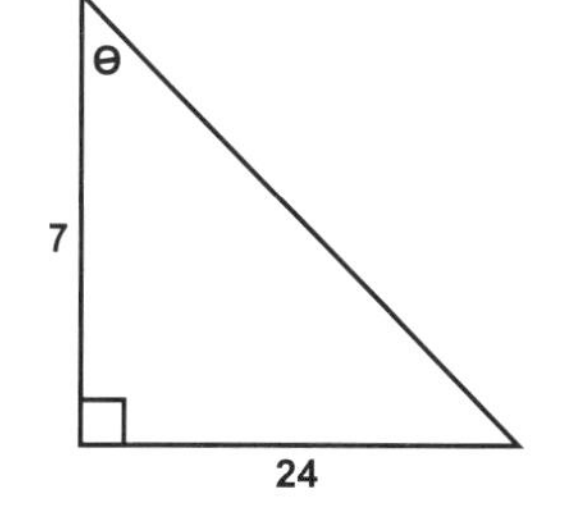

11. 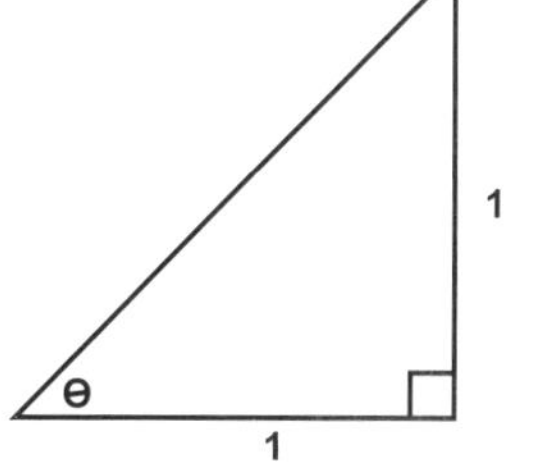

12. 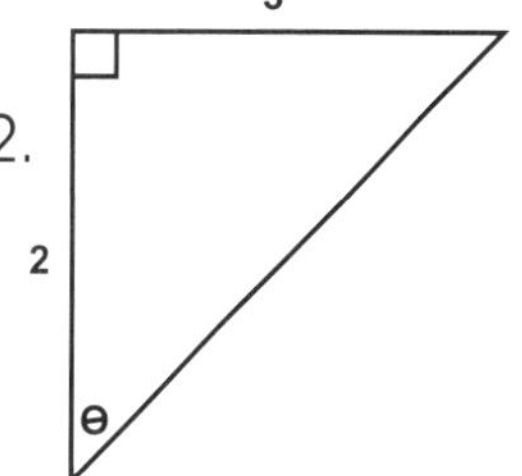

Name________________________________ HSG-SRT.C.6, HSG-SRT.C.7

Solving Any Triangle

$$\tan\theta = \frac{\sin\theta}{\cos\theta}$$

$\sin(^{-}\theta) = -\sin\theta$ $\quad\quad$ $\cos(^{-}\theta) = \cos\theta$

$\cos(\theta + \beta) = \cos\theta\cos\beta - \sin\theta\sin\beta$

$\sin(\theta + \beta) = \sin\theta\cos\beta + \cos\theta\sin\beta$

$\cos\frac{\pi}{2} = 0$ $\quad\quad$ $\sin\frac{\pi}{2} = 1$ $\quad\quad$ $\tan\frac{\pi}{2}$ is undefined

$\cos\pi = {}^{-}1$ $\quad\quad$ $\sin\pi = 0$ $\quad\quad$ $\tan\pi = 0$

$$\begin{aligned}\cos(\theta - \beta) &= \cos[\theta + (^{-}\beta)]\\ &= \cos(\theta)\cos(^{-}\beta) - \sin(\theta)\sin(^{-}\beta)\\ &= \cos\theta\cos\beta - \sin\theta(-\sin\beta)\\ \cos(\theta - \beta) &= \cos\theta\cos\beta + \sin\theta\sin\beta\end{aligned}$$

Evaluate the following using the above identities.

1. $\sin(\theta - \beta)$
2. $\tan(\theta + \beta)$
3. $\tan(\theta - \beta)$
4. $\cos(\theta + \frac{\pi}{2})$
5. $\cos(\theta + \pi)$
6. $\sin(\theta + \frac{\pi}{2})$
7. $\sin(\theta + \pi)$
8. $\sin(2\theta)$
9. $\cos(2\theta)$
10. $\tan(2\theta)$

$$\begin{aligned}\sin\theta &= \sin\theta\cos^2\theta + \sin^3\theta\\ &= \sin\theta(\cos^2\theta) + \sin\theta(\sin^2\theta)\\ &= \sin\theta(\cos^2\theta + \sin^2\theta)\\ &= \sin\theta(1)\\ \sin\theta &= \sin\theta\end{aligned}$$

Verify the following identities.

1. $\csc^2\theta = 1 + \cot^2\theta$
2. $\sec^2\theta = 1 + \tan^2\theta$
3. $\cos\theta = \sec\theta - \tan\theta\sin\theta$
4. $\sin\theta = \csc\theta - \cot\theta\cos\theta$

Name________________________________ HSG-SRT.C.6, HSG-SRT.C.7

Right Triangle Trigonometry Review

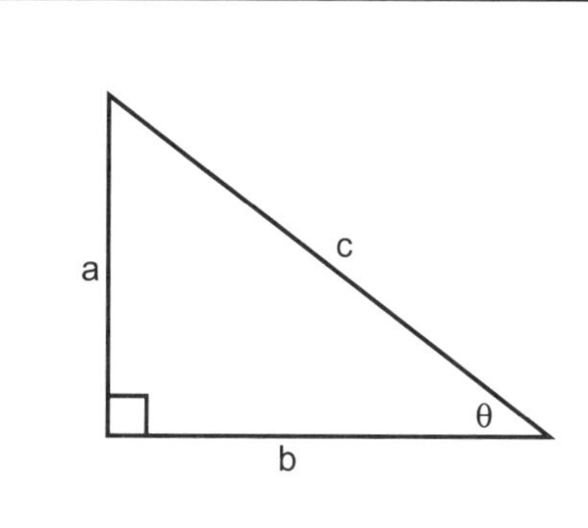

sine	$\sin\theta$	$= \frac{\text{opposite}}{\text{hypotenuse}}$	$= \frac{a}{c}$
cosine	$\cos\theta$	$= \frac{\text{adjacent}}{\text{hypotenuse}}$	$= \frac{b}{c}$
tangent	$\tan\theta$	$= \frac{\text{opposite}}{\text{adjacent}}$	$= \frac{a}{b}$
cosecant	$\csc\theta$	$= \frac{1}{\sin\theta}$	$= \frac{c}{a}$
secant	$\sec\theta$	$= \frac{1}{\cos\theta}$	$= \frac{c}{b}$
cotangent	$\cot\theta$	$= \frac{1}{\tan\theta}$	$= \frac{b}{a}$

Find the six trigonometric functions for the angles below.

1. 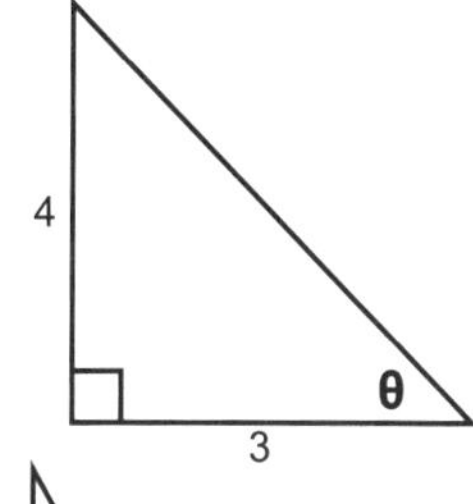

2. 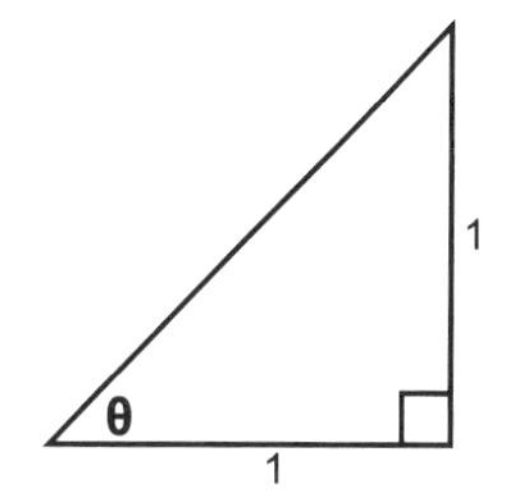

3. 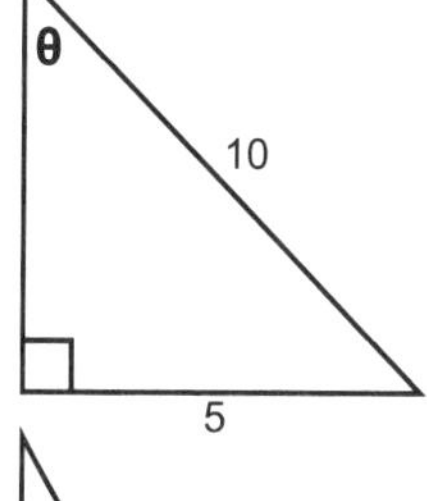

4. 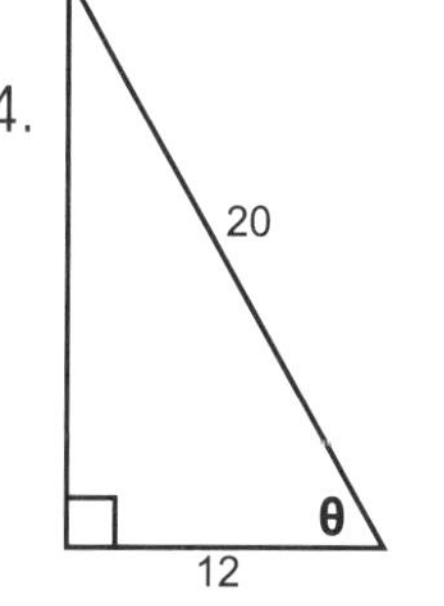

5. 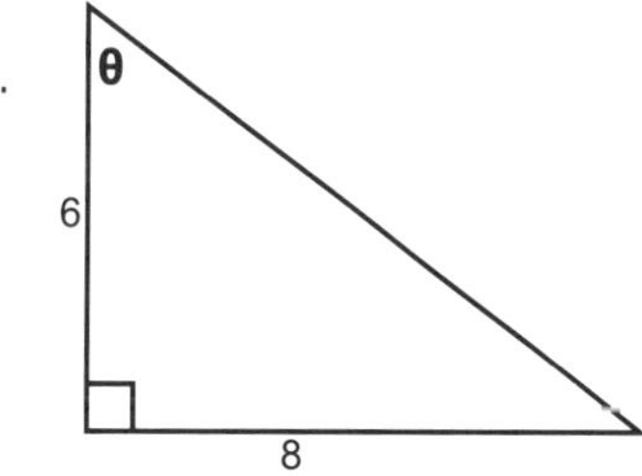

6. 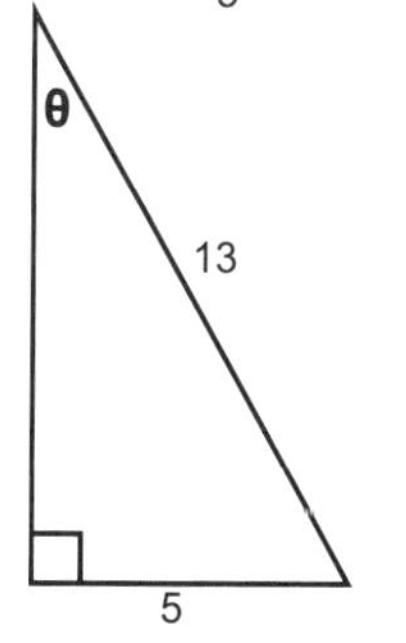

7. 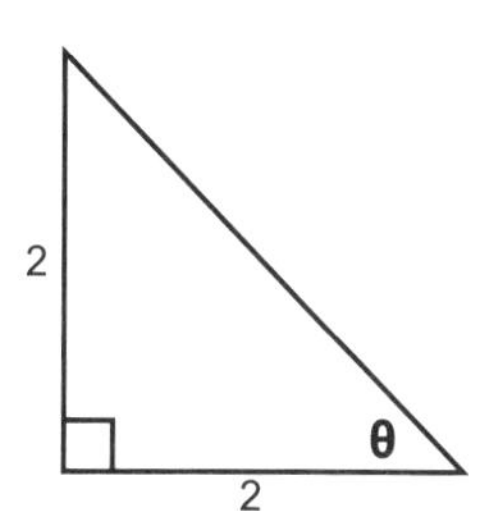

8. 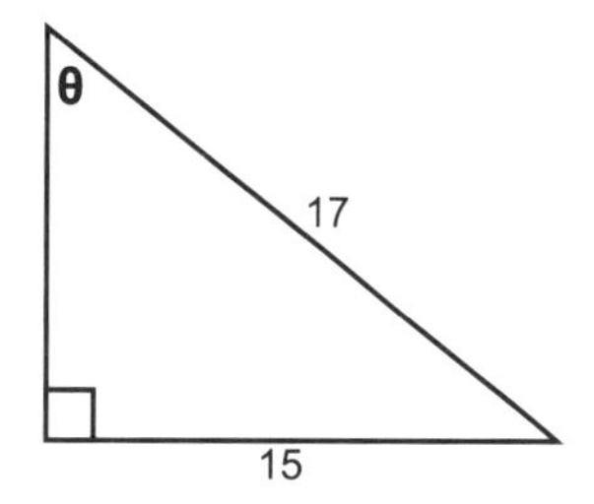

9. 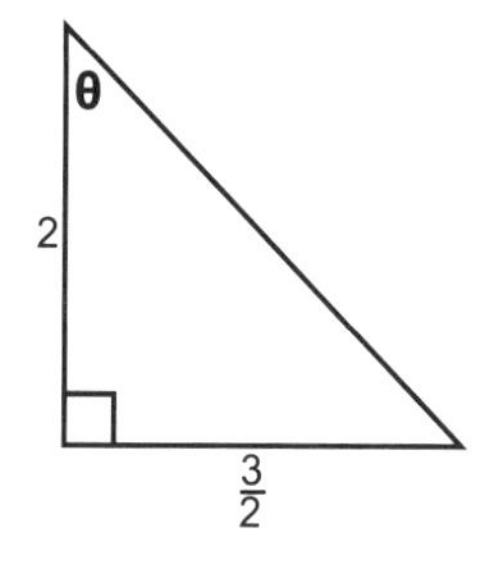

10. 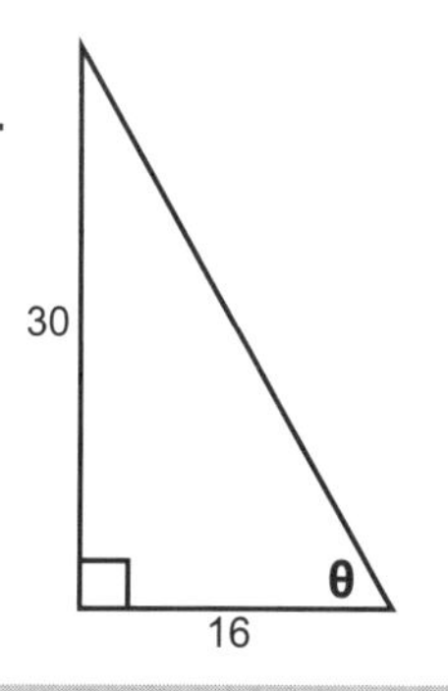

11. 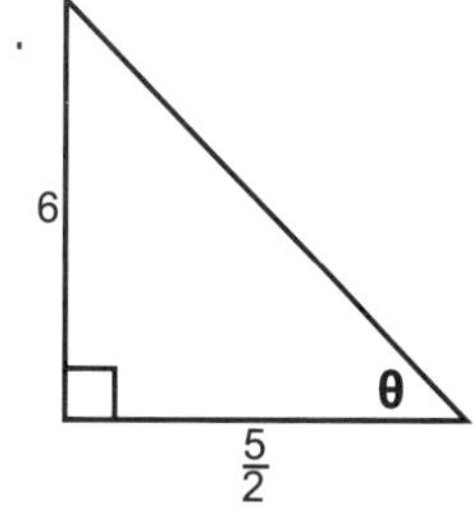

Name________________________________

HSG-SRT.C.8

Solving Any Triangle

Law of cosines

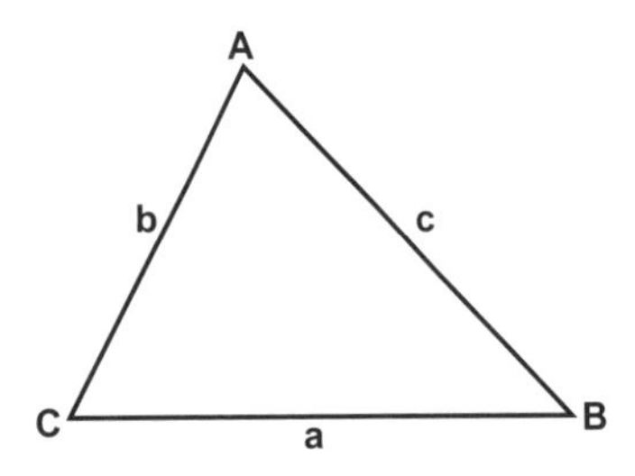

In any triangle ABC.

$a^2 = b^2 + c^2 - 2bc \cos A$

$b^2 = a^2 + c^2 - 2ac \cos B$

$c^2 = a^2 + b^2 - 2ab \cos C$

Use the law of cosines to state an equation to find the missing part, x.

1.

2.

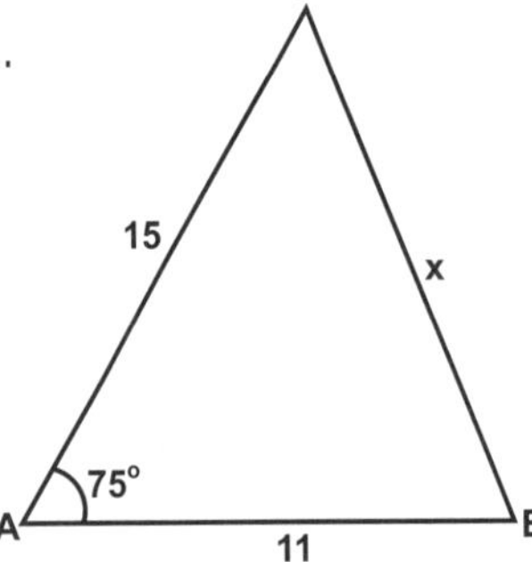

3.

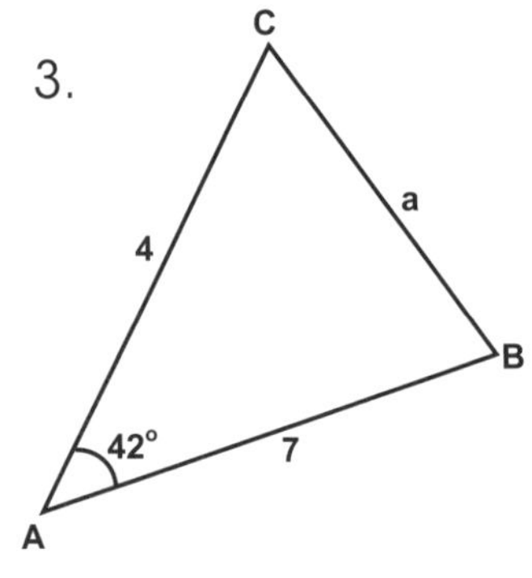

4.

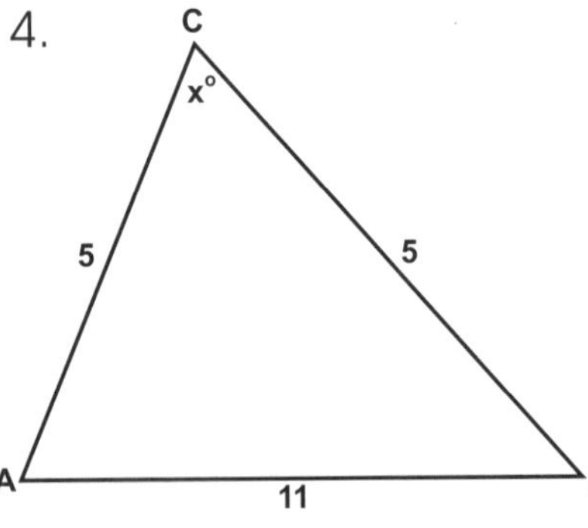

5.

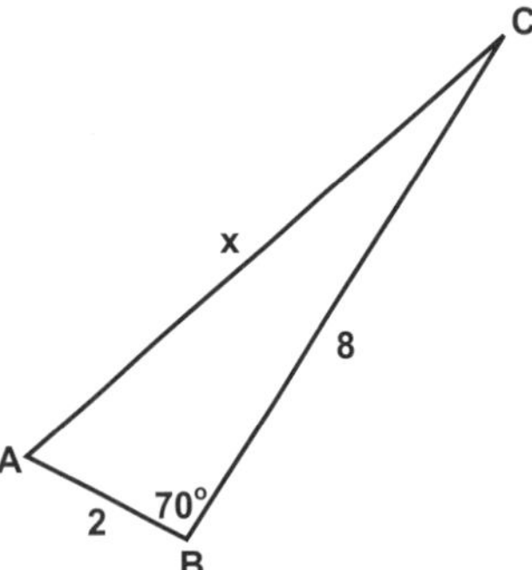

Find the indicated part of ΔABC. Round angles to the nearest tenth and lengths to three significant digits.

6. b = 12, c = 10, ∠A = 38°, a = ______
7. a = 14, b = 15, c = 18, ∠A = ______
8. a = 12, c = 11, ∠B = 81°, b = ______
9. a = 8, b = 9, c = 15, ∠C = ______
10. a = 5, b = 7, ∠C = 40°, c = ______
11. c = 20, b = 30, ∠A = 140°, a = ______
12. b = 2, a = 4, ∠C = 20°, c = ______
13. a = 5, b = 9, c = 11, ∠C = ______
14. a = 1.5, b = 3, c = 2, ∠B = ______
15. a = .6, b = .8, c = 1.2, ∠A = ______

Name____________________________________

HSG-SRT.C.8

Law of Sines

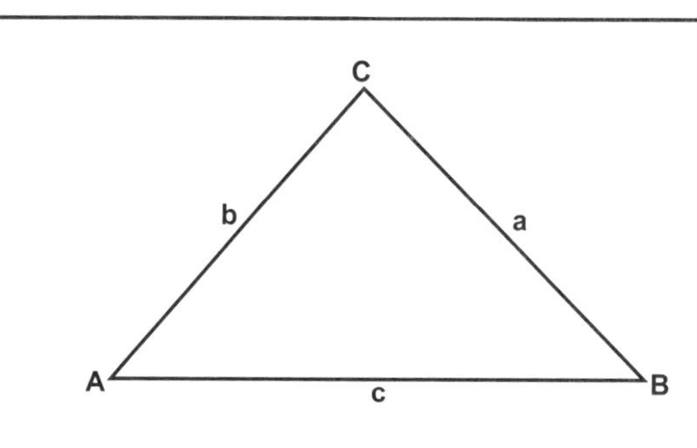

In any triangle ABC,

$$\frac{\sin A}{a} = \frac{\sin B}{b} = \frac{\sin C}{c}$$

Use the law of sines to state an equation for each triangle.

1.

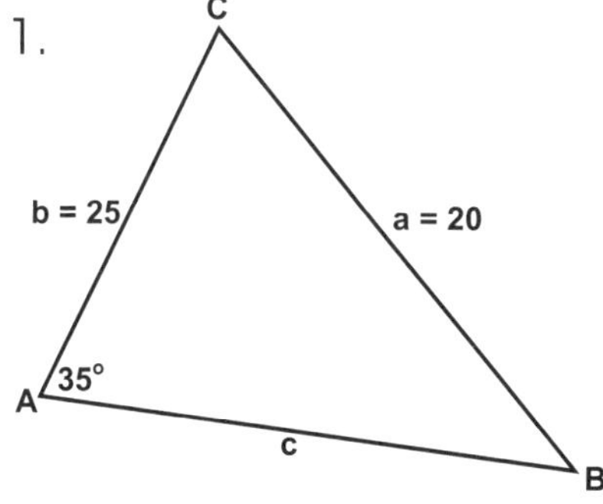

2.

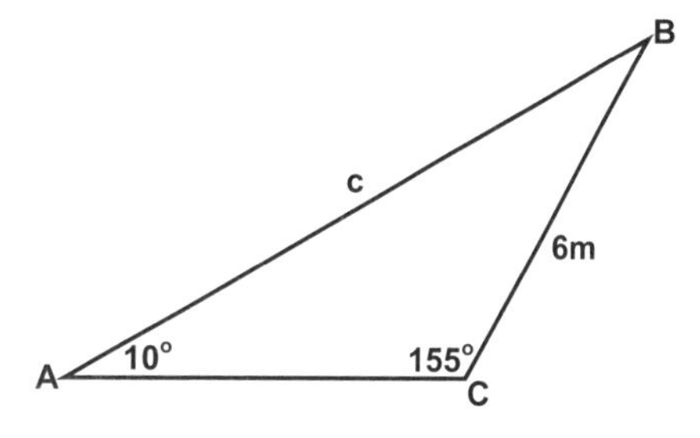

3.

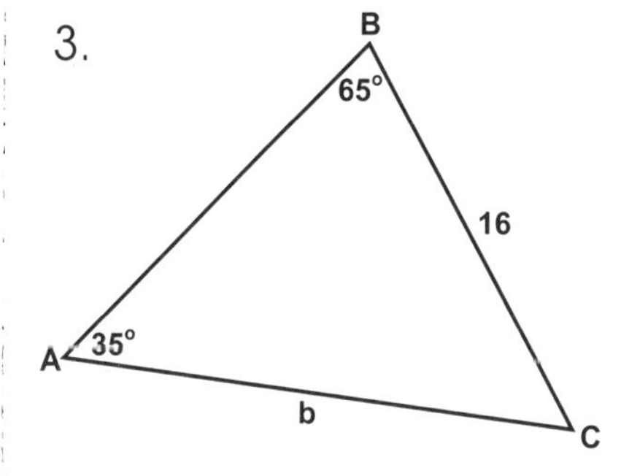

4.

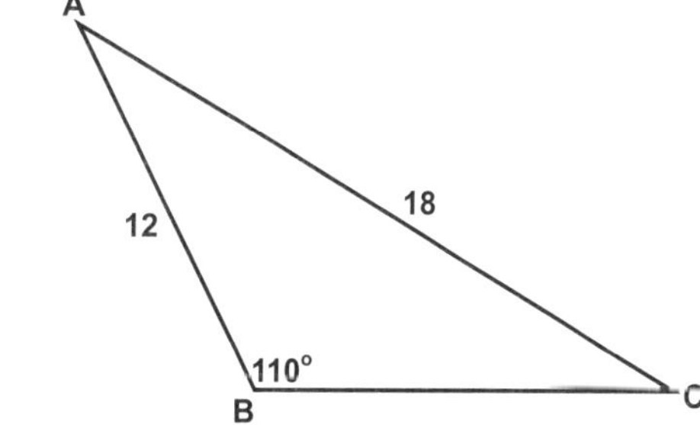

Find the indicated part of ΔABC. Round angles to the nearest tenth and lengths to three significant digits.

5. $c = 10$, $\angle A = 48°$, $\angle C = 63°$, $a =$ ______
6. $a = 20$, $b = 15$, $\angle A = 40°$, $\angle B =$ ______
7. $a = 40$, $b = 50$, $\angle A = 37°$, $\angle B =$ ______
8. $a = 11$, $c = 15$, $\angle A = 40°$, $\angle C =$ ______
9. $c = 30$, $\angle A = 42°$, $\angle C = 98°$, $a =$ ______
10. $a = 1.5$, $b = 2.0$, $\angle B = 35°$, $\angle A =$ ______
11. $a = 16$, $\angle A = 35°$, $\angle C = 65°$, $c =$ ______
12. $b = 18$, $c = 32$, $\angle C = 100°$, $\angle B =$ ______

Name________________________________

8.F.A.1, HSF-TF.A.3, HSG-SRT.D.10

Graphs of Sine and Cosine Functions I

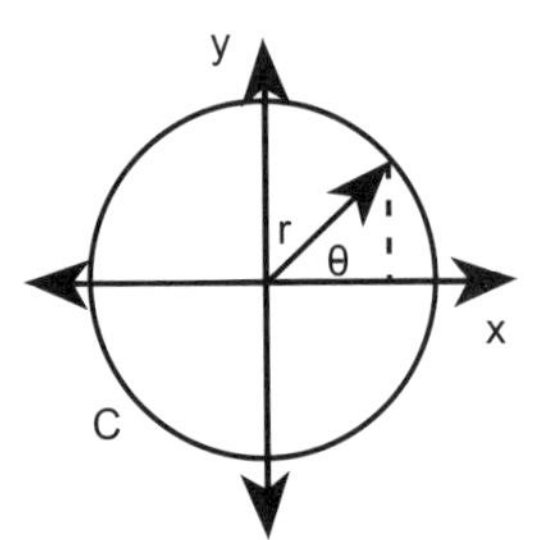

Let C be a unit circle (r = 1) with its center at the origin. The angle θ is positive if the ray is rotated counterclockwise and negative if the ray is rotated clockwise.

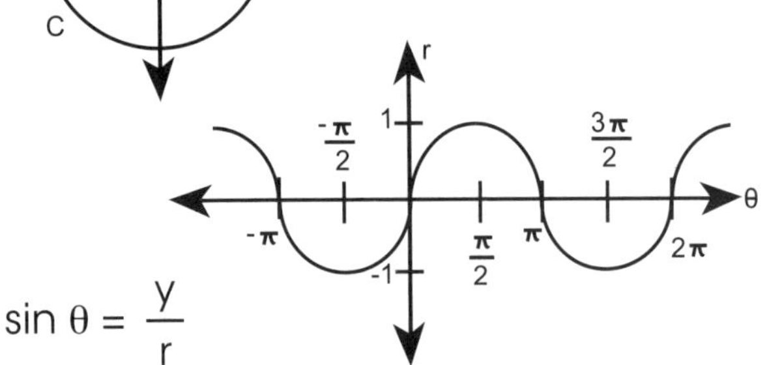

$\sin \theta = \frac{y}{r}$

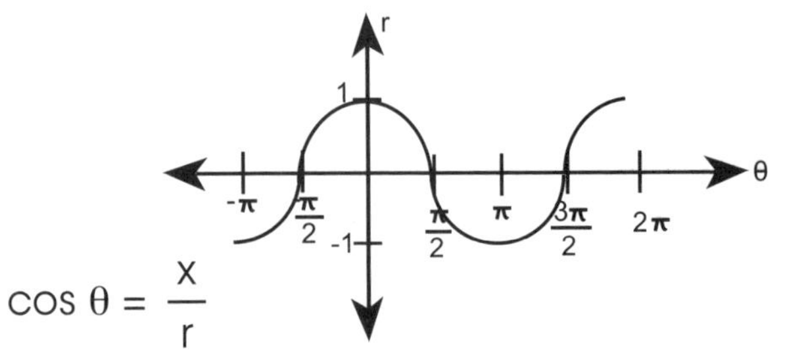

$\cos \theta = \frac{x}{r}$

Complete the following table of values.

y = sin x

	x	y
1.	0	______
2.	$\frac{\pi}{2}$	______
3.	π	______
4.	$\frac{3\pi}{2}$	______
5.	2π	______
6.	$\frac{-\pi}{2}$	______

y = cos x

	x	y
7.	0	______
8.	$\frac{\pi}{2}$	______
9.	π	______
10.	$\frac{3\pi}{2}$	______
11.	2π	______
12.	$\frac{-\pi}{2}$	______

13. What is the value of y when $\sin \theta = 0$?

14. At what values of θ is $\sin \theta = 0$?

15. What is the greatest value sin θ may assume?

16. Name two values of θ that make sin θ a maximum.

17. When is $\cos \theta = 0$?

18. What range of values may cos θ assume?

19. What is the value of sin θ when $\theta = 90°$?

20. $\theta = \pi$ radian. What is sin θ? What is cos θ?

21. $\theta = {}^{-}4\pi$ radian. What is sin θ? What is cos θ?

22. $\theta = \frac{3\pi}{r}$ radian $\cos \theta = \frac{x}{r}$ What is the value of x?

Name____

Graphs of Sine and Cosine Functions II

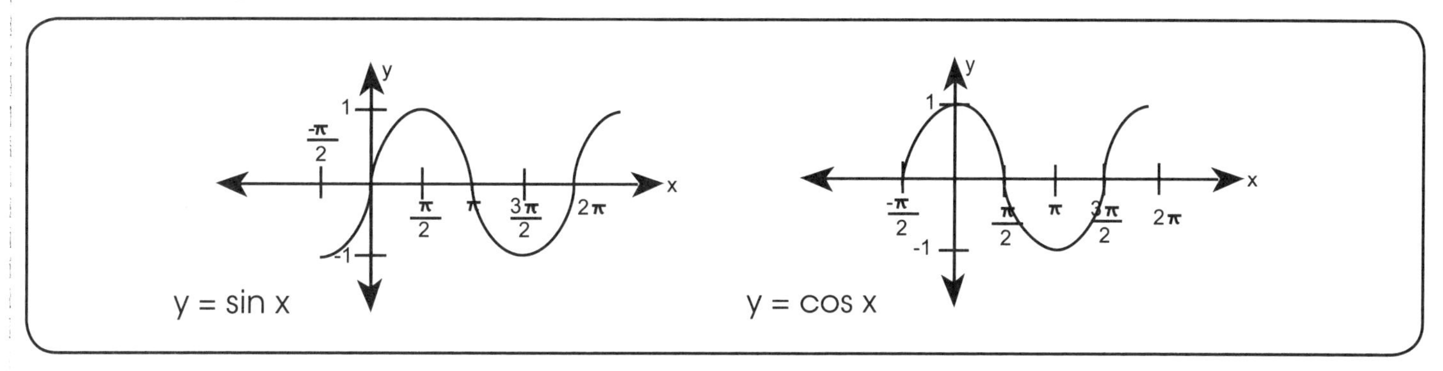

Consider the graph y = 2 sin x.

1. Where will it cross the x-axis?
2. Graph y = 2 sin x.
3. What is the range of y = 2 sin x?

In the equations y = a sin x and y = a cos x, $|a|$ is called the **amplitude** of the curve.

4. What is the amplitude of y = 3 cos x?
5. Graph y = 3 cos x.
6. Where does y = 3 cos x cross the x-axis?
7. How does y = 3 cos x compare to y = cos x?

y = sin x and y = cos x range over the same values each time x runs through an interval of 2π units. The curve is said to be periodic and its period is 2π.

8. Graph y = sin 2x. (Include at least x = 0, $x = \frac{\pi}{4}$, $x = \frac{\pi}{2}$.)
9. Where does y = sin 2x cross the x-axis?
10. What is the period of y = sin 2x?

Name________________________________

HSG-C.A.2

Parts of a Circle

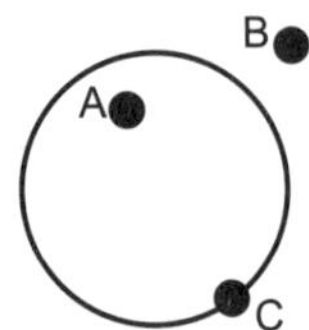

Point A is in the interior of the circle.
Point B is in the exterior of the circle.
Point C lies on the circle.
A circle is named by its center.

Radius—segment joining the center to a point on the ⊙.
Chord—segment joining two points on the ⊙.
Diameter—chord containing the center.

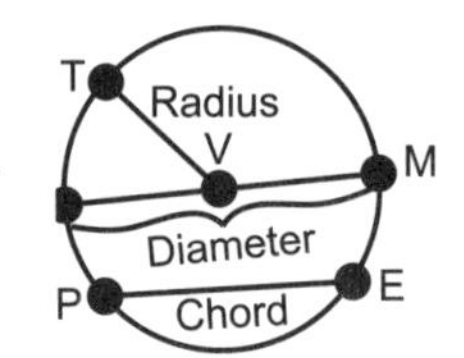

Identify each segment as a radius, chord, or diameter.

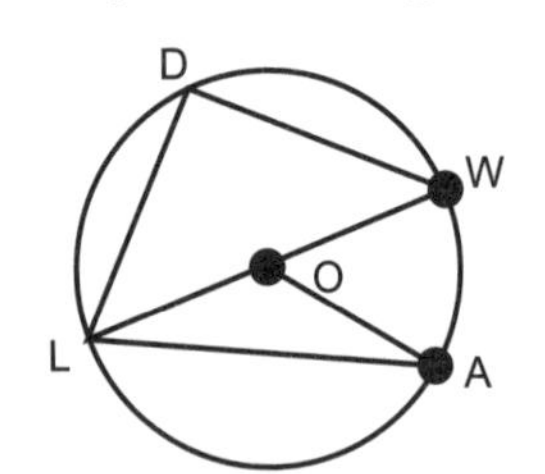

1. $\overline{OA}$
2. $\overline{DL}$
3. $\overline{LW}$
4. $\overline{WO}$
5. $\overline{DW}$
6. $\overline{LA}$
7. $\overline{AW}$
8. $\overline{OL}$

Name all examples of each term shown.

Figure 1

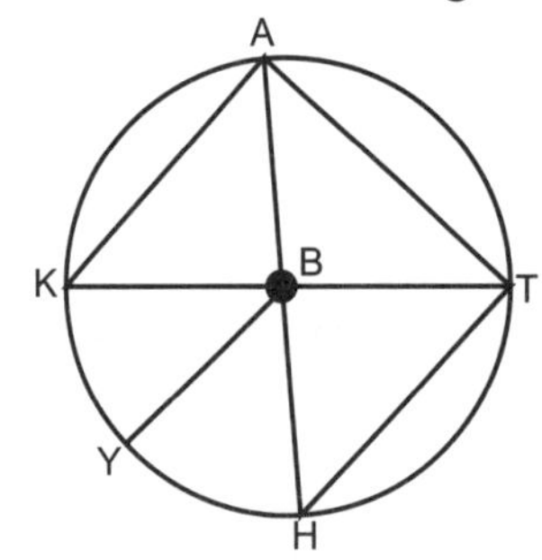

Figure 1

9. Radius
10. Chord
11. Center
12. Diameter

Figure 2

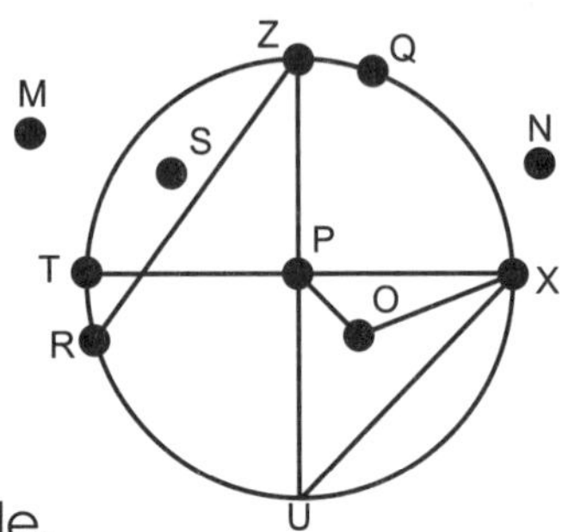

Figure 2

13. Name all the points in the interior of the circle.
14. Name all the points in the exterior of the circle.
15. Name all the points on the circle.
16. Name the center of the circle.
17. Name all segments that are a radius of the circle.
18. Name all segments that are a diameter of the circle.
19. Name all segments that are neither a radius nor a diameter of the circle.
20. Name all segments that are chords of the circle.

Name____ HSG-C.A.2, HSG-C.A.4

Secants and Tangents

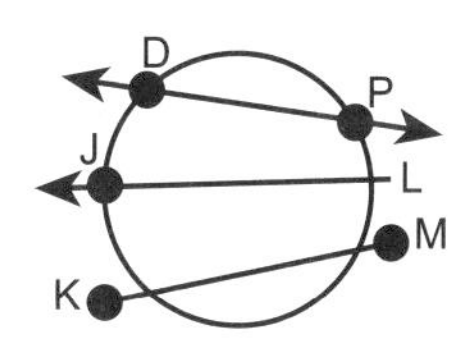

A **secant** is a line, ray, or segment that contains a chord. $\overleftrightarrow{DP}$, $\overrightarrow{LJ}$, and $\overline{KM}$ are secant.

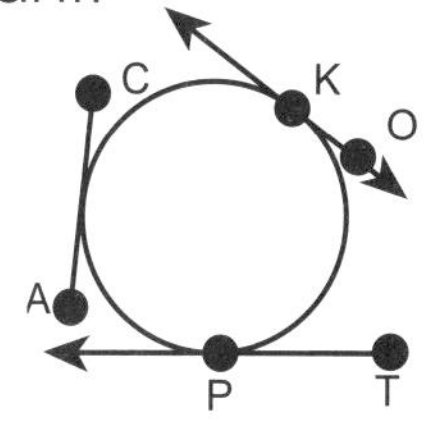

A **tangent** is a line, ray, or segment that contains exactly one point of a circle. $\overline{CA}$, $\overrightarrow{TP}$, and $\overleftrightarrow{KO}$ are tangents.

Name the indicated parts of circle T.

1. 3 radii
2. 1 diameter
3. 1 tangent ray
4. 1 secant line
5. 2 chords

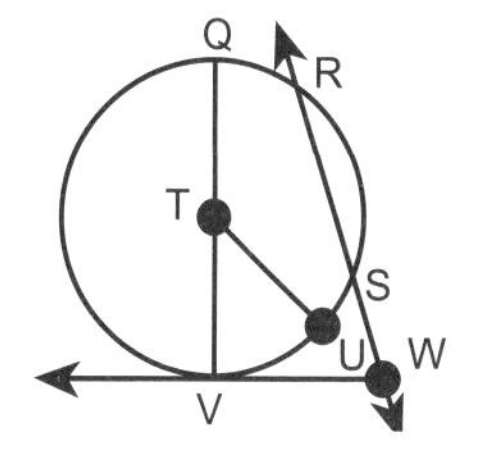

Circles may be internally tangent

or externally tangent.

Find the indicated lengths.

6.

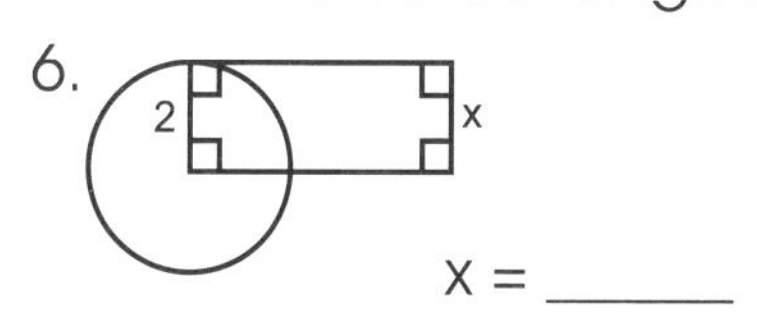

X = ______

9. DE = 9.1

EF = ______

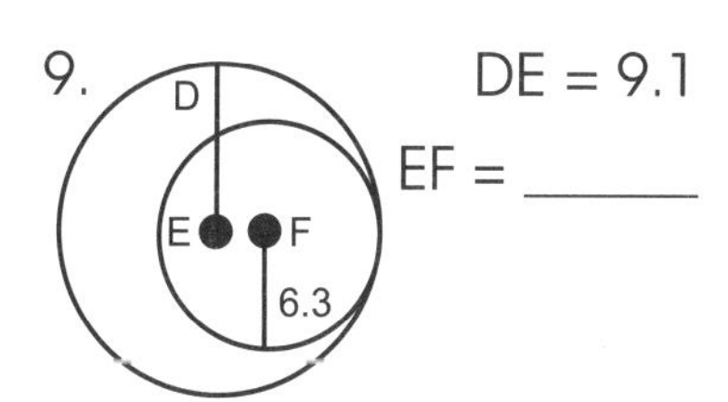

12.

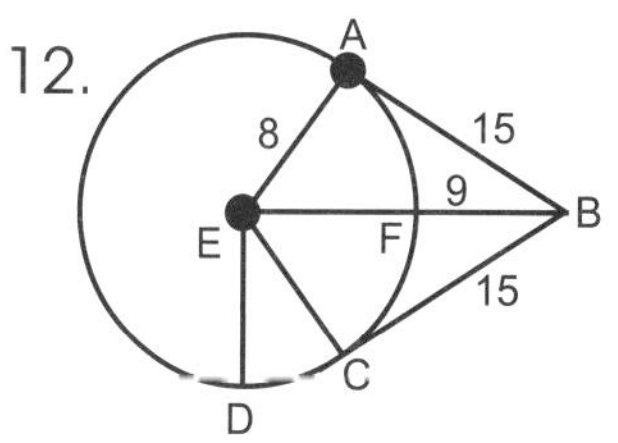

EF = ______

7. OP = ______

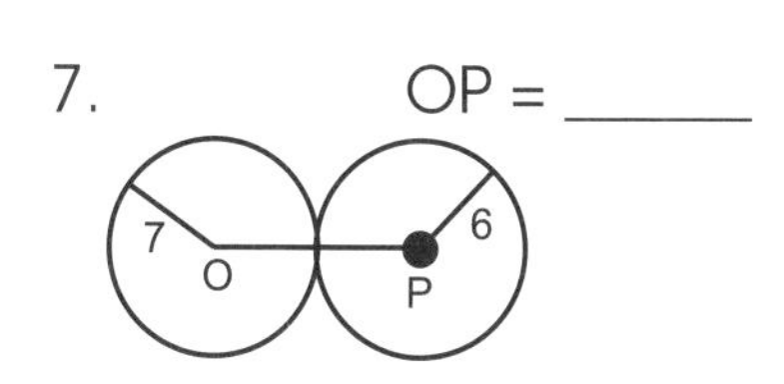

10.

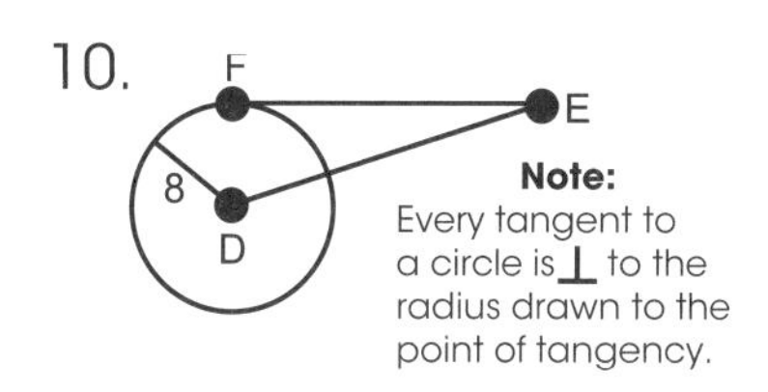

Note: Every tangent to a circle is ⊥ to the radius drawn to the point of tangency.

DE = 17 FE = ______

13.

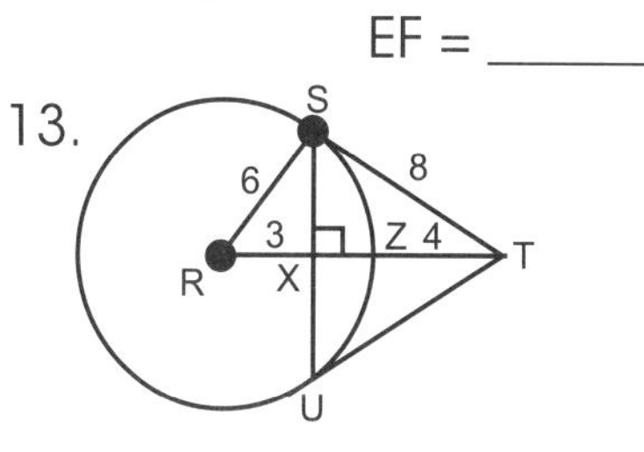

RZ = 6 ST = 8

RX = 3 SU = ______

8. AC = 14

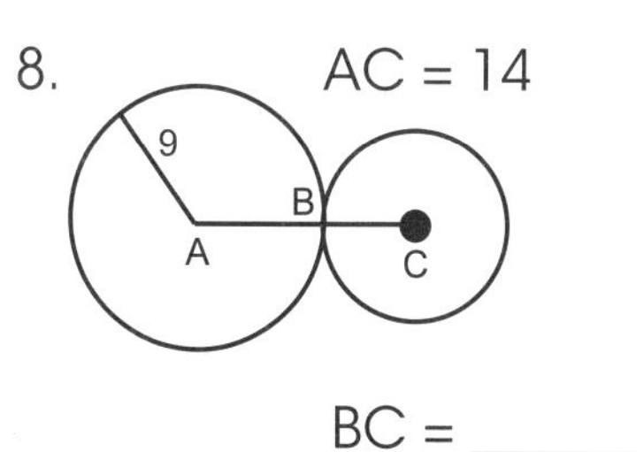

BC = ______

11.

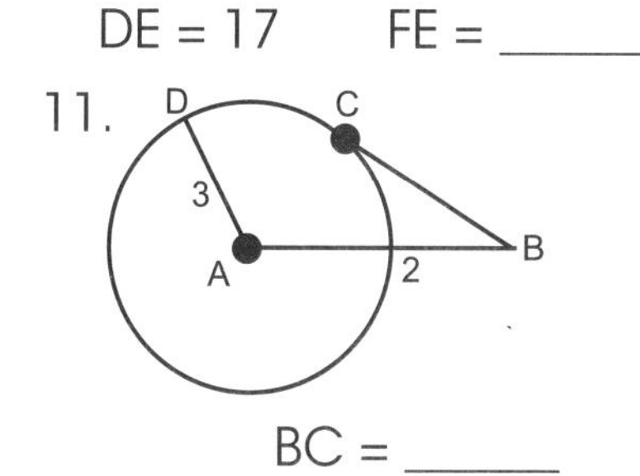

BC = ______

14.

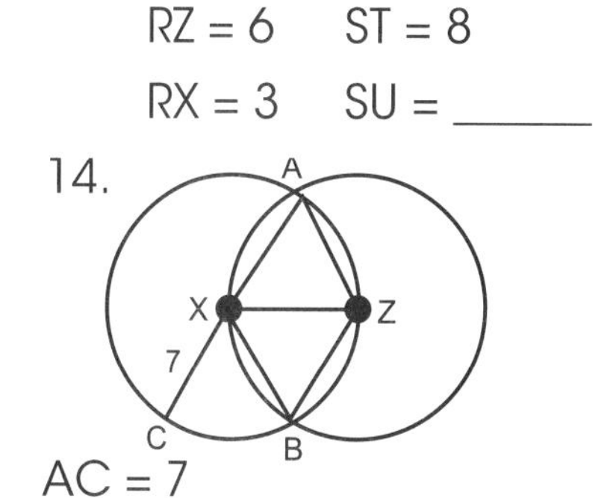

AC = 7

perimeter of AZBX = ______

Name______________________________ HSG-C.A.2, HSG-C.A.4

Tangents, Secants, and Chords

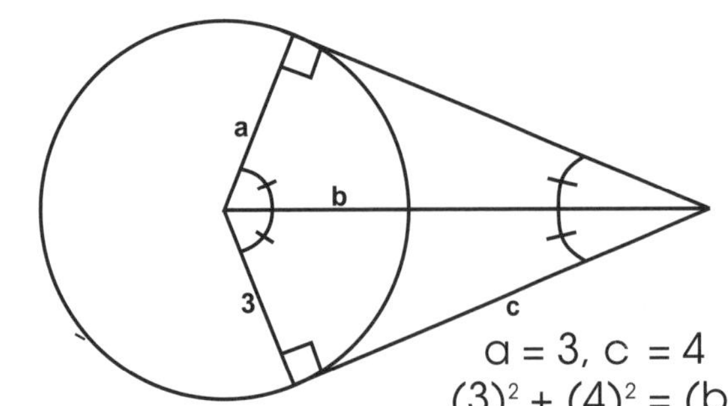

$a = 3,\ c = 4$

$(3)^2 + (4)^2 = (b)^2$

$9 + 16 = b^2$

$b = 5$

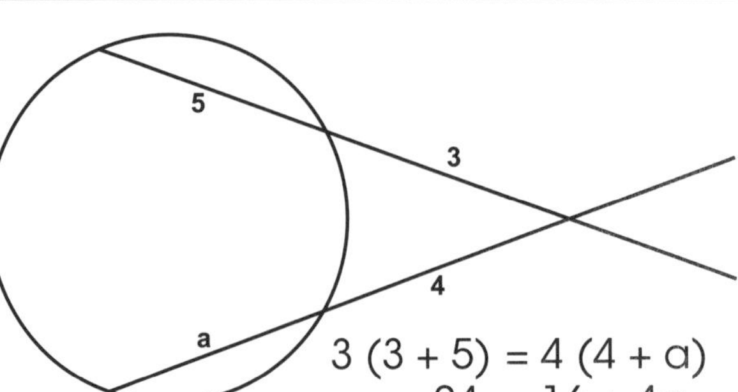

$3\ (3 + 5) = 4\ (4 + a)$

$24 = 16 + 4a$

$8 = 4a$

$a = 2$

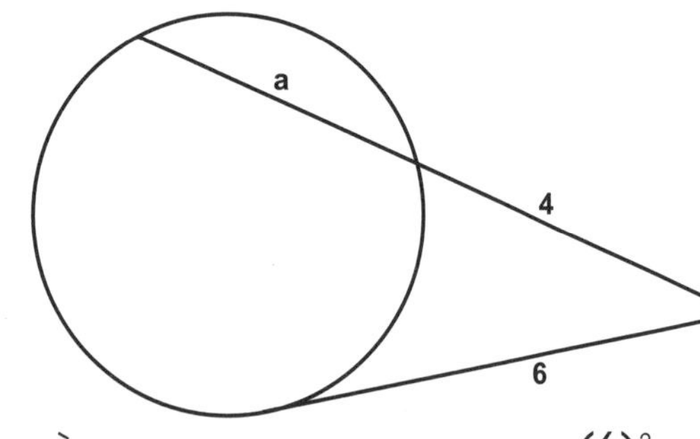

$(6)^2 = 4\ (4 + a)$

$36 = 16 + 4a$

$20 = 4a$

$a = 5$

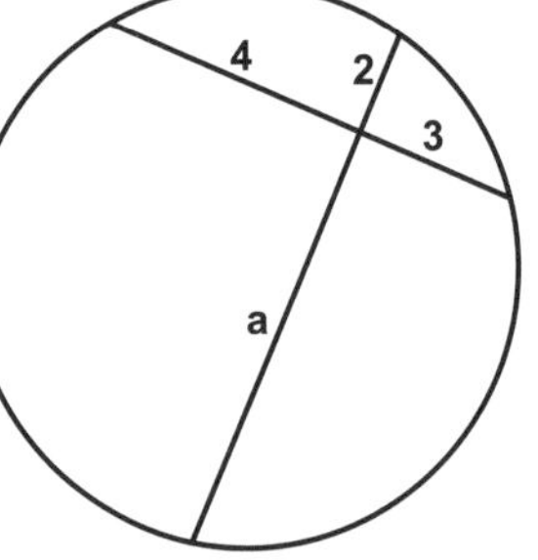

$2\ (a) = 3\ (4)$

$2a = 12$

$a = 6$

Find the labeled lengths.

1.

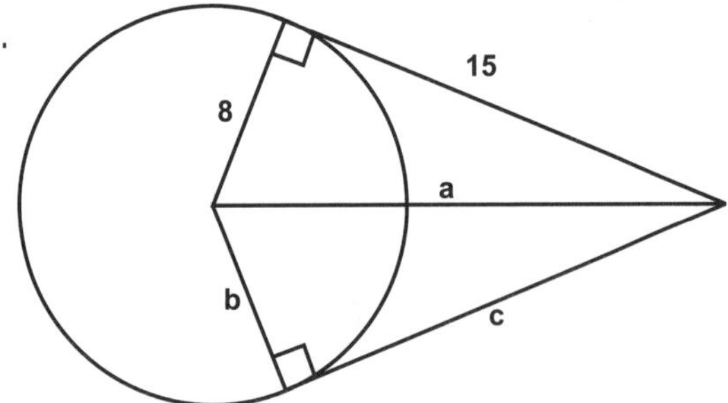

2.

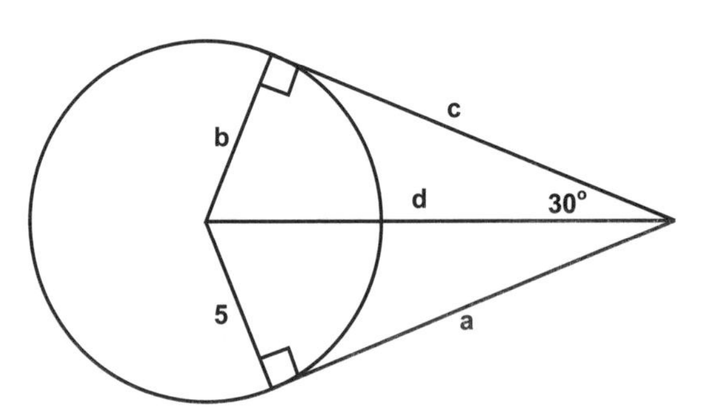

3.

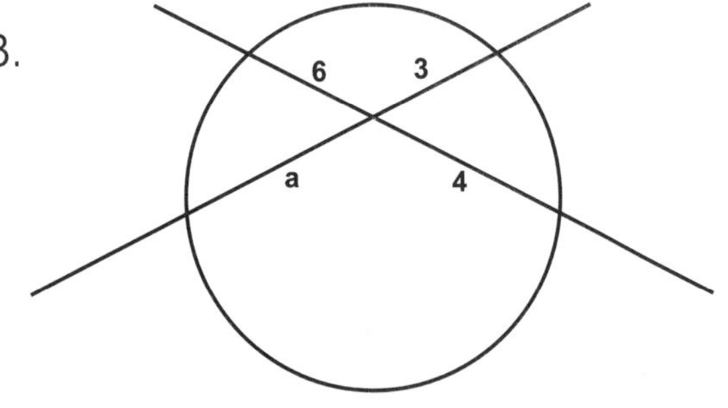

4.

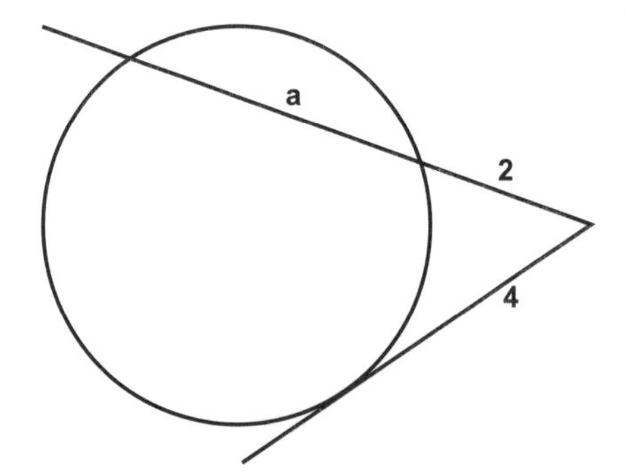

Arcs and Angles

Minor Arc	Semicircle	Major Arc
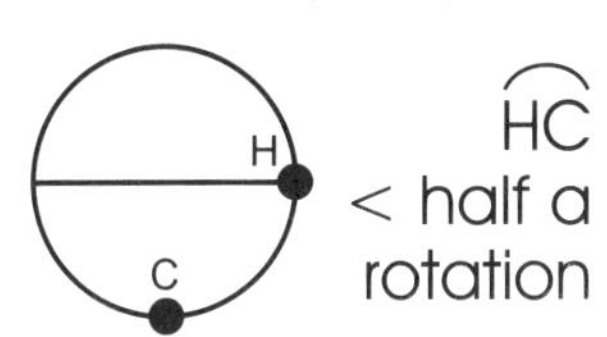 $\overset{\frown}{HC}$ < half a rotation	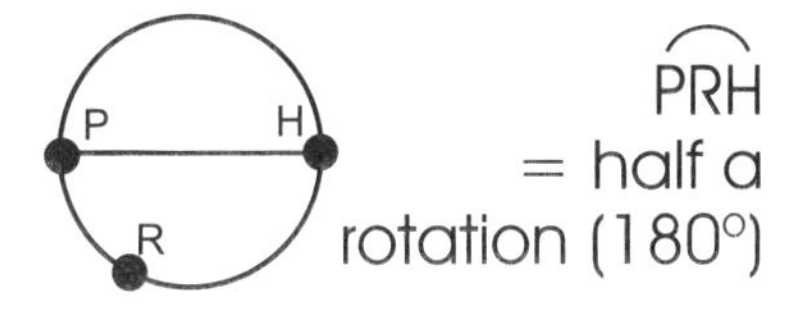 $\overset{\frown}{PRH}$ = half a rotation (180°)	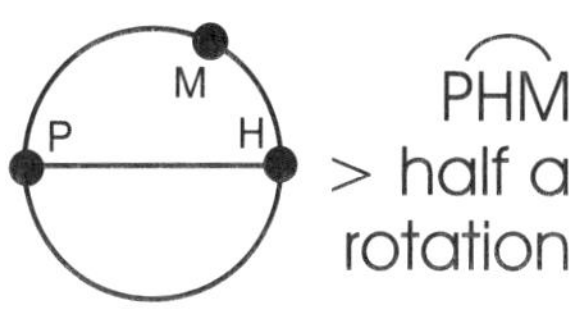 $\overset{\frown}{PHM}$ > half a rotation

Note: A complete rotation measures 360°.

A central angle has its vertex at the center of a circle and its measure is equal to the measure of its minor arc.

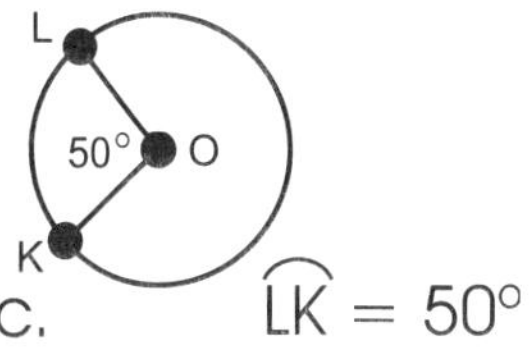

$\overset{\frown}{LK}$ = 50°

Name the indicated parts of circle B.

1. 3 central angles
2. 1 semicircle
3. 2 minor arcs
4. 2 major arcs

Find the measure of each angle or arc listed.

5. 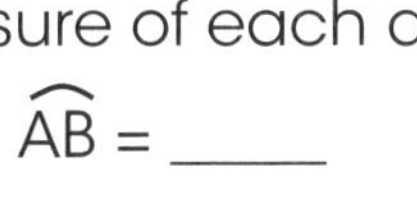 $\overset{\frown}{AB}$ = _____ $\overset{\frown}{BC}$ = _____

$\overline{AC}$ is a diameter.

6. 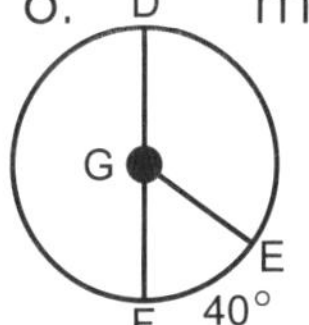 m∠FGE = _____ 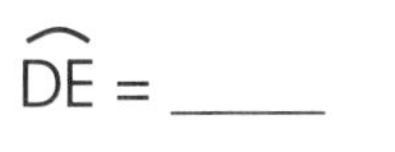 $\overset{\frown}{DE}$ = _____

$\overline{DF}$ is a diameter.

7. 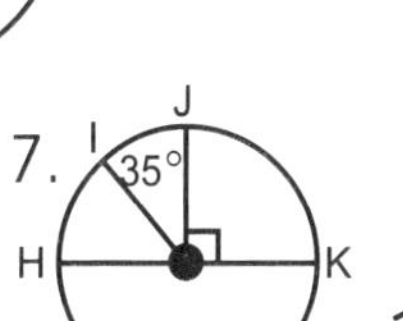 $\overset{\frown}{HI}$ = _____ $\overset{\frown}{HKJ}$ = _____

$\overline{HK}$ is a diameter.

8. 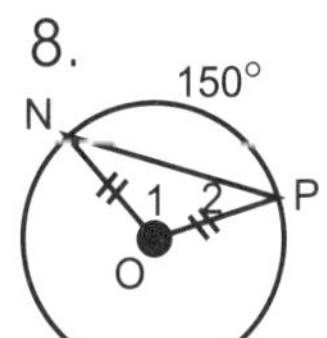 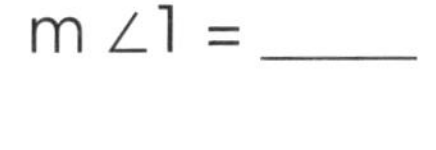 m∠1 = _____ 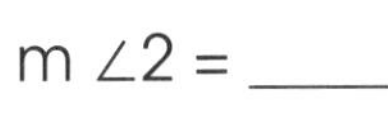 m∠2 = _____

Hint: △NOP is isosceles; ∠N≅∠P

9. 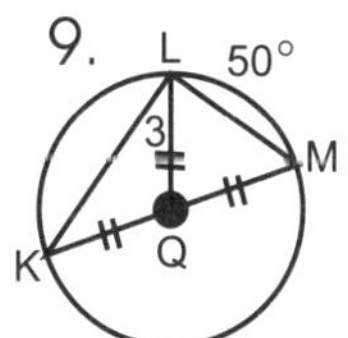 $\overset{\frown}{KL}$ = _____ 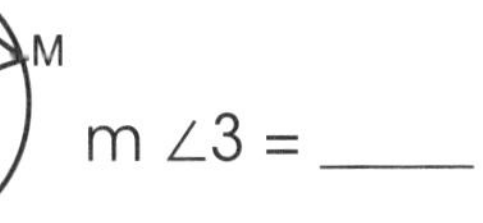 m∠3 = _____

$\overline{KM}$ is a diameter.

10. 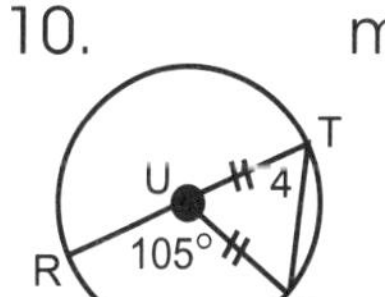 m∠4 = _____ 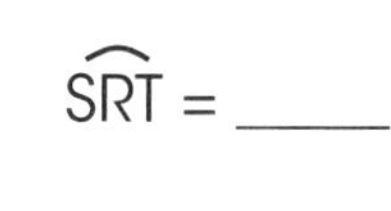 $\overset{\frown}{SRT}$ = _____

$\overline{RT}$ is a diameter.

11. 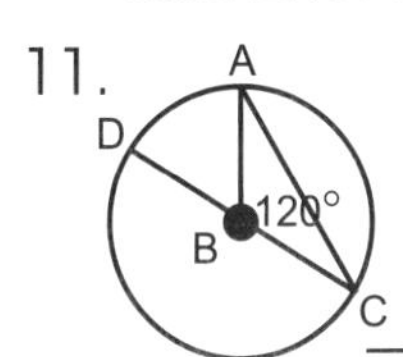 $\overset{\frown}{AC}$ = _____ $\overset{\frown}{DA}$ = _____

$\overline{DC}$ is a diameter.

12. 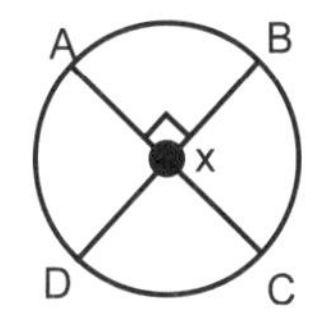 m∠AXB = _____ $\overset{\frown}{AC}$ = _____

13. 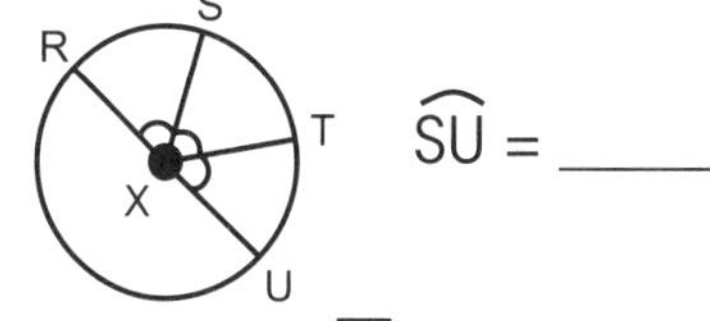 m∠RXS = _____ $\overset{\frown}{SU}$ = _____

$\overline{RU}$ is a diameter.

14. m∠5 = m∠6, m∠BED = 90°

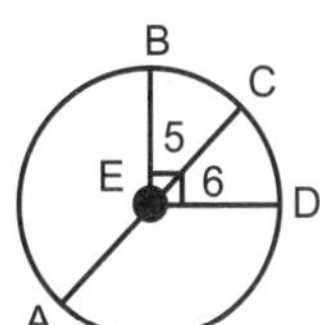

m∠5 = _____

$\overset{\frown}{AB}$ = _____ $\overline{AC}$ is a diameter.

$\overset{\frown}{BDA}$ = _____

15. m∠7 = m∠8 = 70°

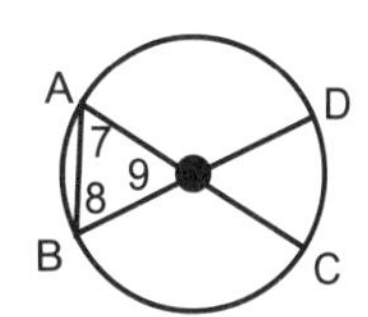

m∠9 = _____

$\overset{\frown}{CD}$ = _____

$\overline{AC}$ and $\overline{BD}$ are diameters.

Name____________________________________ HSG-C.A.2, HSG-C.B.5

Arcs and Angles

Fill in the blanks.

16. The measure of a minor arc ________________ the measure of the central angle that intercepts it.
17. A complete rotation has a measure of ________ degrees.
18. A ___________ measures 180°.
19. The measure of a major arc equals 360° minus the measure of the ___________ angle that intercepts it.

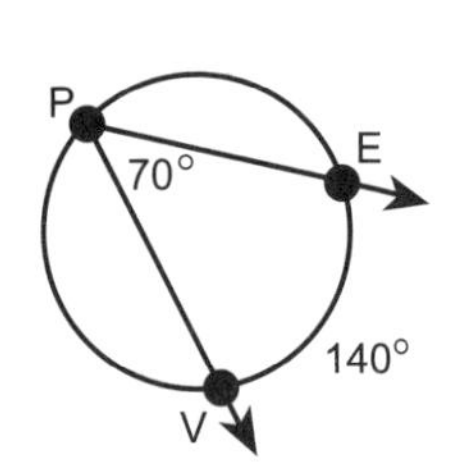

An **inscribed** angle has its vertex on the circle and its sides form chords of the circle. Inscribed ∠VPE intercepts $\overset{\frown}{VE}$.

m ∠VPE $= \frac{1}{2}$ m $\overset{\frown}{VE}$

Name the arc intercepted by the given angle.

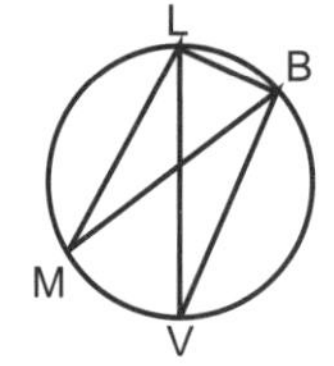

20. ∠LMB
21. ∠MBV
22. ∠LBV
23. ∠BVL
24. ∠MLB
25. ∠MBL
26. ∠BLV
27. ∠MLV

Find the measures of the indicated arcs and angles.

28. 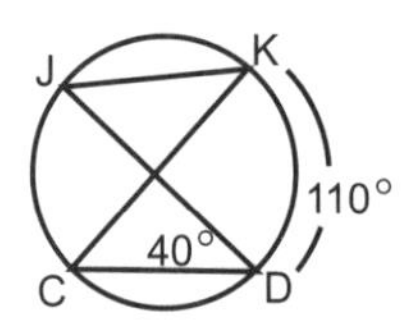

$\overset{\frown}{JC}$ = ______

m ∠JKC = ______

m ∠KJD = ______

29. 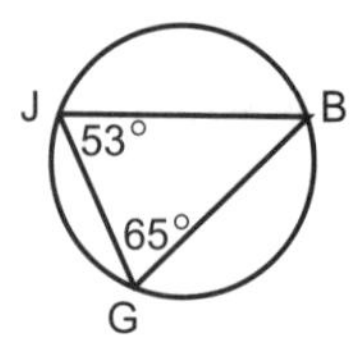

$\overset{\frown}{JB}$ = ______

$\overset{\frown}{GB}$ = ______

$\overset{\frown}{JG}$ = ______

30. 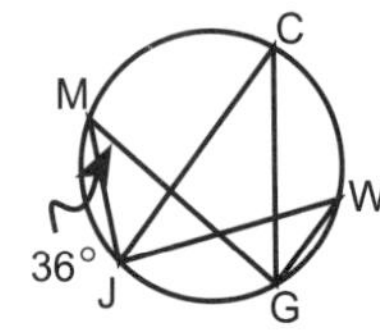

$\overset{\frown}{JG}$ = ______

m ∠C = ______

m ∠W = ______

31. 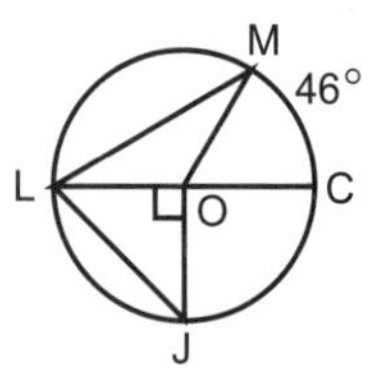

$\overset{\frown}{JM}$ = ______

m ∠JLM = ______

$\overset{\frown}{LM}$ = ______

32.  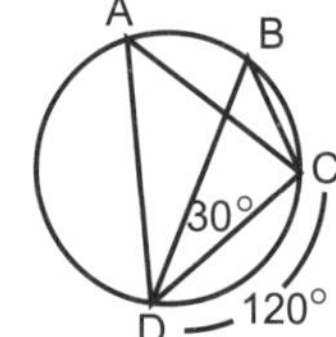

m ∠CBD = ______

m ∠CAD = ______

$\overset{\frown}{BC}$ = ______

33.  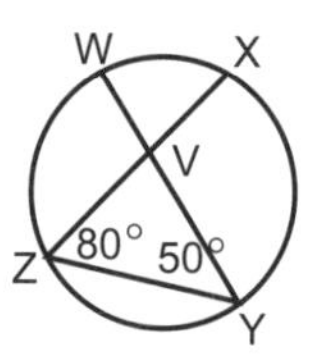

m ∠YVZ = ______

$\overset{\frown}{XY}$ = ______

$\overset{\frown}{WZ}$ = ______

34.  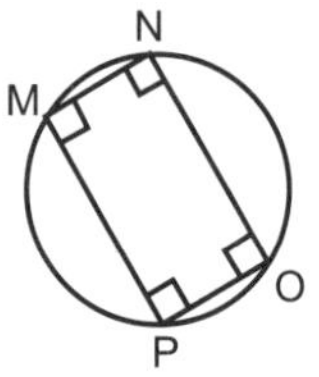

$\overset{\frown}{MPO}$ = ______

$\overline{NP}$ is a___________

35. 

m ∠KVE = ______

m ∠KIE = ______

Name________________________________ HSG-C.B.5

Sectors and Arcs

length of an arc (l) = $\frac{a}{180}\pi r$

where q is the measure of the arc

area of a sector (A) = $\frac{a}{360}\pi r^2$

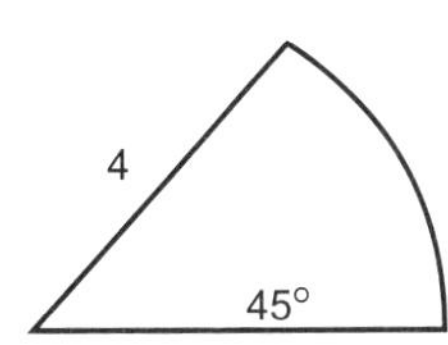

$l = \frac{45}{180}\pi(4)$ $A = \frac{45}{360}\pi(4)^2$

$= \frac{1}{4}\pi(4)$ $= \frac{1}{8}\pi(16)$

$l = \pi$ units $A = 2\pi$ square units

Find the length of each arc and the area of each sector.

1. 2.

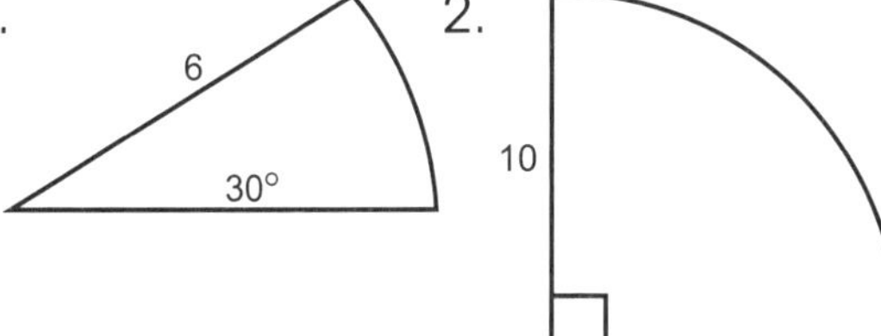

3.

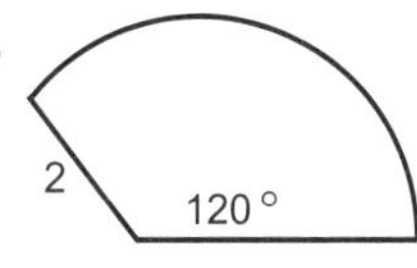

4.

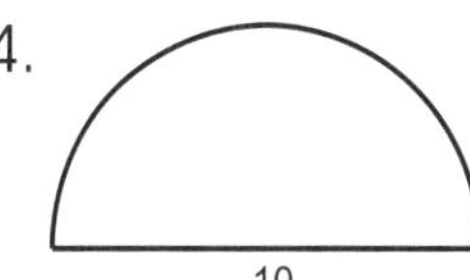

5.

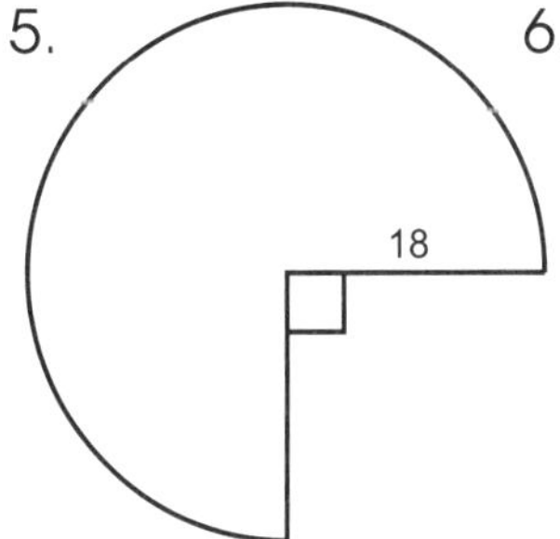

6.

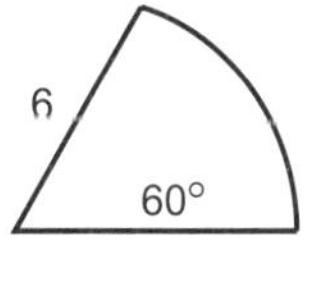

7.

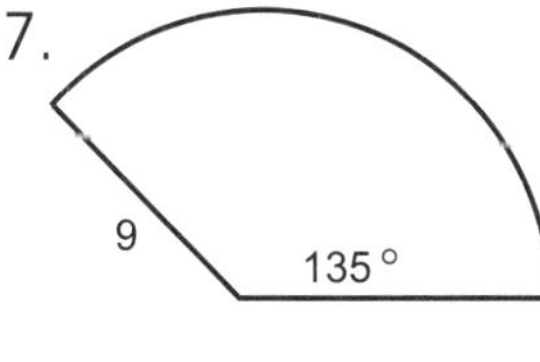

8.

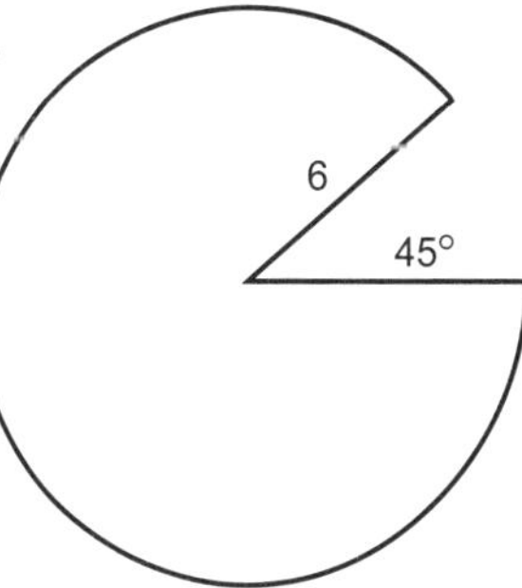

9.

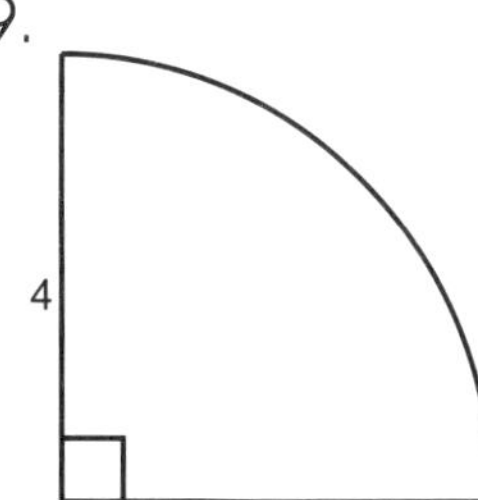

10.

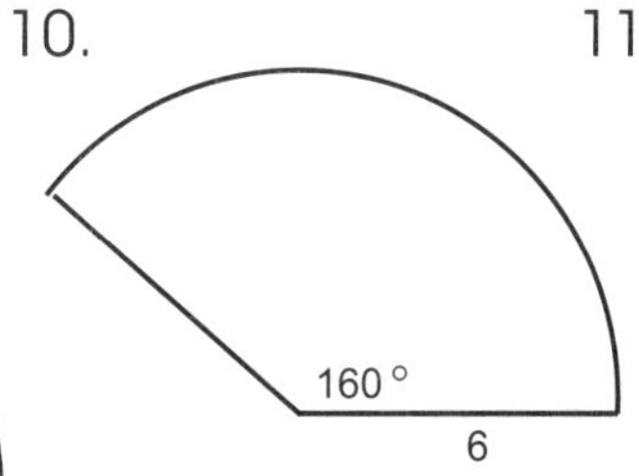

11.

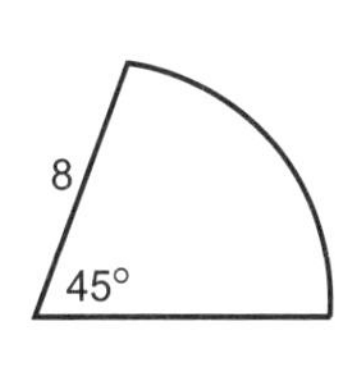

12.

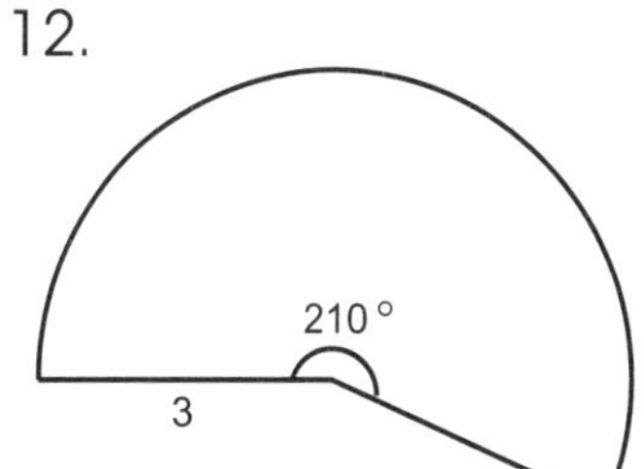

Name________________________________ HSG-C.A.2, HSG-C.B.5

Arcs and Chords

Two chords of a circle are congruent if they intercept congruent arcs.

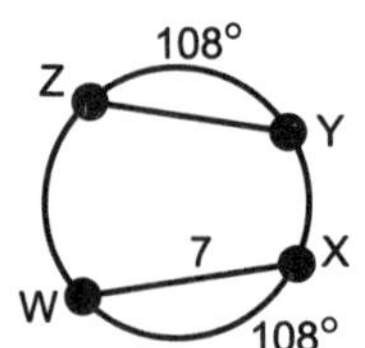

$\overline{ZY} \cong \overline{WX}$
$\overline{ZY} = 7$

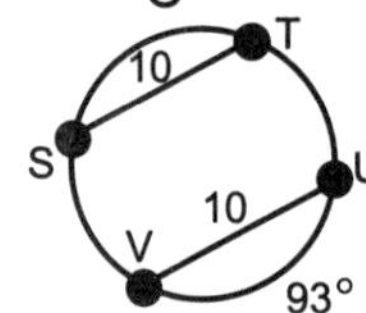

Two arcs of a circle are congruent if their chords are congruent.

The perpendicular bisector of a chord of a circle contains the center of the circle and bisects the arcs of chord.

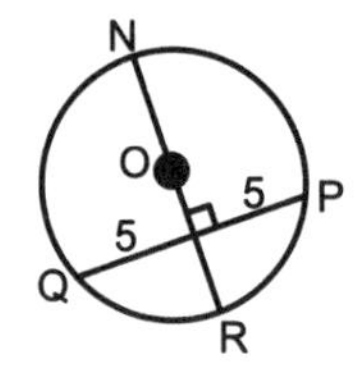

$\widehat{QR} \cong \widehat{PR}$
$\widehat{QN} \cong \widehat{PN}$

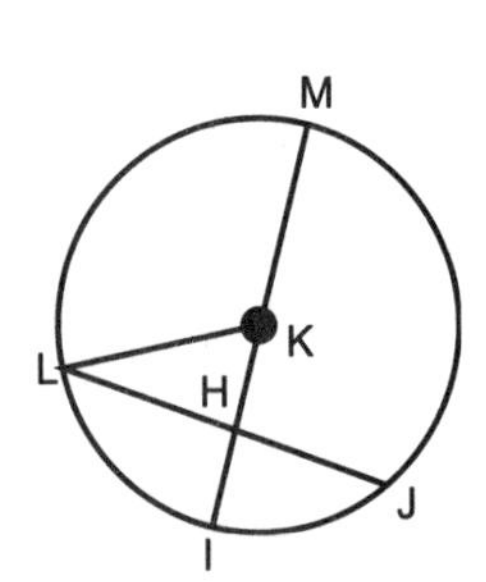

In ⊙ K, MI __ LJ; KH = 6, LJ = 16.

1. $\overline{LH} \cong$ ______
2. $\widehat{LM} \cong$ ______
3. $\overline{MI}$ bisects ______, ______ and ______.
4. The midpoint of $\overline{LJ}$ is ______.
5. LH = ______
6. LK = ______ **Remember:** ΔLHK is a right triangle.

Find the arc measures and segment lengths indicated.

7.

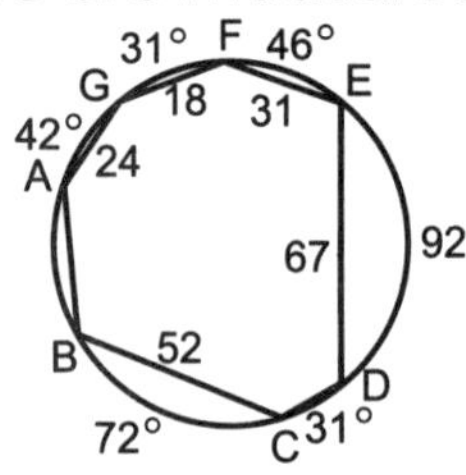

CD = ______
$\widehat{AB}$ = ______
AB = ______

8.

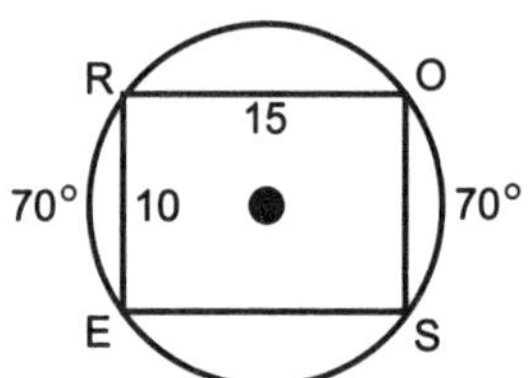

OS = ______
$\widehat{RO}$ = ______
ES = ______

9.

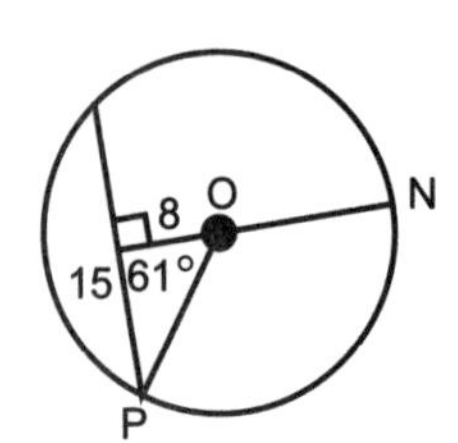

OP = ______
ON = ______

10.

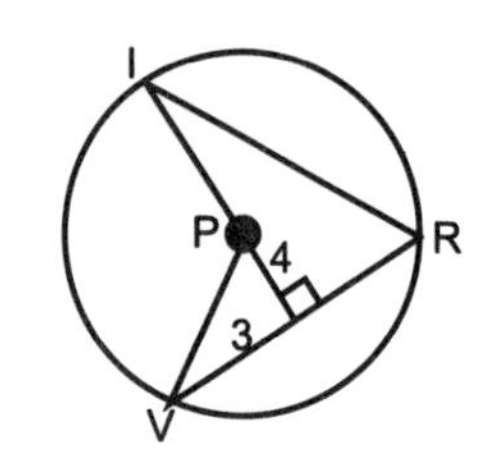

VP = ______
IR = ______

11.

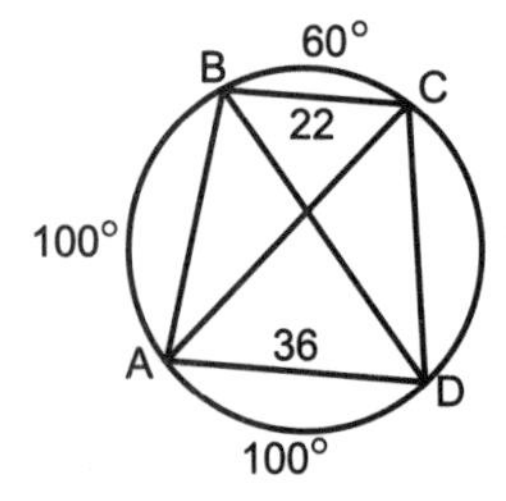

$\widehat{ADC}$= ______
AB = ______
CD = ______
$\widehat{BCD}$= ______

12.

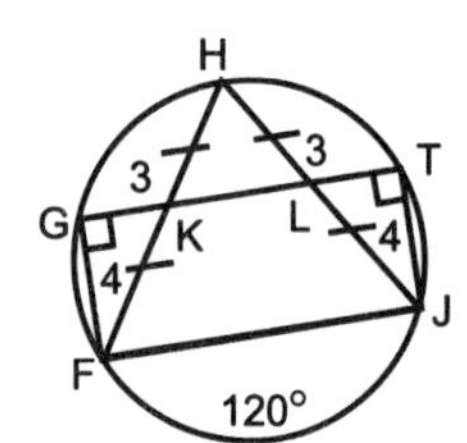

LJ = ______
HJ = ______
HF = ______
FJ = ______
KL = ______

Name________________________________ HSG-C.A.2

Lengths of Segments in a Circle

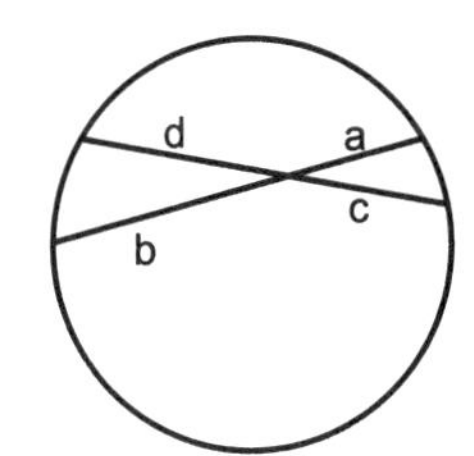

When 2 chords intersect, the product of the lengths along one chord equals the product of the lengths along the other chord.

$a \bullet b = c \bullet d$

Find the indicated lengths.

1.
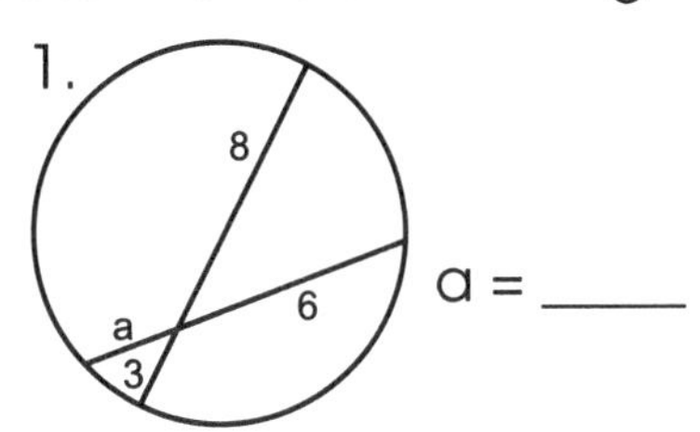

a = _____

5.
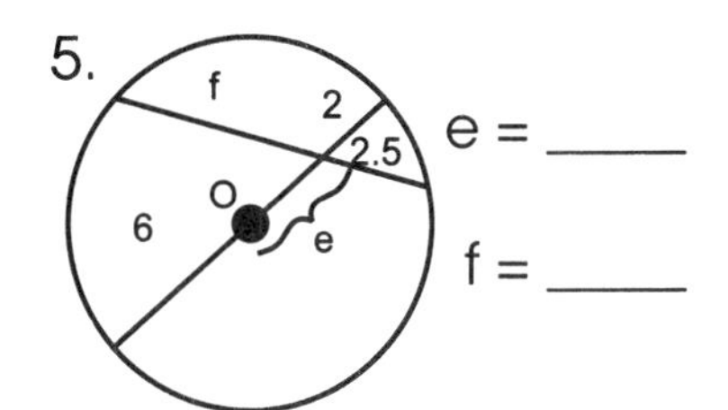

e = _____

f = _____

9.
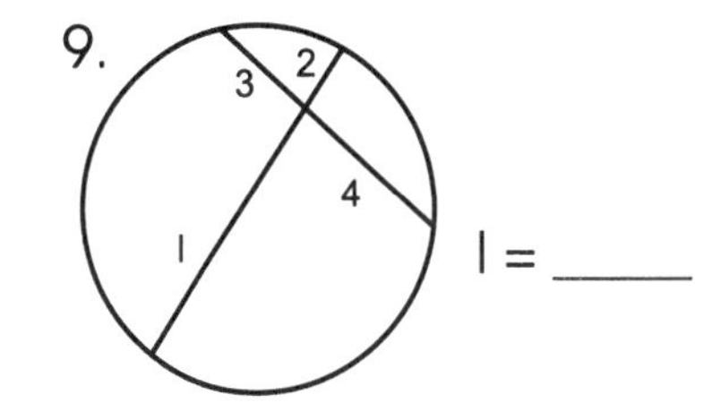

l = _____

2.
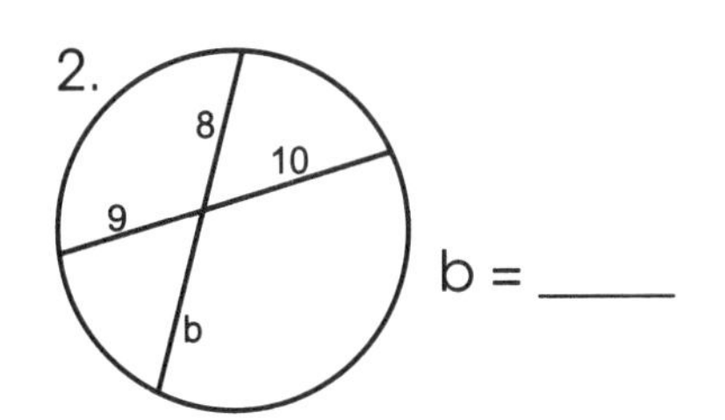

b = _____

6.
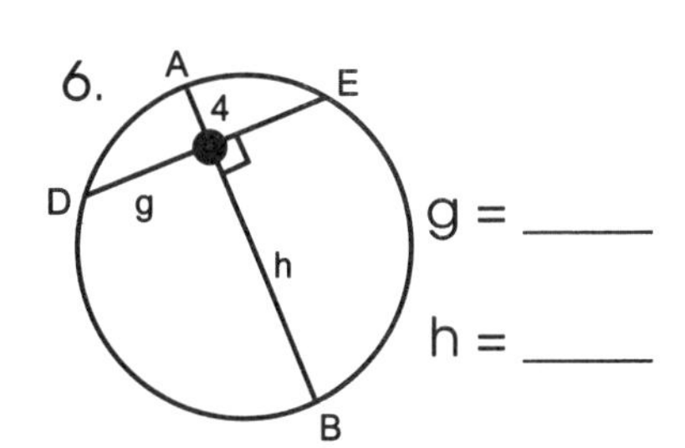

g = _____

h = _____

10.
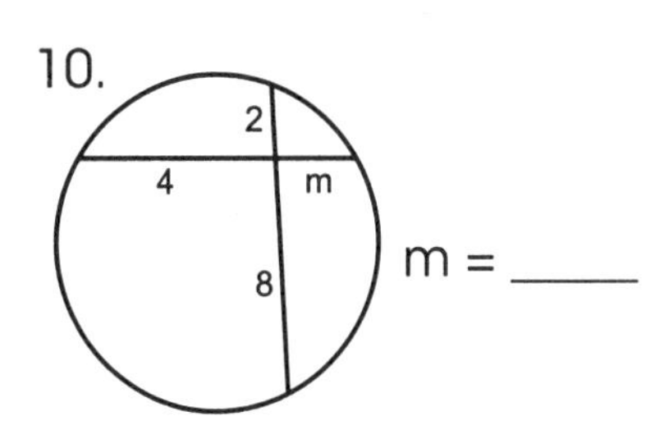

m = _____

3.
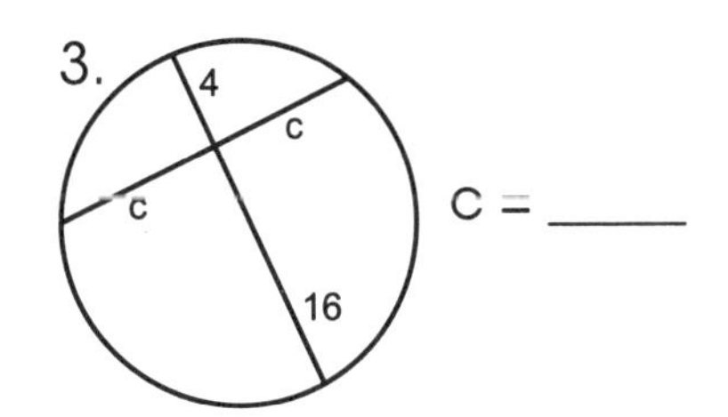

c = _____

7.
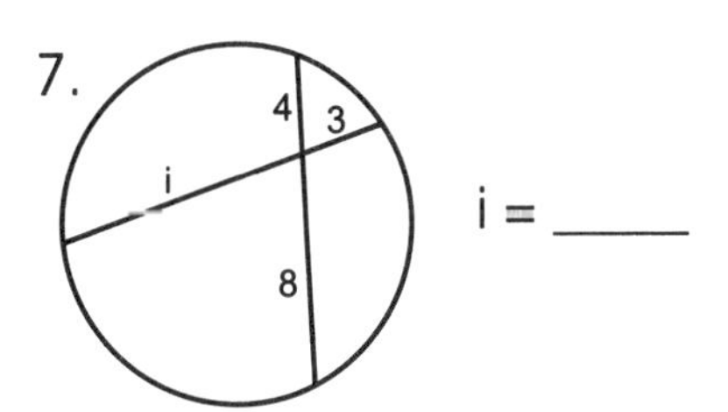

i = _____

11.
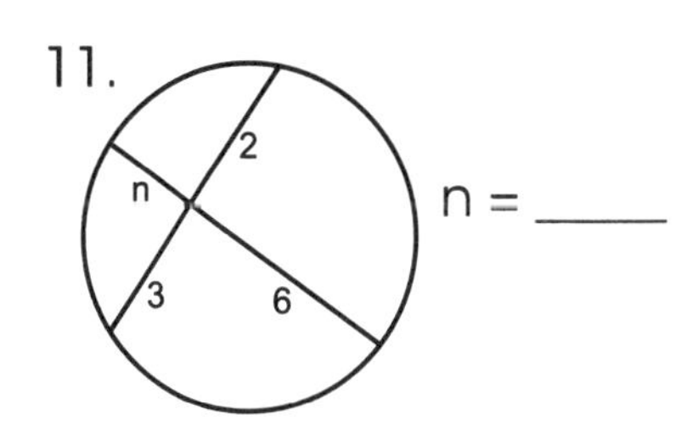

n = _____

4.
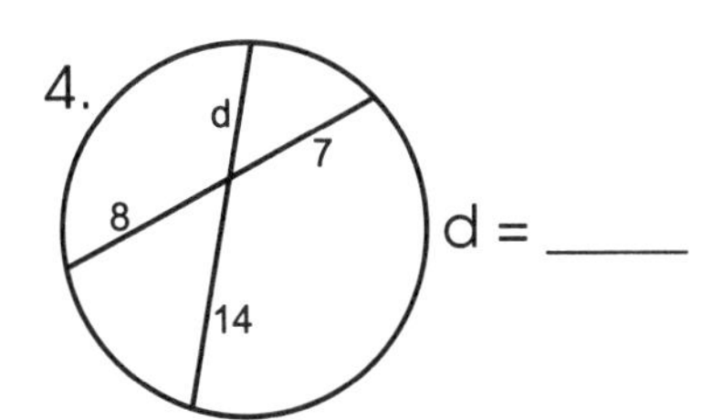

d = _____

8.
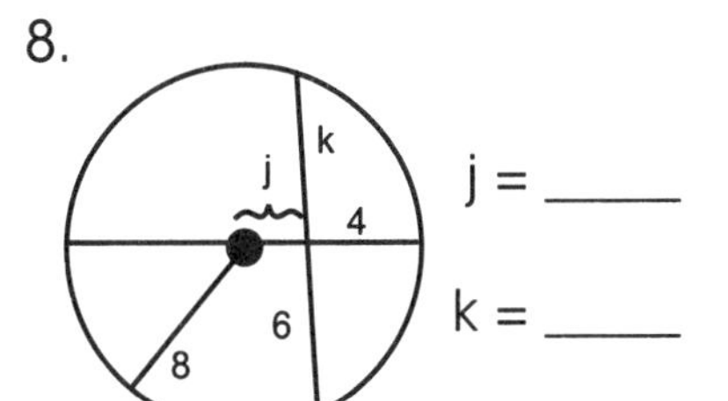

j = _____

k = _____

12.
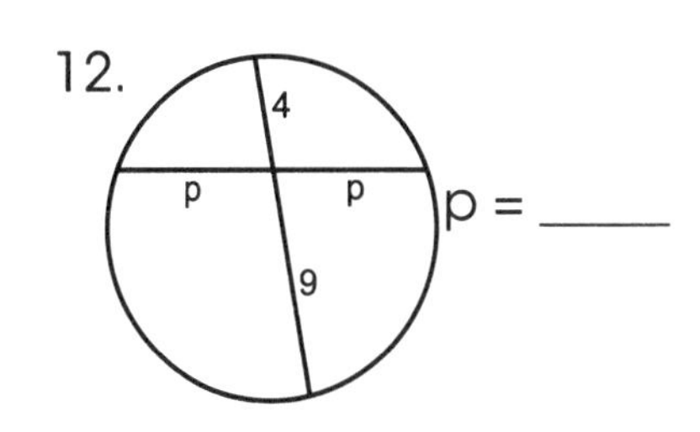

p = _____

The Earth's equator is a great circle. The diameter of the Earth at the equator is approximately 7926 miles. Substitute the values of the letters into the following to check your work.

(A x B x C) x [(D x E) + F] – (G x H x I) – [(J + K) x (L + M + N)] – P = 7926

(__x__x__) x [(__x__) +__] - (__x__x__) - [(__+__) x (__+__+__)] -__ = 7926

Name______________________________

HSG-C.A.2, HSG-C.A.3, HSG-C.B.5

Mixed Practice with Circles

Find the indicated parts.

1.

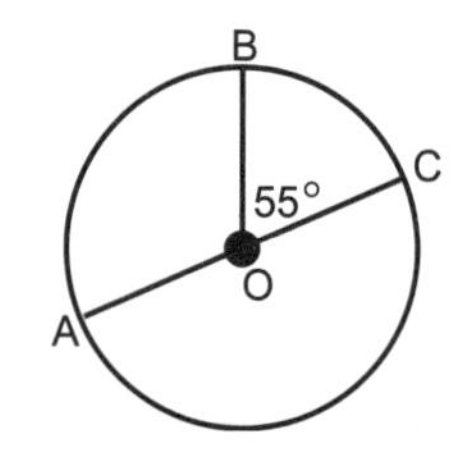

$\overset{\frown}{ABC}$ = ______

m ∠AOB = ______

$\overset{\frown}{AB}$ = ______

2.

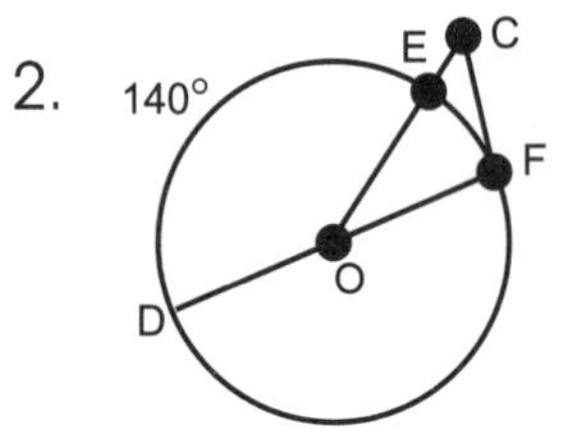

$\overset{\frown}{EF}$ = ______

m ∠EOF = ______

m ∠CFO = ______

3.

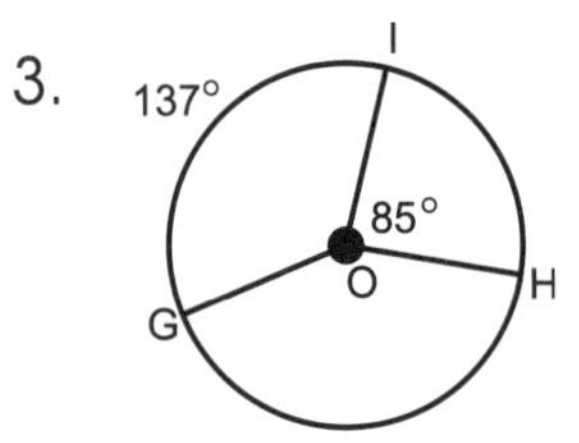

$\overset{\frown}{IH}$ = ______

$\overset{\frown}{GH}$ = ______

m ∠GOI = ______

4.

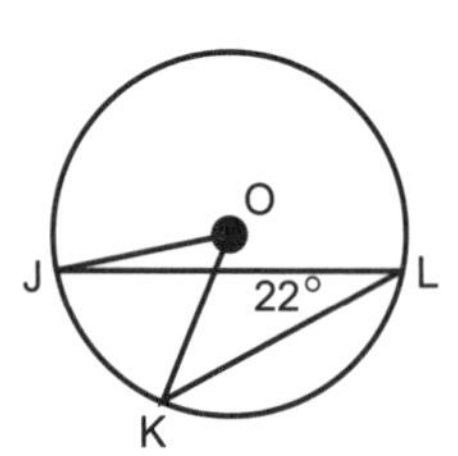

$\overset{\frown}{JK}$ = ______

m ∠JOK = ______

5.

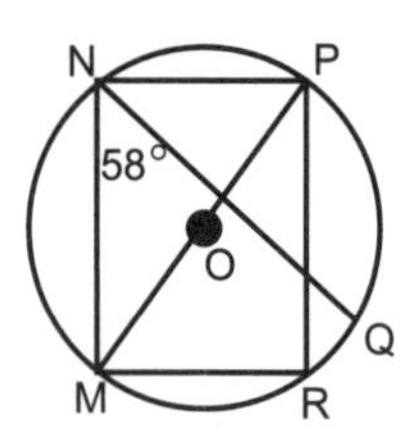

$\overset{\frown}{MQ}$ = ______

$\overset{\frown}{QP}$ = ______

m ∠PNQ = ______

m ∠MRP = ______

6.

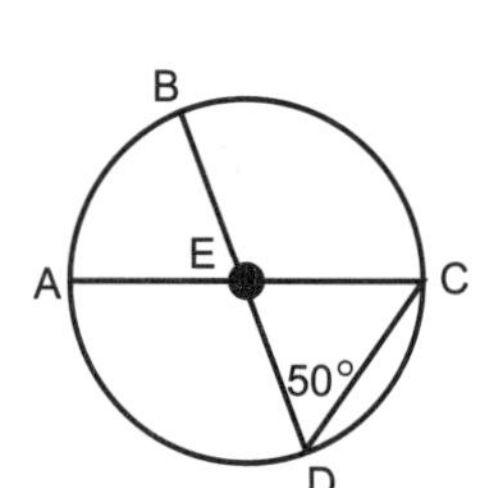

$\overset{\frown}{BC}$ = ______

m ∠BEC = ______

m ∠CED = ______

m ∠ECD = ______

7.

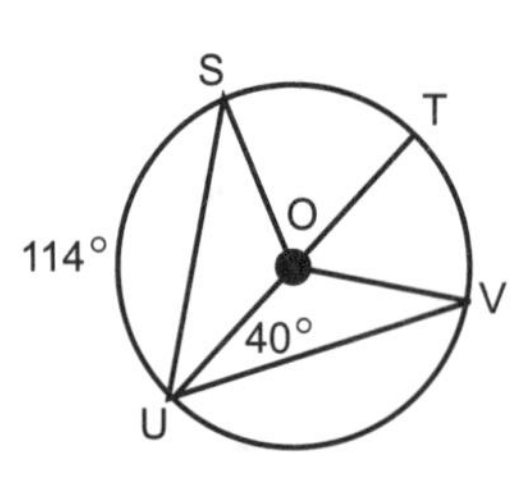

$\overset{\frown}{TV}$ = ______

m ∠UOV = ______

m ∠SUT = ______

8.

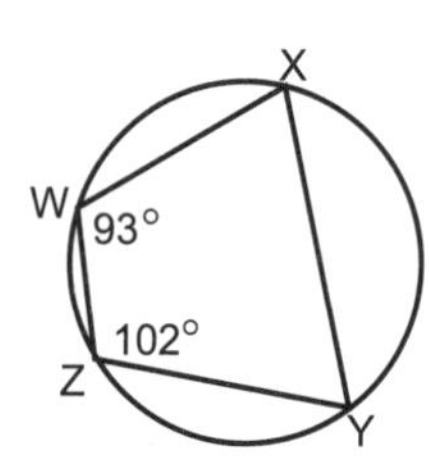

$\overset{\frown}{XYZ}$ = ______

m ∠Y = ______

$\overset{\frown}{WZY}$ = ______

9.

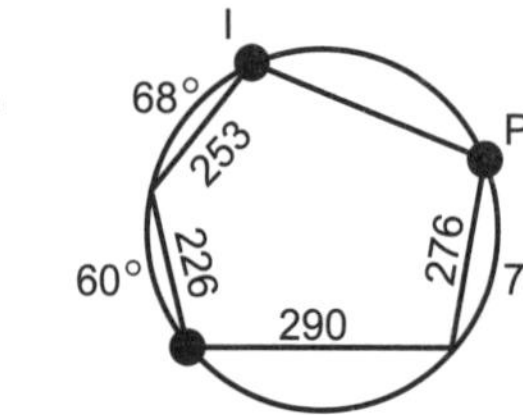

$\overset{\frown}{IP}$ = ______

IP = ______

10.

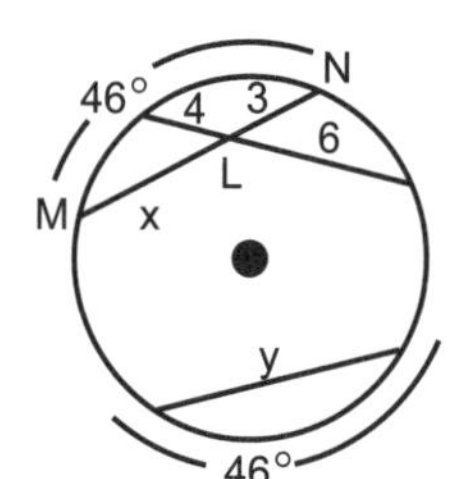

x = ______

y = ______

11.

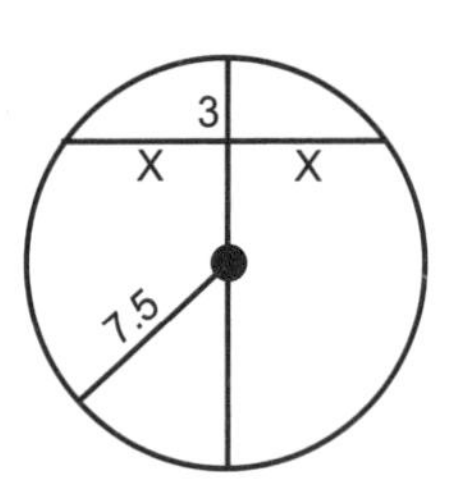

x = ______

12.

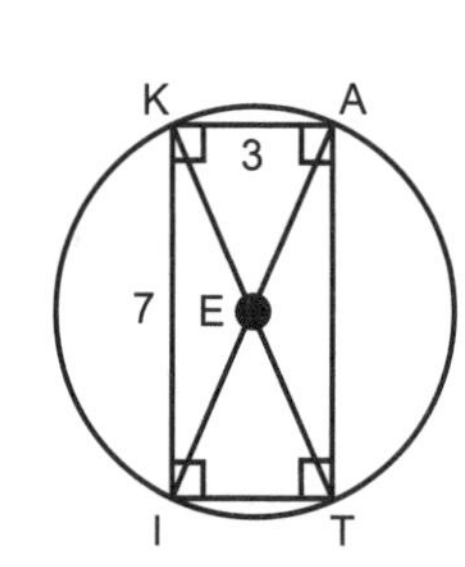

AT = ______

IT = ______

$\overset{\frown}{AI}$ = ______

Name______________________________

HSG-C.B.5

Radians

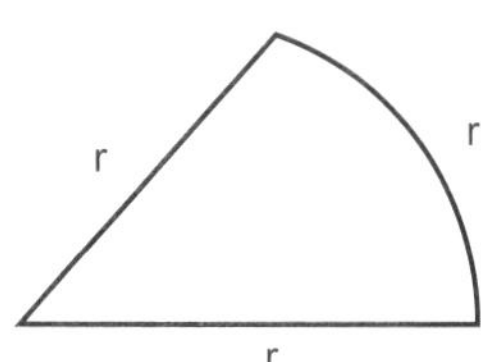

A **radian** is defined to be the measure of an angle which has its vertex at the center of a circle and which intercepts an arc whose length is equal to the radius.

The circumference and the radius are related by the equation $C = 2\pi r$. Thus, there are 2π radians in the complete circle. From this, we can obtain the following:

$$2\pi \text{ radians} = 360°$$

$$1 \text{ radian} = \frac{180°}{\pi} = 57.3°$$

$$1° = \frac{\pi}{180°} \text{ radians} = 0.01745 \text{ radians}$$

Convert the following angle measures from degrees to radians or from radians to degrees.

$\text{degrees} \times \frac{\pi}{180} = \text{radians}$

$\text{radians} \times \frac{180}{\pi} = \text{degrees}$

1. 180°
2. $\frac{\pi}{2}$ radians
3. 27°
4. 45°
5. 6.2832 radians
6. 4.7 radians
7. 2 radians
8. 90°
9. 0.05235 radians
10. $\frac{\pi}{3}$ radians
11. 1.0472 radians
12. 36°

Name________________________________

HSG-C.B.5

Application of Radians

$\theta = \frac{\pi}{2}$ rad., r = 2

$S = \theta r$

$= (\frac{\pi}{2})(2)$

$S = \pi$ units

$\theta = 45°$, r = 8

$\theta = 45° \times \frac{\pi}{180°} = \frac{\pi}{4}$ radians

$S = (\frac{\pi}{4})(8)$

$S = 2\pi$ units

The length of an arc is directly proportional to the size of the central angle. In other words, the greater the angle, the greater the arc. Since there are 2π radians in the complete circle, then the length of the arc can be expressed as $S = \theta r$, where θ is the measure of the central angle and S is the length of the arc.

Complete the following table.

	θ	r	S
1.	180°	1	
2.		1	3π
3.	$\frac{\pi}{4}$ rad.	5	
4.		2	4π
5.	2 rad.	3	
6.	$\frac{\pi}{2}$ rad.		2π
7.	45°	4	
8.	3 rad.		4
9.	1°	1	
10.	270°	6	

Name______________________________________

HSG-CO.D.12

Draw, Fold, But Don't Spindle

Several geometric shapes can be drawn or folded using simple materials.

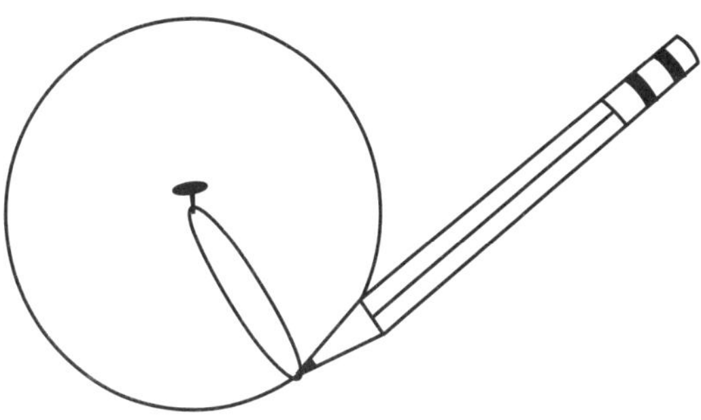

Circle: Thumbtack, string, cardboard, pencil

1. Place the thumbtack in the center of the cardboard.
2. Tie a string in a loop that when pulled taut is the length of the radius.
3. Place the loop around the thumbtack and pull the loop taut with the pencil.
4. Draw the circle keeping the loop taut.

Why: A circle is a set of points a given distance (radius) from a point (center).

Ellipse: 2 thumbtacks, string, cardboard, pencil

1. Place 2 thumbtacks (foci) two inches apart on the cardboard.
2. Tie a string in a loop that when pulled taut is four inches in length.
3. Place the loop around the thumbtacks and pencil. Pull the loop taut with the pencil.
4. Draw the ellipse keeping the loop taut.

Why: An ellipse is the set of points in which the sum of the distances from the foci is a constant.

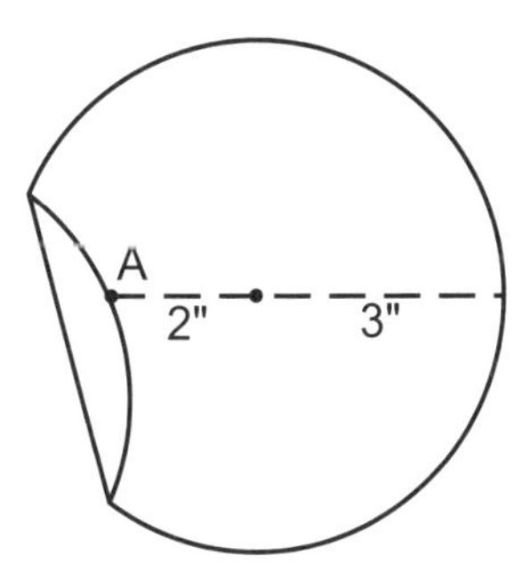

Ellipse: Thin paper or wax paper, ruler, pencil

1. Draw a three-inch radius circle. Mark the center.
2. Draw a point A two inches from the center.
3. Fold and crease the paper so a point on the circle touches point A.
4. Make about 40 folds around the circle.

Why: The sum of the distance from the fold to the center and the fold to point A is constant.

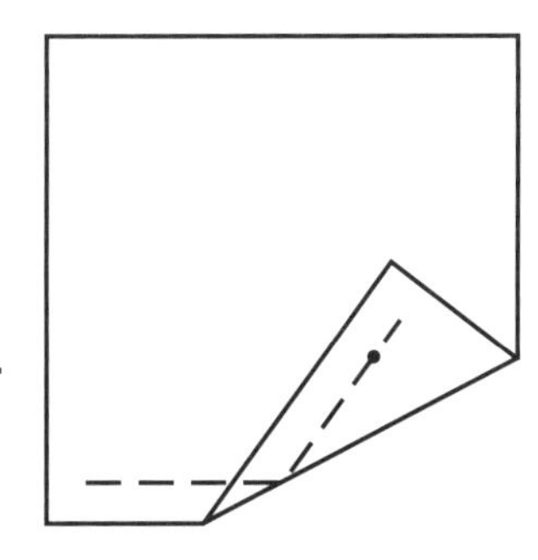

Parabola: Thin paper or wax paper, ruler, pencil

1. Draw a point (focus).
2. Draw a line parallel to the bottom of the paper.
3. Fold and crease the paper about 40 times so the line touches the point.

Why: A parabola is the set of points equidistant from a point and a line.

Name________________________________

8.EE.B.5, HSG-GPE.B.4

The Coordinate Plane

Each point is designated by two coordinates (x, y).
Point A (⁻3, 5)

The quadrants of the plane are numbered counterclockwise as shown.

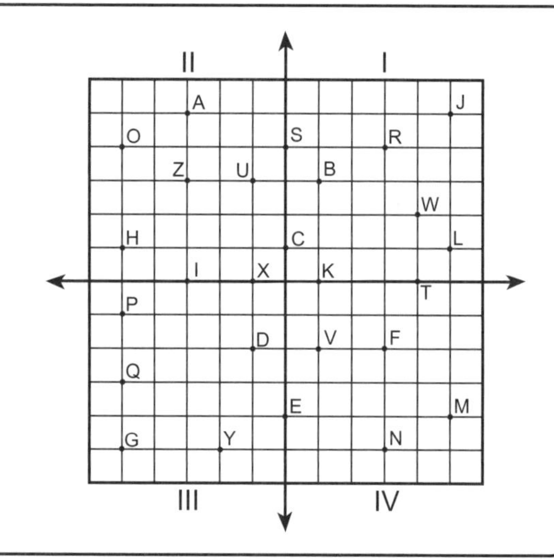

Give the coordinates of the following points.

C _______ G _______ P _______ U _______ V _______ Z _______

Use the coordinates to locate the correct letter on the graph.

1. Where is Rutherford B. Hayes buried?

(3, ⁻2) (3, 4) (0, ⁻4) (5, ⁻4) (⁻5, 4) (3, ⁻5) (4, 0)   (⁻5, 4) (⁻5, 1) (⁻3, 0) (⁻5, 4)

2. Darwin, MN claims to have the largest what?

(1, 3) (⁻3, 5) (5, 1) (5, 1)   (⁻5, 4) (3, ⁻2) (4, 0) (4, 2) (⁻3, 0) (3. ⁻5) (0, ⁻4)

3. Who was the tenth president of the United States?

(5, 5) (⁻5, 4) (⁻5, 1) (3, ⁻5)   (4, 0) (⁻2, ⁻5) (5, 1) (0, ⁻4) (3, 4)

4. A line is a simple figure in the coordinate plane. Name three points on the line.

_______ _______ _______

5. The line passes through which quadrants?

____

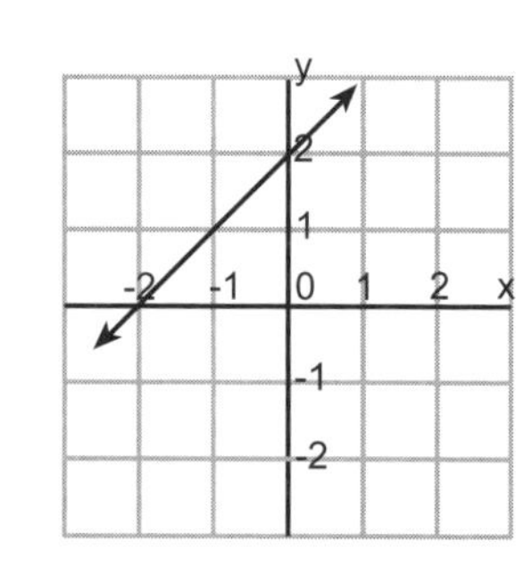

6. Give the location by quadrant(s) of the following points.

(⁻2, ⁻5) _______   (3, ⁻1) _______

Equal x- and y-coordinates. _______
Opposite x- and y-coordinates. _______

Name________________________

Lines and Their Equations

I. Plotting Points

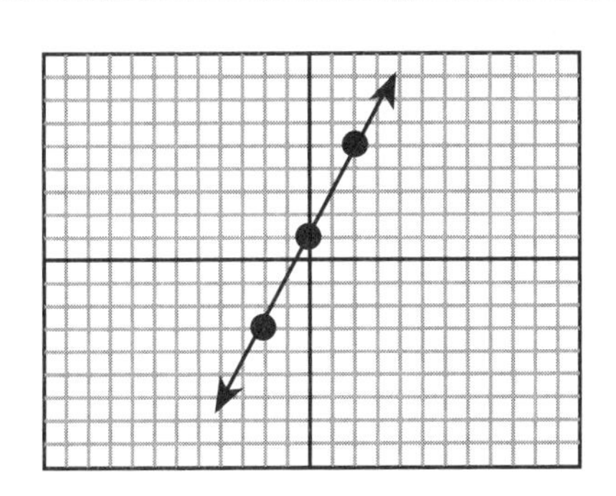

x	y
$^-2$	$^-3$
0	1
2	5

$y = 2x + 1$

$y = 2(^-2) + 1 = {^-4} + 1 = {^-3}$
$y = 2(0) + 1 = 0 + 1 = 1$
$y = 2(2) + 1 = 4 + 1 = 5$

Graph the following lines by plotting points. Use your own graph paper.

1. $y = -x + 6$
2. $y = \frac{1}{2}x - 5$
3. $2x + 3y = 6$
4. $8x - 2y = {^-6}$
5. $3x - 2y = {^-6}$
6. $x + 6 = {^-3}y$

II. Slope-Intercept

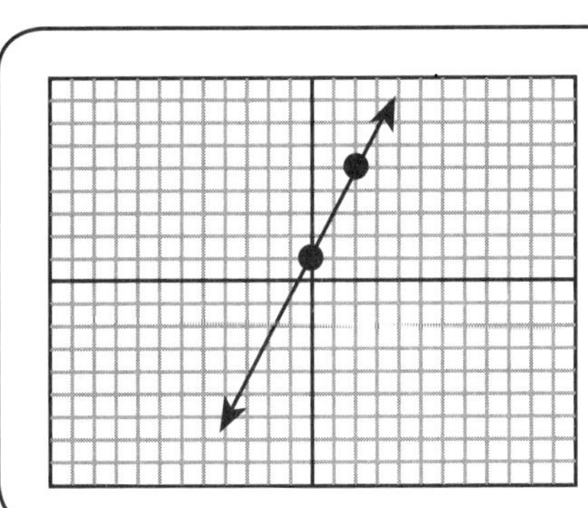

slope-intercept form is $\mathbf{y = mx + b}$
slope = m   y-intercept = b

$$\text{slope} = \frac{\text{vertical change}}{\text{horizontal change}} = \frac{\text{rise}}{\text{run}}$$

$y = 2x + 1$ slope = $\frac{2}{1}$
y -intercept = 1
Start at (0, 1). Then, move up 2 and right 1.

Graph the following lines by using the slope and y-intercept. Use your own graph paper.

7. $y = {^-2}x + 3$
8. $y = \frac{2}{3}x - 1$
9. $y = 4x - 3$
10. $y = \frac{1}{3}x$
11. $y = -\frac{1}{4}x + 3$
12. $y = 2$

Name______________________________ 8.EE.B.5, HSG-GPE.B.5

Lines and Their Equations

III. Standard Form

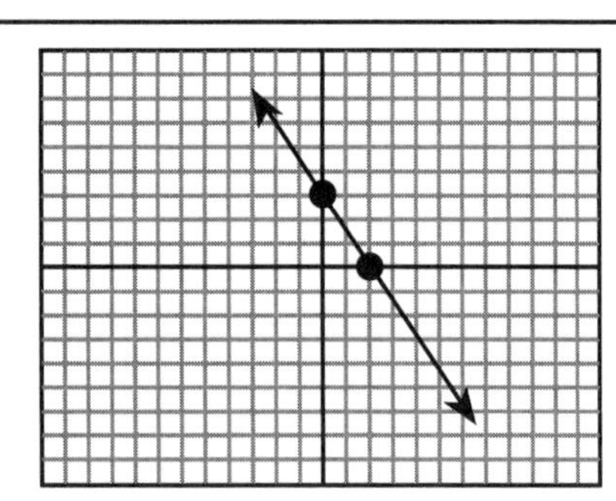

Standard form is $Ax + By = C$

$3x + 2y = 6$

$2y = {}^-3x + 6$

$y = -\frac{3}{2}x + 3$

$m = -\frac{3}{2}$ y-intercept = 3

Start at (0, 3). Then, move down 3 and right 2.

Graph the following lines by using the slope and y-intercept. Use your own graph paper.

13. $2x + 5y = 10$
14. $3x - 4y = 12$
15. ${}^-2x - 3y = 6$
16. $3x + 4y = 1$
17. ${}^-2x + y = 4$
18. $x - 3y = 2$

IV. Segments

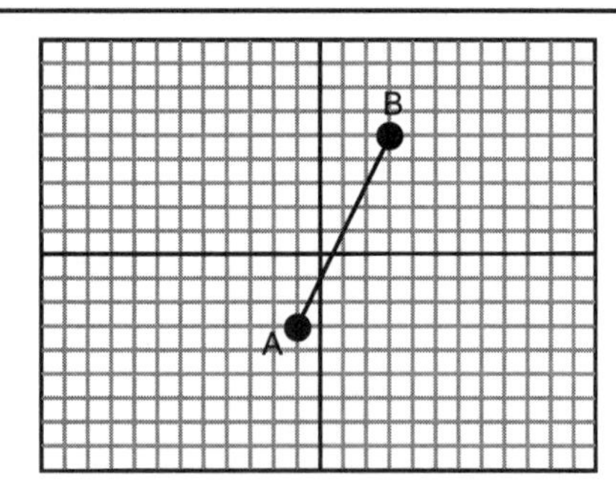

A $({}^-1, {}^-3)$ B $(3, 5)$
and
$y = 2x - 1$ for ${}^-1 \le x \le 3$
are two different ways to express $\overline{AB}$.

Graph the following segments. Use your own graph paper.

19. $\overline{CD}$ C $({}^-2, 2)$ D $(4, {}^-1)$
20. $\overline{EF}$ E $(0, {}^-4)$ F $(5, 2)$
21. $\overline{GH}$ G $({}^-3, {}^-5)$ H $(2, 5)$
22. $\overline{IJ}$ $y = \frac{1}{2}x + 2$ for ${}^-4 \le x \le 2$
23. $\overline{KL}$ $y = {}^-2x + 3$ for ${}^-1 \le x \le 3$
24. $\overline{MN}$ $2x - 3y = {}^-6$ for ${}^-3 \le x3 \le 3$

Name____ 8.EE.B.5, HSG-GPE.B.5

Equation of a Line in Standard Form: Ax + By = C

I. Given the slope and a point.

$m = \frac{1}{4}$, (⁻4, 3)

Use slope-intercept form and solve for b.

$y = mx + b$	$y = \frac{1}{4}x + 4$
$3 = \frac{1}{4}(^-4) + b$	$4y = x + 16$
$3 = {}^-1 + b$	$-x + 4y = 16$
$4 = b$	$x - 4y = {}^-16$

Match the equation of the line to the given conditions.

1. m = ⁻2, (3, 1)	A. 3x + 2y = 5
2. m = 2, (1, ⁻2)	B. 2x + y = 7
3. $m = -\frac{3}{2}$, (1, 1)	C. x + y = 1
4. m = ⁻1, (⁻1, 2)	D. 2x – y = 4
5. $m = , \frac{1}{3}$ (6, 3)	E. x – 3y = ⁻3

II. Given two points

(1, 4), (⁻1, ⁻2)

Use slope formula to find m. Use slope-intercept form to find b.

$m = (y_2 - y_1)/(x_2 - x_1) = (^-2 - 4)/(^-1 - 1) = {}^-6/{}^-2 = 3$

$y = mx + b$	$y = 3x + 1$
$4 = 3(1) + b$	$^-3x + y = 1$
$4 = 3 + b$	$3x - y = {}^-1$
$1 = b$	

Match the equation of the line to the given conditions.

6. (1, 3), (⁻1, ⁻1)	F. x + 3y = 6
7. (3, 1), (6, 0)	G. 2x – 3y = ⁻6
8. (⁻3, 0), (0, 2)	H. 2x – y = ⁻1
9. (4, 2), (2, 4)	I. x + 2y = 5
10. (1, 2), (⁻1, 3)	J. x + y = 6

Name________________________________ 8.EE.B.5, HSG-GPE.B.5

Equation of a Line in Standard Form: Ax + By = C

III. Given a parallel line and a point

Parallel to $y = 2x - 1$ through (3, 5)
Remember: Parallel lines have the same slope.
$m = 2 \quad (3, 5)$

$y = mx + b$
$5 = 2(3) + b$
$5 = 6 + b$
$^-1 = b$

$y = 2x - 1$
$^-2x + y = {}^-1$
$2x - y = 1$

Match the equation of the line to the given conditions.

11. Parallel to $y = 3x + 4$ through (0, -2)
12. Parallel to $y = \frac{1}{2}x - 3$ through (4, 2)
13. Parallel to $2x + 3y = 6$ through (4, -3)
14. Parallel to $3x - 4y = 1$ through (2, 2)
15. Parallel to $3x + 2y = 10$ through $(1, \frac{1}{2})$

K. $3x + 2y = 4$
L. $2x + 3y = {}^-1$
M. $3x - 4y = {}^-2$
N. $3x - y = 2$
O. $x - 2y = 0$

IV. Given a perpendicular line and a point

Perpendicular to $y = 2x - 1$ through (3, 5)
Remember: Perpendicular lines have slopes whose product is negative one.
$y = 2x - 1$ has a slope of 2, so m will equal $-\frac{1}{2}$ because $-\frac{1}{2} \bullet 2 = -1$

$y = mx + b$
$5 = -\frac{1}{2}(3) + b$
$5 = \frac{-3}{2} + b$
$\frac{13}{2} = b$

$y = -\frac{1}{2}x + \frac{13}{2}$
$2y = -x + 13$
$x + 2y = 13$

Match the equation of the line to the given conditions.

16. Perpendicular to $y = 3x + 4$ through (0, $^-2$)
17. Perpendicular to $y = \frac{1}{2}x - 3$ through (1, 4)
18. Perpendicular to $3x + 2y = 6$ through (1, 1)
19. Perpendicular to $2x - 5y = 2$ through (2, 3)
20. Perpendicular to $x + y = 4$ through ($^-3$, $^-3$)

P. $x + 3y = {}^-6$
Q. $2x - 3y = {}^-1$
R. $2x + y = 6$
S. $x - y = 0$
T. $5x + 2y = 16$

V. Summary

Write the equation of the line with the following conditions.

21. $m = 4$ through $(\frac{1}{2}, {}^-2)$________________
22. Through (2, $^-1$) and (8, 1)________________
23. Parallel to $2x - y = -3$ through (2, $^-1$)________
24. Perpendicular to $3x + 6y = 5$ through (4, 1) ________________

Name________________________________ HSG-GPE.A.1

Nonlinear Equations— Circles

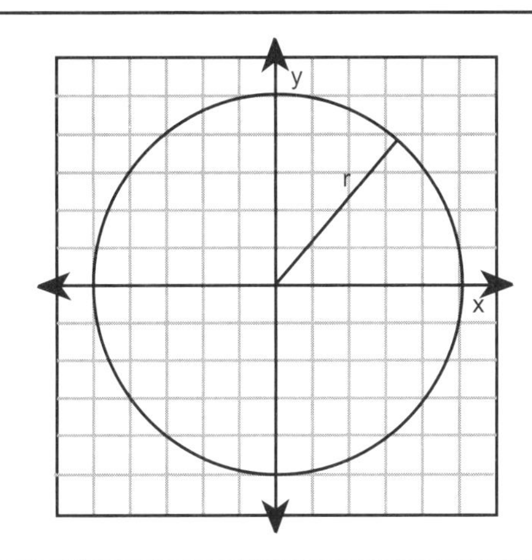

The graph of $x^2 + y^2 = r^2$ is a **circle** with radius r and center at the origin. A more general equation can be derived from the distance formula.

$$\sqrt{(x-a)^2 + (y-b)^2} = r \rightarrow (x-a)^2 + (y-b)^2 = r^2$$

This is an equation for a circle with radius r and center at (a, b).

1. Graph the following equations by plotting points.
   a. $x^2 + y^2 = 4$
   b. $x^2 + y^2 = 25$
   c. $(x-1)^2 + (y+1)^2 = 16$
   d. $(x+2)^2 + (y-2)^2 = 9$

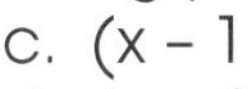

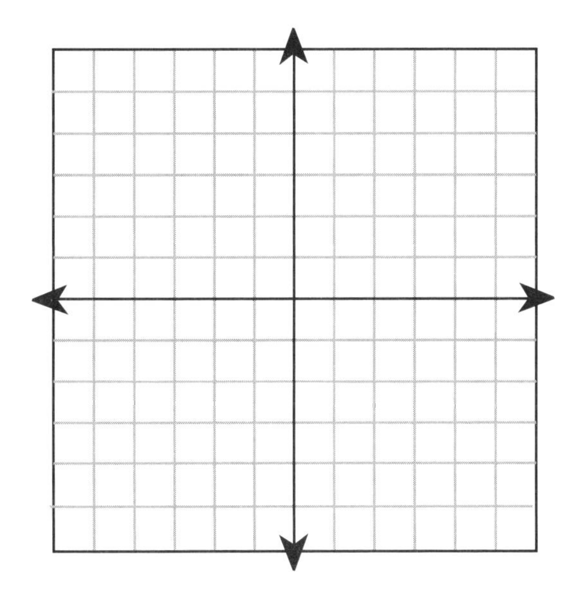

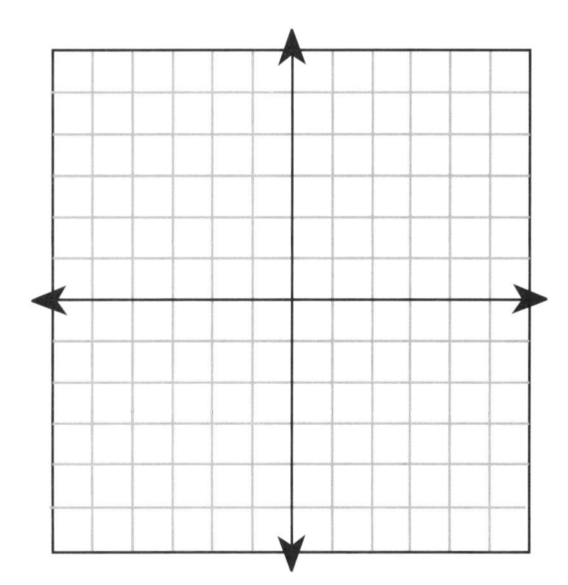

2. Give the center and radius for the circles below.
   a. $x^2 + y^2 = 9$
   b. $(x-3)^2 + (y-4)^2 = 16$
   c. $(x+2)^2 + (y-1)^2 = 4$
   d. $x^2 + (y+3)^2 = 25$
   e. $(x-1)^2 + (y+2)^2 = 4$
   f. $(x+5)^2 + (y-3)^2 = 81$
   g. $(x-7)^2 + (y+5)^2 = 24$
   h. $(x-3)^2 + (y-3)^2 = 18$

3. Write the equations of the following circles.
   a. $r = 1$ $(2, {}^{-}3)$
   b. $r = 2$ $(3, 4)$
   c. $r = 3$ $(2, 2)$
   d. $r = 6$ $(0, 0)$
   e. $r = 5$ $({}^{-}1, {}^{-}3)$
   f. $r = 2$ $({}^{-}2, 4)$
   g. $r = 1$ $(3, {}^{-}2)$
   h. $r = 4$ $({}^{-}2, {}^{-}3)$

Name________________________________

HSG-GPE.A.2

Nonlinear Equations— Parabolas

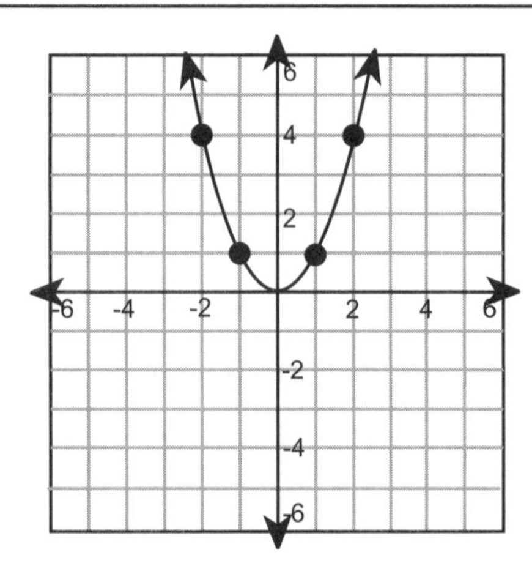

$y = x^2$ is the equation of a **parabola**.
It is graphed at the left.

1. Looking at the curve $y = x^2$, what do you think the graph of $y = 2x^2$ would look like?

   ____

   Sketch the graph of $y = 2x^2$.

2. Sketch the graph of $y = \frac{1}{2}x^2$.

3. Sketch the graph of $y = x^2 + 2$.

4. Sketch the graph of $y = (x + 2)^2$.

5. Sketch the graph of $y = -x^2$.

6. What do you expect $y = (x - 3)^2 + 1$ to look like?________________________
   Check your answer by graphing the equation.

7. What do you expect $y = (x + 1)^2 - 2$ to look like?________________________
   Check your answer by graphing the equation.

8. What do you expect $y = 3x^2 - 2$ to look like?________________________
   Check your answer by graphing the equation.

Name________________________________

HSG-GPE.A.3

Nonlinear Equations—Ellipses

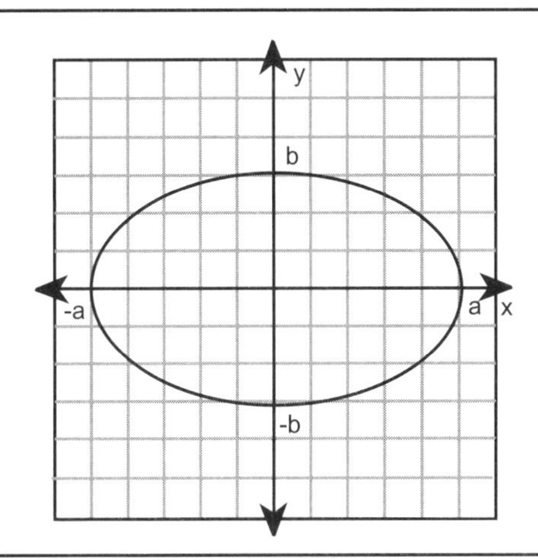

The standard form for the equation of an **ellipse** is

$$\frac{x^2}{a^2} + \frac{y^2}{b^2} = 1$$

This gives an ellipse with its center at the origin and its major axis along the x-axis with a length of 2a and its minor axis along the y-axis with a length of 2b. (a > b)

1. Graph the following equations.

a. $\frac{x^2}{9} + \frac{y^2}{4} = 1$

b. $\frac{x^2}{25} + \frac{y^2}{16} = 1$

*c. $\frac{3x^2}{2} + \frac{3y^2}{1} = 6$

*d. $4x^2 + 9y^2 = 36$

***Hint:** Put these in standard form before graphing.

2. The following are not in standard form.
Identify major and minor axes and graph them.

a. $4x^2 + y^2 = 16$

b. $\frac{5x^2}{4} + \frac{4y^2}{5} = 20$

*c. $\frac{(x-3)^2}{25} + \frac{(y-4)^2}{16} = 1$

*d. $4(x+2)^2 + 9y^2 = 36$

Hint: You've seen something similar to this with circles.

3. Write the equation for an ellipse with the following perimeters.
   a. a = 3, b = 2, center (0, 0)
   b. a = 4, b = 3, center (1, 2)
   c. a = 6, b = 3, center ($^-3$, 4)
   d. a = 12, b = 5, center ($^-2$, $^-4$)

Name_______________________________________ HSG-GPE.A.3

Nonlinear Equations—Hyperbola

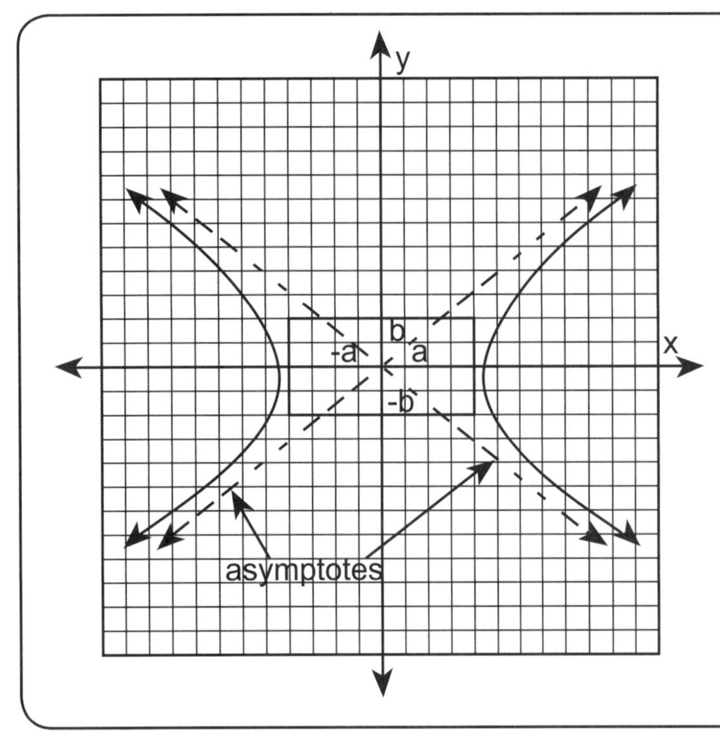

The standard form for the equation of a **hyperbola** is

$$\frac{x^2}{a^2} - \frac{y^2}{b^2} = 1$$

This gives a hyperbola with its center at the origin and asymptotes $y = \pm \frac{b}{a}x$.

The vertices are at (a, 0) and (-a, 0).

1. Graph the following equations.

   a. $\frac{x^2}{9} - \frac{y^2}{4} = 1$

   b. $\frac{x^2}{4} - \frac{y^2}{9} = 1$

   c. $\frac{y^2}{9} - \frac{x^2}{4} = 1$

   d. $\frac{y^2}{4} - \frac{x^2}{9} = 1$

2. The following are not in standard form. Identify asymptotes and graph them.

   a. $x^2 - 4y^2 = 4$

   b. $4x^2 - 9y^2 = 36$

   c. $(x + 1)^2 - 9(y - 1)^2 = 9$

   d. $(x - 1)^2 - (y - 2)^2 = 4$

3. Write the equation for a hyperbola with the following perimeters.*

   a. a = 2, b = 4, center (2, $^-1$)

   b. a = 3, b = 5, center (3, 2)

   c. a = 1, b = 3, center ($^-2$, 4)

   d. a = 4, b = 2, center ($^-1$, $^-3$)

   (*All open in the x direction.)

Name________________________________

HSG-CO.A.2

Polar Coordinates

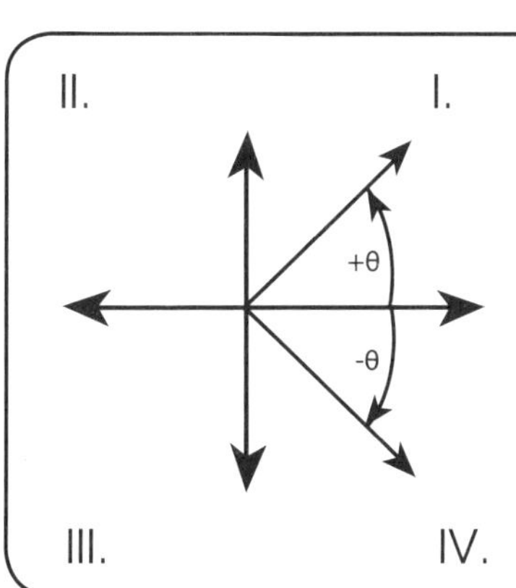

A point in a plane can also be given a unique representation by using polar coordinates. A positive angle is generated by the counterclockwise rotation of the positive x-axis about the origin. Similarly, a negative angle is generated by the clockwise rotation of the positive x-axis about the origin. A point can be named by giving the distance from the origin and the measure of the angle in the form (r, θ).

Graph the following polar coordinates.

1. $(2, \frac{\pi}{2})$
2. $(3, \pi)$
3. $(1, 45°)$
4. $(5, \frac{-\pi}{2})$
5. $(4, 270°)$
6. $(2, -\frac{3}{2}\pi)$
7. $(2, 540°)$
8. $(2, 2\pi)$

In which quadrant would you expect to find the following points?

9. $(2, \frac{\pi}{4})$
10. $(4, \frac{3}{4}\pi)$
11. $(3, \frac{-\pi}{3})$
12. $(1, \pi)$
13. $(2, \frac{15}{3}\pi)$
14. $(4, -\frac{4\pi}{3})$
15. $(3, -700°)$
16. $(-1, 45°)$

Name________________________________ HSG-CO.A.2, HSG-CO.A.3, HSG-GPE.B.7

Converting Polar and Rectangular Coordinates

The equations below can be used to convert between polar coordinates and rectangular coordinates.

$x = r \cdot \cos \theta$ $\tan \theta = \frac{y}{x}$

$y = r \cdot \sin \theta$ $r = \sqrt{x^2 + y^2}$

Find the polar coordinates corresponding to the following rectangular coordinates.

1. (2, 0)
2. (3, 4)
3. (0, 5)
4. (4, 4)
5. ($^-2$, $^-2$)
6. ($^-6$, 8)

Find the rectangular coordinates corresponding to the following polar coordinates.

7. $(2, \frac{\pi}{2})$
8. $(3, \frac{\pi}{4})$
9. $(4, \pi)$
10. (2, 0)
11. $(5, \frac{\pi}{3})$
12. $(\sqrt{2}, \frac{\pi}{4})$

Express the following equations in polar notation.

13. $x = 2$
14. $x^2 + y^2 = 4$
15. $y = 2x + 1$
16. $y = 4$
17. $x^2 + y^2 - 3 = 13$
18. $y = 3x$

Express the following equations in rectangular notation.

19. $r = 5$
20. $r = 3 \csc \theta$
21. $\tan \theta = 3$
22. $r = 2 \sec \theta$
23. $r = 7$
24. $\tan \theta = 2$

Name________________________________

HSG-GPE.B.7

The Distance Formula

The distance between 2 points: (x_1, y_1), (x_2, y_2)

$$D = \sqrt{(x_2 - x_1)^2 + (y_2 - y_1)^2}$$

Find the distance between $(^-3, 2)$ and $(1, ^-2)$

$$D = \sqrt{(1 - ^-3)^2 + (^-2 - 2)^2}$$
$$= \sqrt{(4)^2 + (^-4)^2} = \sqrt{16 + 16}$$
$$= \sqrt{32} = \sqrt{2 \cdot 2 \cdot 2 \cdot 2 \cdot 2}$$
$$= 4\sqrt{2}$$

Find the distance between these pairs of points.

1. $(6, 4)$ and $(2, 1)$
2. $(^-2, ^-4)$ and $(3, 8)$
3. $(0, 0)$ and $(5, 10)$
4. $(^-5, 2)$ and $(7, ^-7)$
5. $(0, ^-8)$ and $(8, 7)$
6. $(^-2, 11)$ and $(4, 3)$
7. $(2, 1)$ and $(4, 0)$
8. $(6, 4)$ and $(6, ^-2)$
9. $(^-2, 2)$ and $(4, ^-1)$
10. $(^-3, ^-5)$ and $(2, 5)$
11. $(^-4, 0)$ and $(2, 3)$
12. $(^-1, 5)$ and $(3, ^-3)$
13. $(0, 0)$ and $(3, 4)$
14. $(1, 8)$ and $(3, 10)$
15. $(9, 8)$ and $(^-3, 4)$
16. $(2, 2)$ and $(2, 4)$

Name______________________________ HSG-GPE.B.7

Distance and Midpoint

Distance Formula

$d = \sqrt{(x_2 - x_1)^2 + (y_2 - y_1)^2}$

A(⁻1, ⁻3) B(3, 5)

$d(AB) = \sqrt{(3 - ^-1)^2 + (5 - ^-3)^2}$

$= \sqrt{(4)^2 + (8)^2}$

$= \sqrt{16 + 64}$

$= \sqrt{80}$

$d(AB) = 4\sqrt{5}$

Midpoint Formula

$\left(\frac{x_1 + x_2}{2}, \frac{y_1 + y_2}{2}\right)$

A(⁻1, ⁻3) B(3, 5)

$\left(\frac{^-1 + 3}{2}, \frac{^-3 + 5}{2}\right)$

$\left(\frac{2}{2}, \frac{2}{2}\right)$

(1, 1)

Find the distance and the midpoint between the given points. Cross out the correct answers below. Use the remaining letters to complete the statement.

	Distance	Midpoint
1. (⁻2, 2) and (4, ⁻1)	________	________
2. (⁻3, ⁻5) and (2, 5)	________	________
3. (⁻1, 5) and (3, ⁻3)	________	________
4. (0, 0) and (3, 4)	________	________
5. (1, 2) and (4, 7)	________	________
6. (⁻2, 4) and (3, ⁻5)	________	________
7. (2, 2) and (6, 6)	________	________
8. (3, 6) and (5, ⁻2)	________	________
9. (⁻1, ⁻4) and (3, 5)	________	________

5 S	$(-\frac{1}{2}, 0)$ Q	$(0, -\frac{1}{2})$ P	$4\sqrt{5}$ U	$\sqrt{5}$ Y	10 T	$(1, \frac{1}{2})$ M	$9\sqrt{7}$ H	$(1, \frac{1}{2})$ G
$\sqrt{97}$ B	$3\sqrt{21}$ A	$(\frac{1}{2}, -\frac{1}{2})$ H	(5/2, 9/2) R	21 G	$3\sqrt{5}$ W	(2, 4) O	(4, 2) E	$\sqrt{34}$ U
0 R	(-1, -1) E	$5\sqrt{5}$ S	$2\sqrt{53}$ A	1 N	$(1\frac{1}{2}, 2)$ V	$(2, 1\frac{1}{2})$ T	$4\sqrt{17}$ H	$2\sqrt{17}$ S
(1, 1) R	(0, 0) E	34 O	$\sqrt{106}$ Y	32 R	$4\sqrt{2}$ L	(2, 2) E	(4, 4) T	25 M

10. This Distance Formula is based on the

_ _.

Name________________________________

Just for Fun

Conditional	Statement
$p \rightarrow q$	If p, then q. If it rains, you use an umbrella.
Converse	
$q \rightarrow p$	If q, then p. If you use an umbrella, then it rains.
Inverse	
$\sim p \rightarrow \sim q$	If not p, then not q. If it is not raining, you do not use an umbrella.
Contrapositive	
$\sim q \rightarrow \sim p$	If not q, then not p. If you are not using an umbrella, then it is not raining.

For problems 1–7, write the converse, inverse, and contrapositive statements based upon the given conditional statements.

1. If I own a horse, then I own an animal.
   Converse ________________________________
   Inverse ________________________________
   Contrapositive ________________________________
2. If I study, then I do well in school.
   Converse ________________________________
   Inverse ________________________________
   Contrapositive ________________________________
3. If today is Monday, then yesterday was Sunday.
   Converse ________________________________
   Inverse ________________________________
   Contrapositive ________________________________
4. If it is Saturday, I do not go to school.
   Converse ________________________________
   Inverse ________________________________
   Contrapositive ________________________________
5. If I do not go to bed early, I do not sleep well.
   Converse ________________________________
   Inverse ________________________________
   Contrapositive ________________________________
6. If $6x = 18$, then $x = 3$.
   Converse ________________________________
   Inverse ________________________________
   Contrapositive ________________________________
7. If $AB + BC = AC$, then B is between A and C.
   Converse ________________________________
   Inverse ________________________________
   Contrapositive ________________________________
8. Given a conditional statement, the ______________ statement is **always** true. (Choose from converse, inverse, or contrapositive.)

Name________________________________

Just for Fun

Use the following statements to determine the names of the men playing each position on this baseball team.

1. Andy and the catcher are best friends.
2. Ed's sister is engaged to the second basemen.
3. The center fielder is taller than the right fielder.
4. Harry and the third baseman live in the same building.
5. Paul and Allen each went to a party at the pitcher's house.
6. Ed and the outfielders play basketball during their free time.
7. The pitcher's wife is the third baseman's sister.
8. All the battery and infield, except Allen, Harry and Andy, are shorter than Sam. (battery = catcher and pitcher)
9. Paul, Andy, and the shortstop went to Florida last year.
10. Paul, Harry, Bill, the catcher, and the second baseman work downtown.
11. Sam is undergoing divorce proceedings.
12. The catcher and the third baseman each have two children.
13. Ed, Paul, Jerry, the right fielder, and the center fielder are bachelors. The others are married.
14. The shortstop, the third baseman, and Bill grew up in the Midwest.
15. One of the outfielders is either Mike or Andy.
16. Jerry is taller than Bill. Mike is shorter than Bill. Each of them is heavier than the third baseman.

	C	P	SS	1st	2nd	3rd	LF	CF	RF
Mike									
Ed									
Harry									
Paul									
Allen									
Bill									
Jerry									
Sam									
Andy									

Name________________________________

HSG-GMD.A.1

Circumference and Area

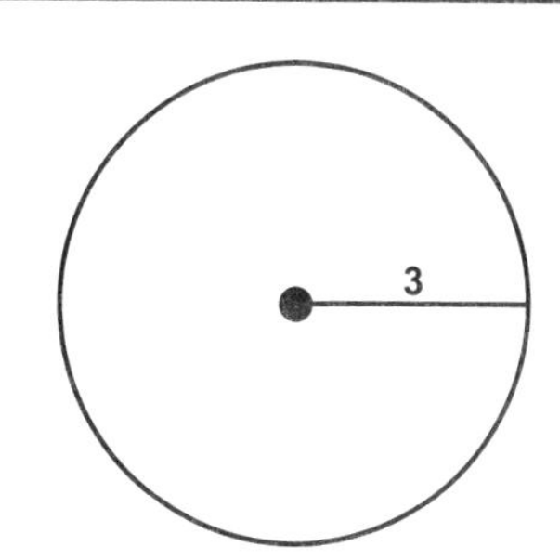

circumference = $2\pi r$
area = πr^2

$C = 2\pi r$
$= 2\pi(3)$
$C = 6\pi$ units

$A = \pi r^2$
$= \pi(3)^2$
$A = 9\pi$ square units

Find the circumference and area of each circle.

1.

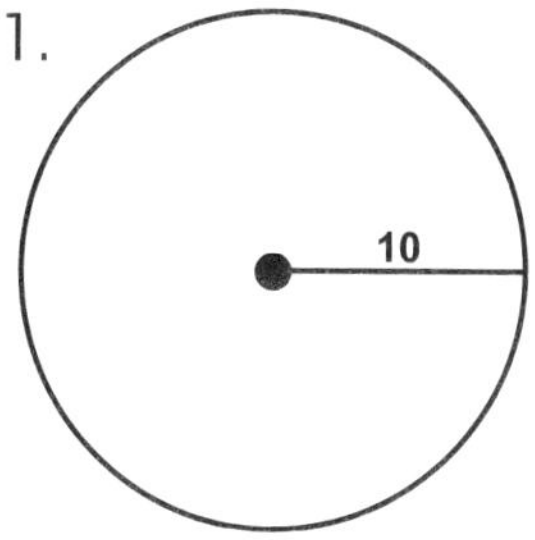

2. 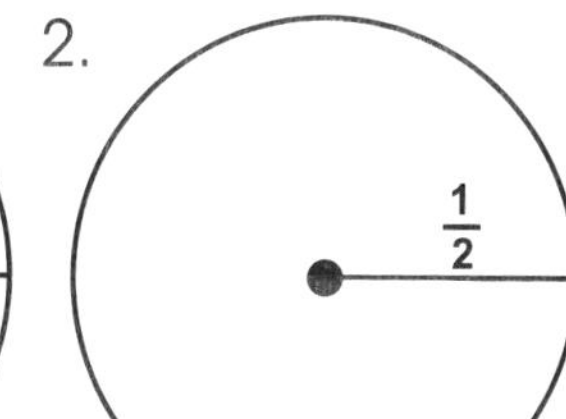

3. 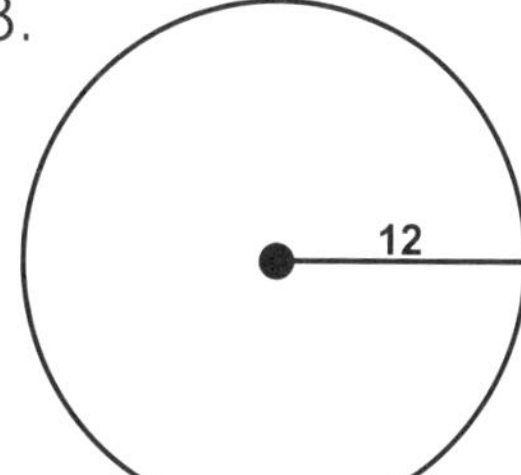

4. 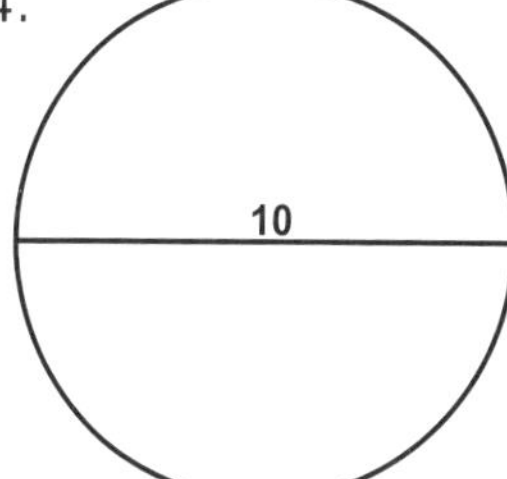

5.

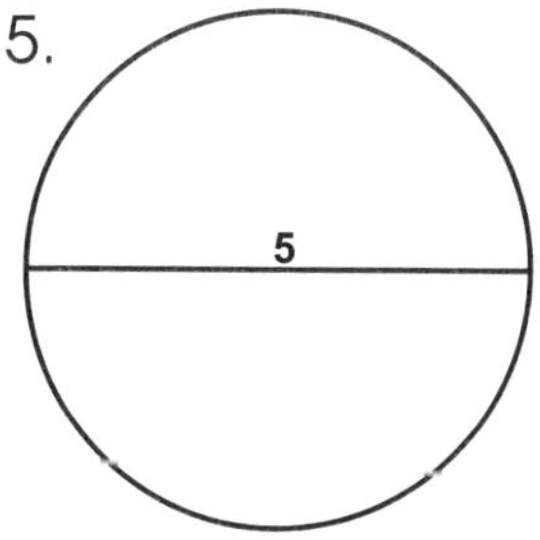

6. 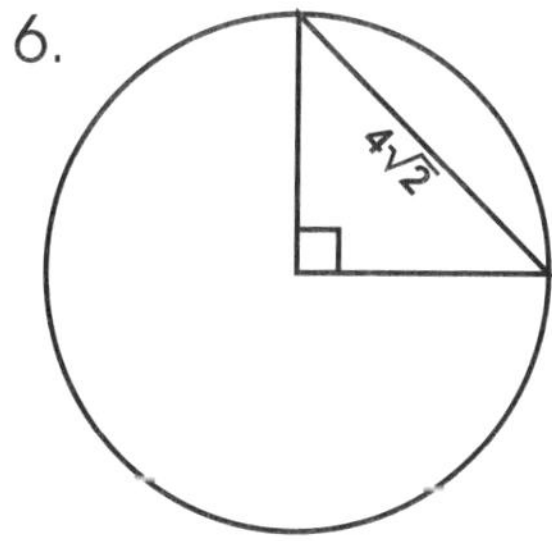

7. 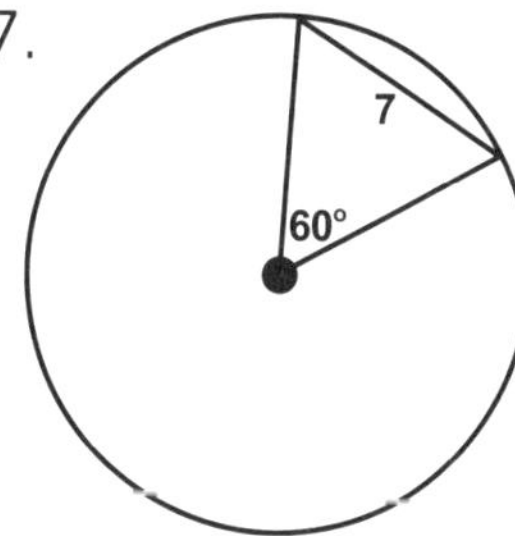

8. 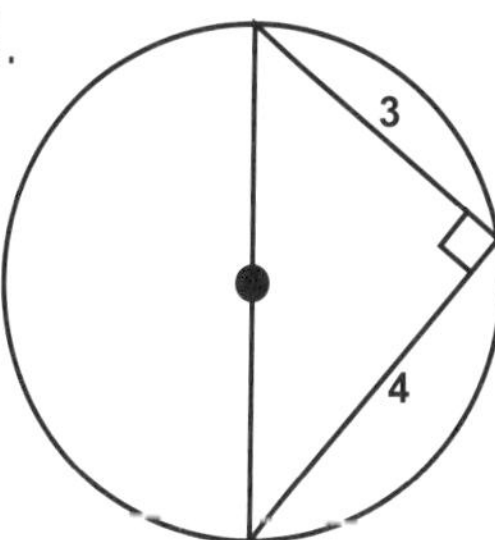

9. 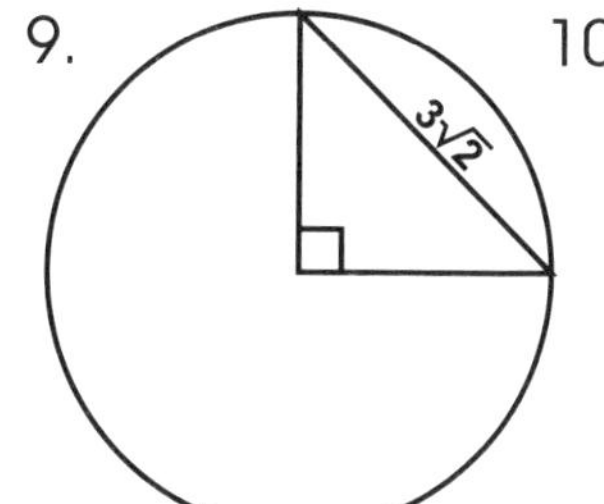

10. 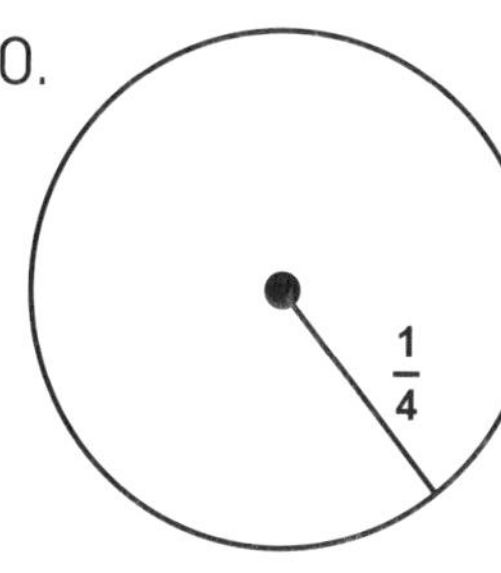

11. 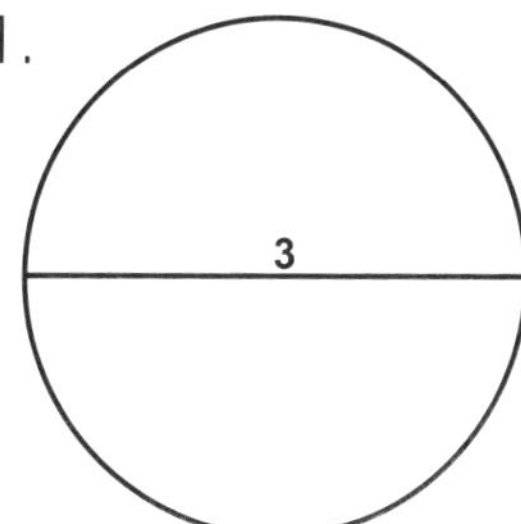

12. 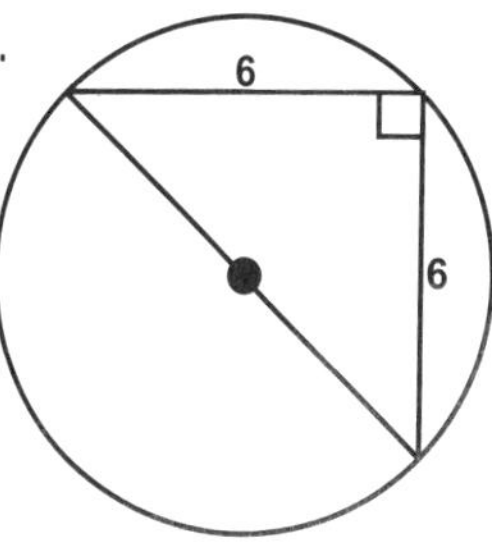

13. 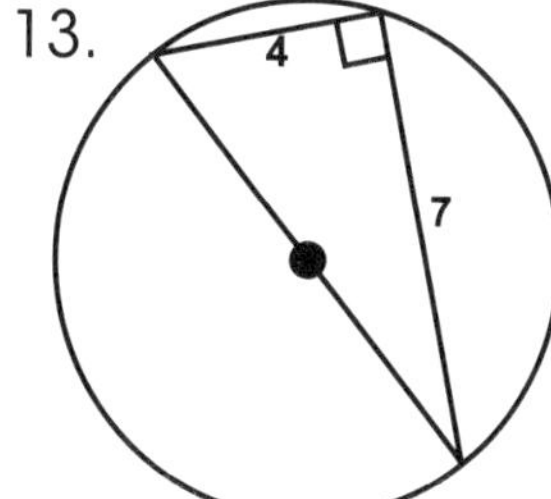

14. 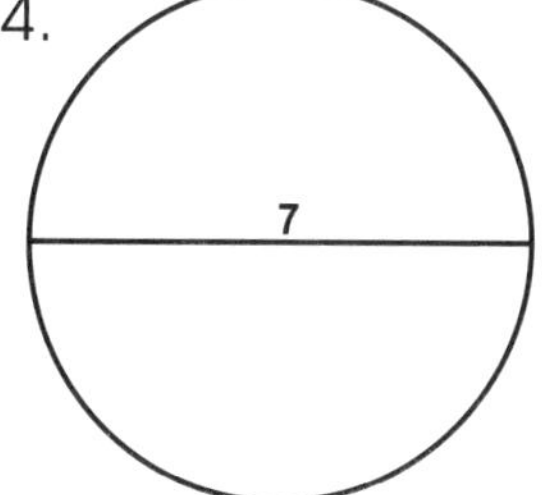

15. 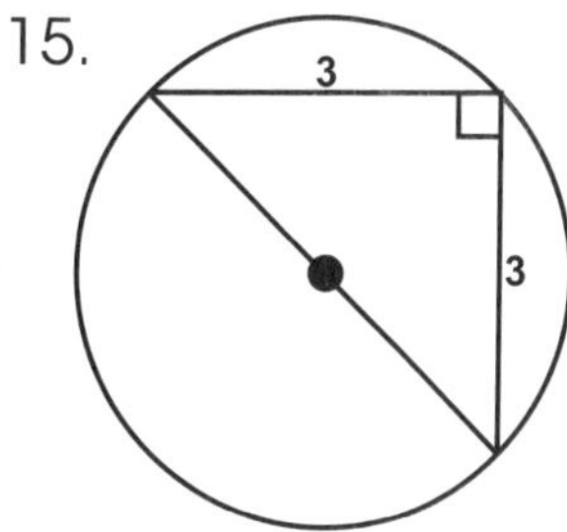

16. 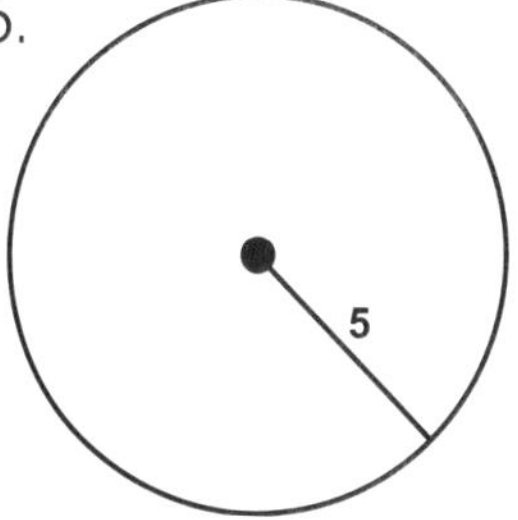

Name________________________________ 8.G.C.9, HSG-GMD.A.3

Prisms

volume = (area of base) • (height)
lateral area = (perimeter of base) • (height)
total area = (lateral area) + 2 • (area of base)

V = (3 • 2) • 4 = 24 cubic units
LA = (3 + 2 + 3 + 2) • 4 = 40 square units
TA = 40 + 2 • (3 • 2) = 52 square units

Find the volume, lateral area, and total area of the following prisms.

1.

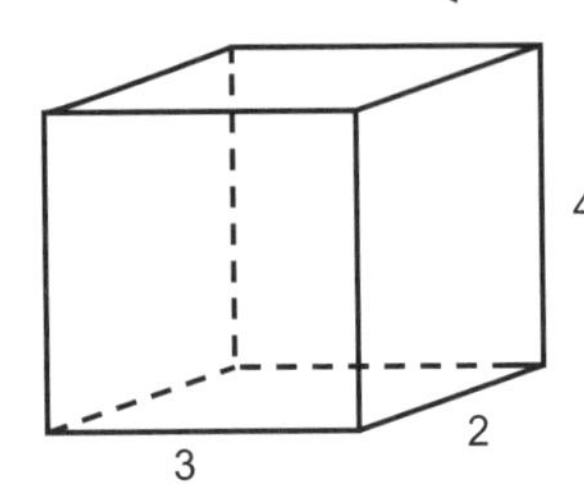

2.

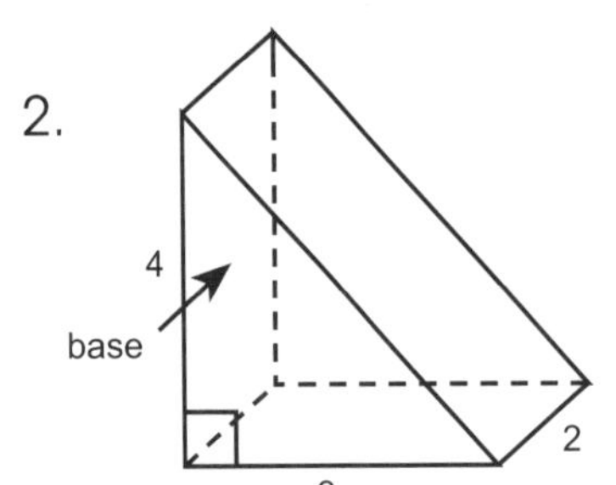

3.

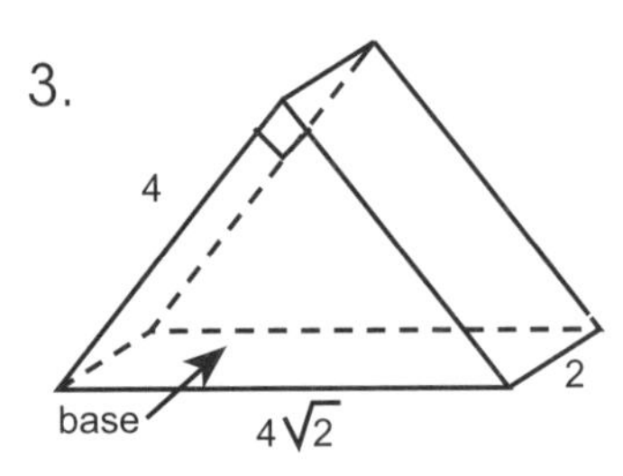

4.

5.

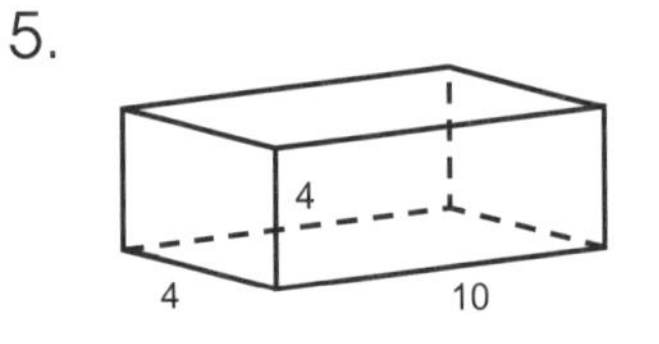

6.

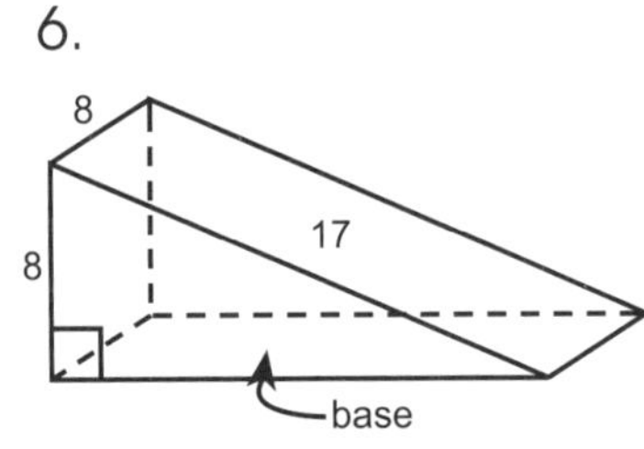

7.

8.

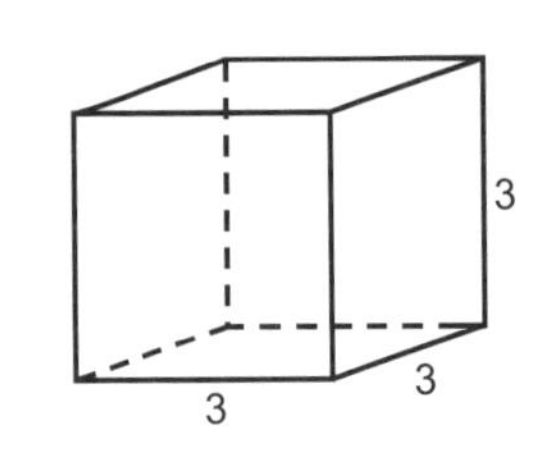

9.

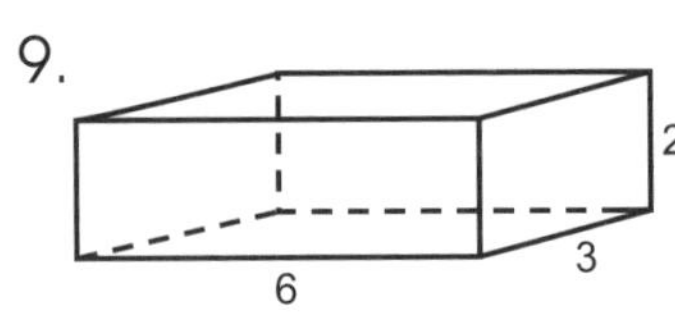

10.

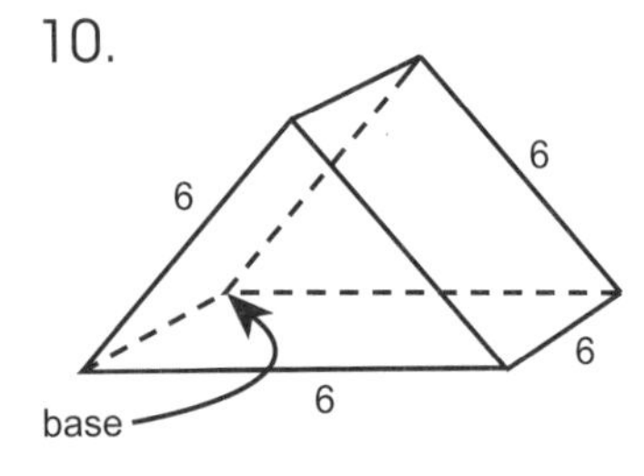

11.

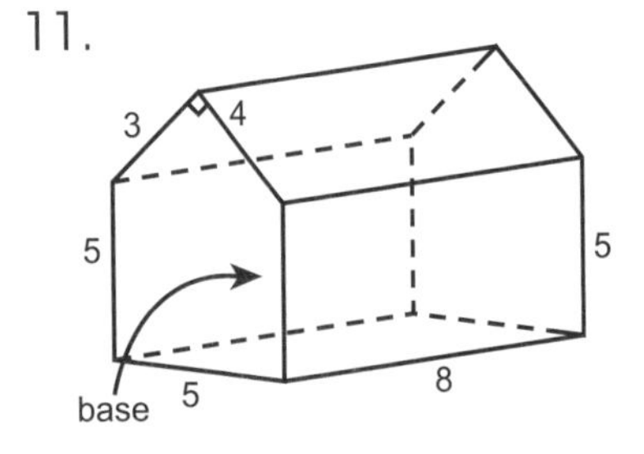

12.

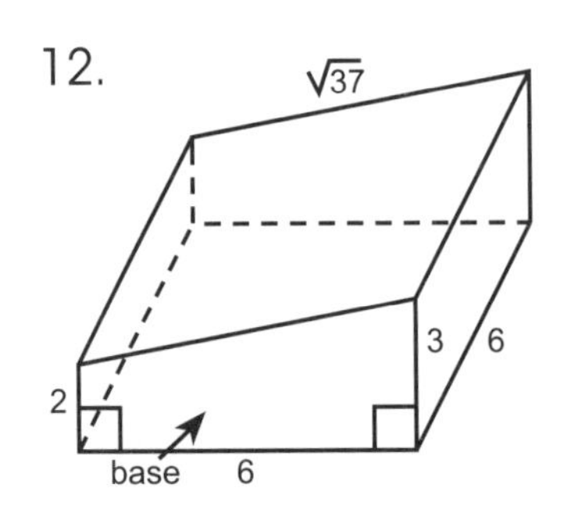

Name____ 8.G.C.9, HSG-GMD.A.3

Right Circular Cylinders

volume = π • (radius)^2 • (height)
lateral area = 2 • π • (radius) • (height)
total area = (lateral area) + 2 • π • (radius)^2

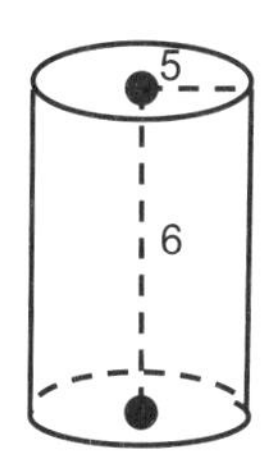

$V = \pi \cdot (5)^2 \cdot (6) = 150\pi$ cubic units
$LA = 2 \cdot \pi \cdot (5) \cdot (6) = 60\pi$ square units
$TA = 60\pi + 2 \cdot \pi \cdot (5)^2 = 110\pi$ square units

Find the volume, lateral area, and total area of the following right circular cylinders.

1. 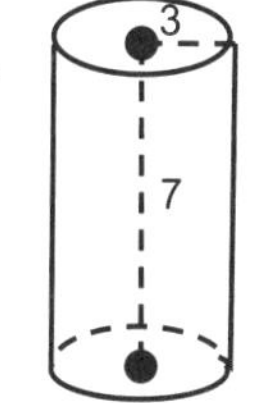

2. 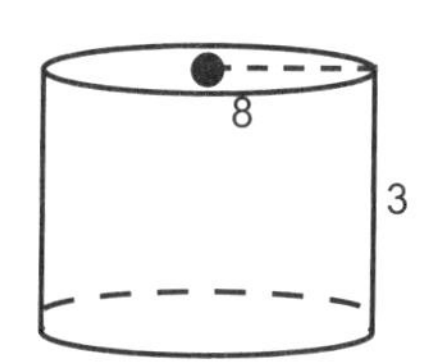

3. 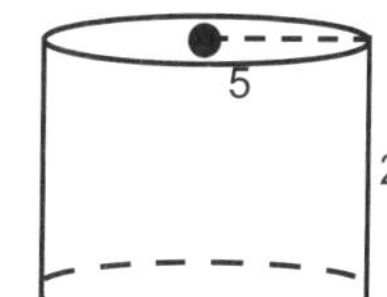

4. 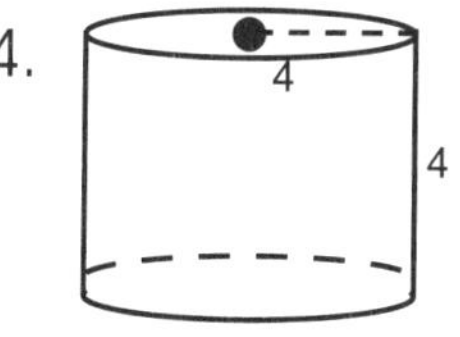

5. 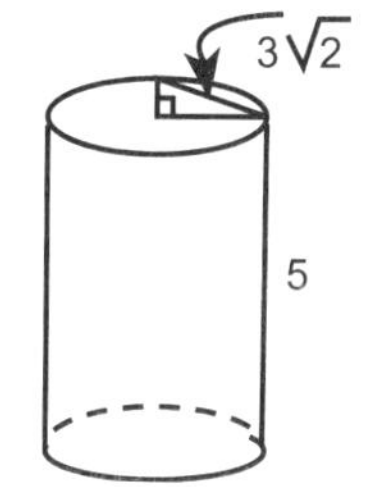

6. 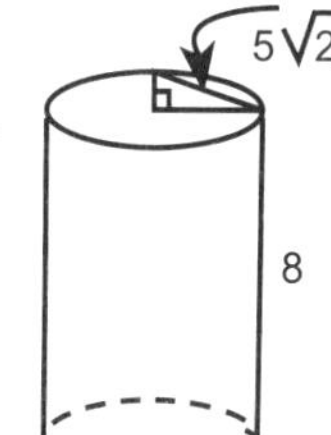

7. 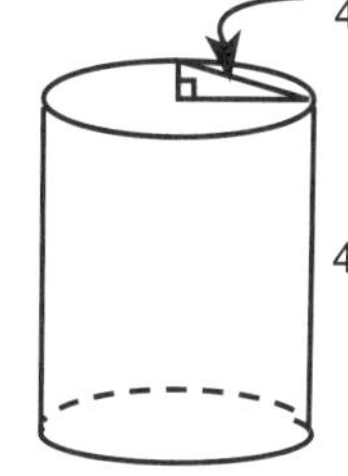

8. 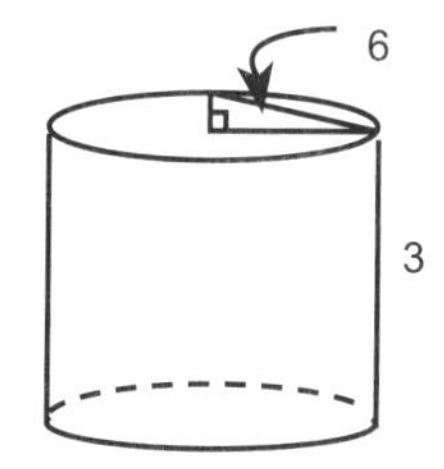

9. 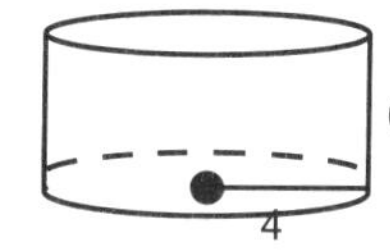

10. 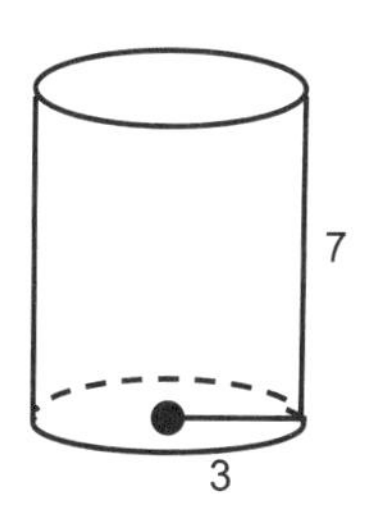

11. 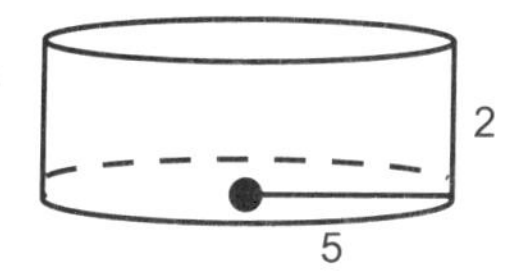

12. 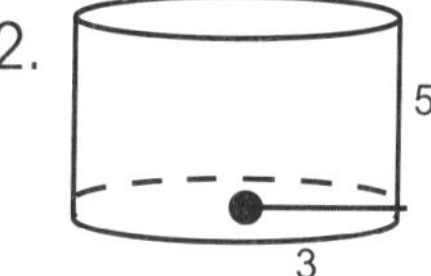

Name________________________________ 8.G.C.9, HSG-GMD.A.3

Pyramids

volume = $\frac{1}{3}$ • (area of base) • (height)

lateral area = $\frac{1}{2}$ • (perimeter) • (slant height)

total area = (lateral area) + (area of base)

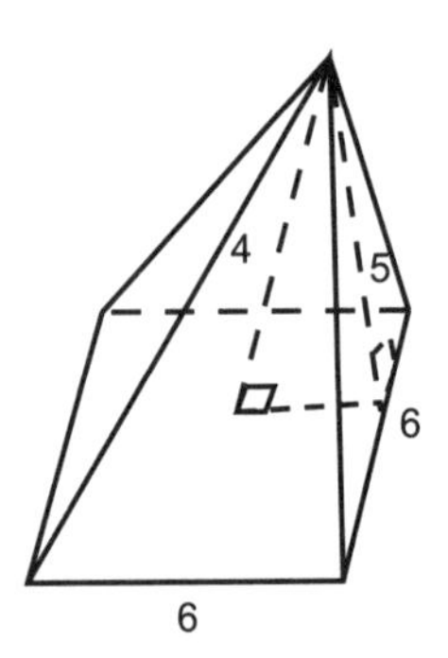

$V = \frac{1}{3} \cdot (6 \cdot 6) \cdot (4) = 48$ cubic units

$LA = \frac{1}{2} \cdot (6 + 6 + 6 + 6) \cdot (5) = 60$ square units

$TA = 60 + (6 \cdot 6) = 96$ square units

Find the volume, lateral area, and total area of the following pyramids.

1. 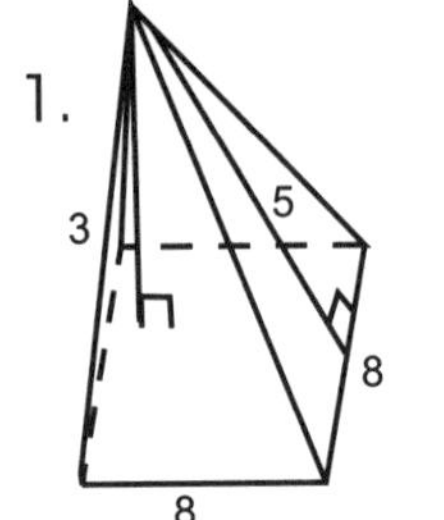

2. 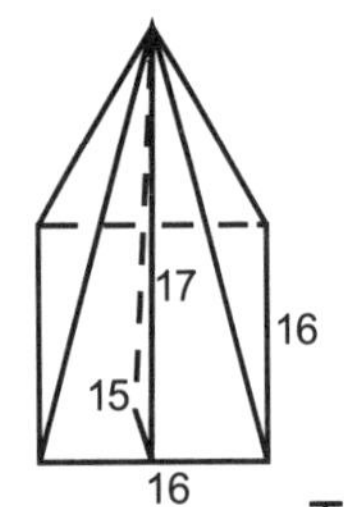

3. 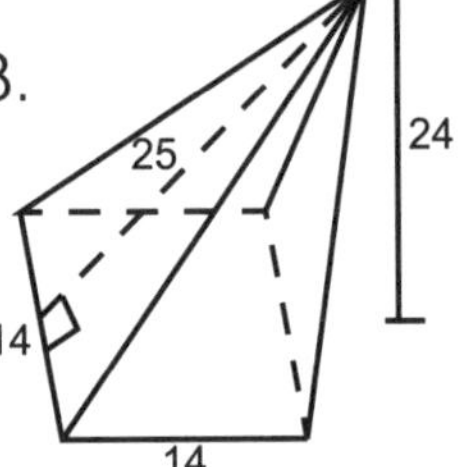

4. 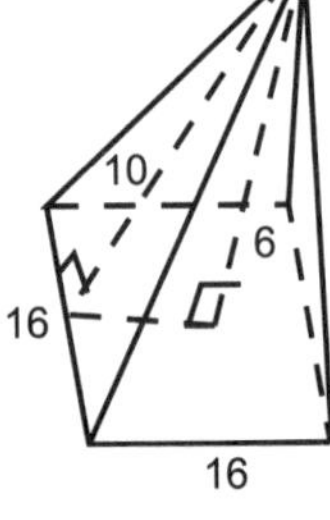

5. 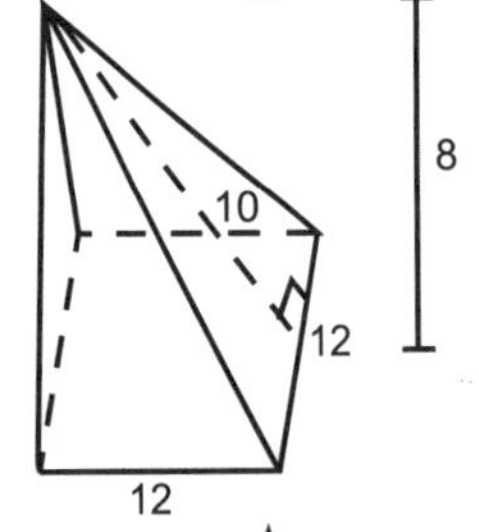

6. 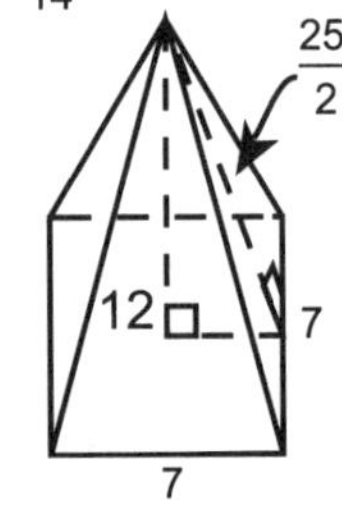

7. 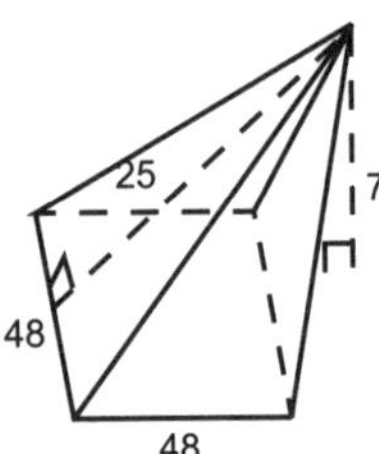

8. 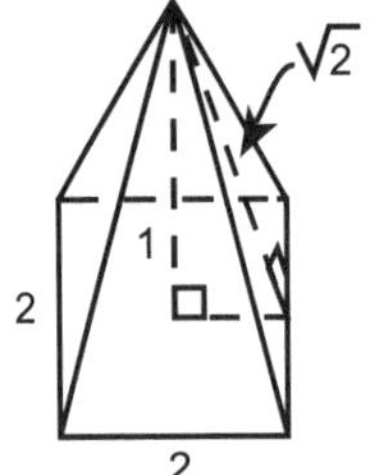

9. 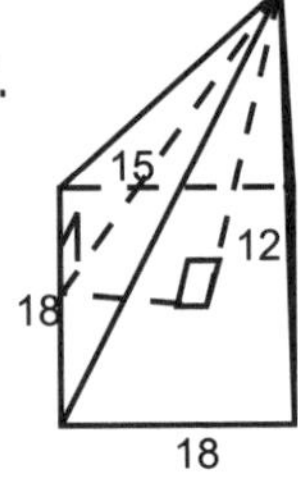

10. 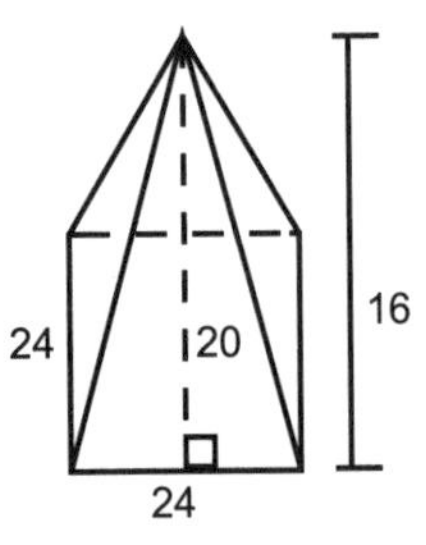

11. 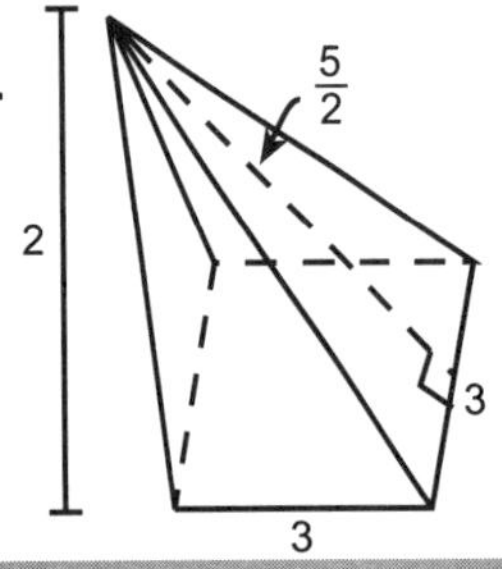

12. 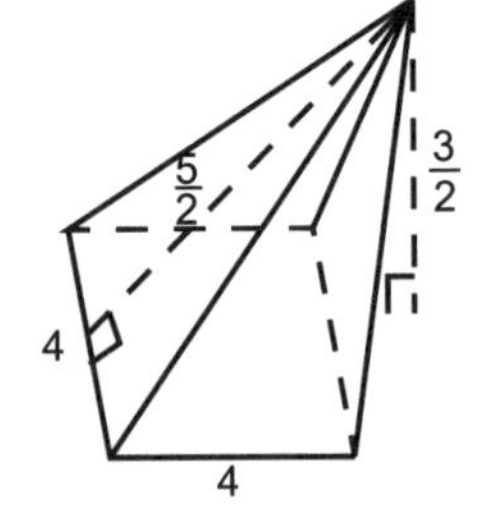

Name____________________________________ 8.G.C.9, G.GMD.A.3

Right Circular Cones

volume = $\frac{1}{3}$ **• π • (radius)² • (height)**

lateral area = π • (radius) • (slant height)

total area = (lateral area) + π • (radius)²

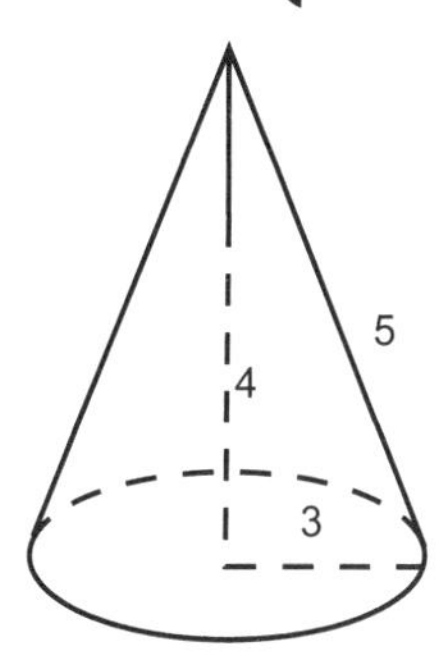

$V = \frac{1}{3} \cdot \pi \cdot (3)^2 \cdot (4) = 12\pi$ cubic units

$LA = \pi \cdot (3) \cdot (5) = 15\pi$ square units

$TA = 15\pi + 15\pi \cdot (3)^2 = 24\pi$ square units

Find the volume, lateral area, and total area of the following right circular cones.

1.
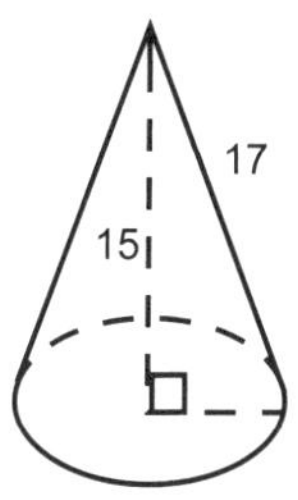

2.
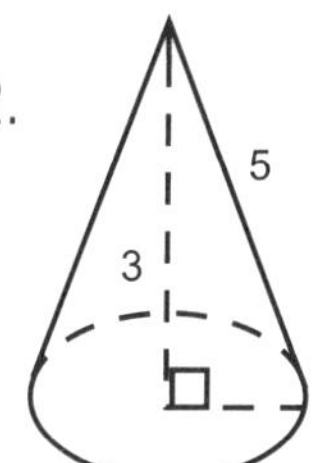

3.
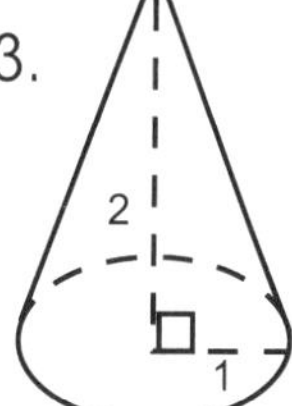

4.
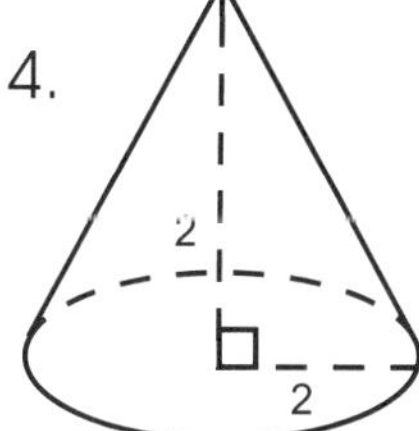

5.
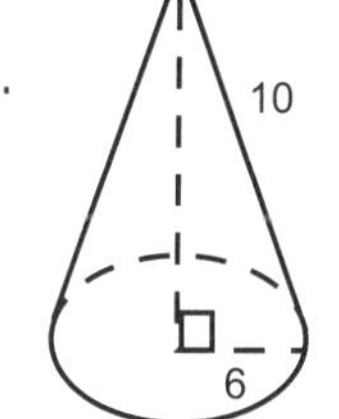

6.

25

7

7.
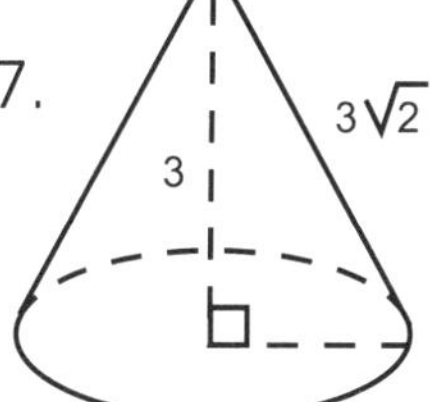

8.
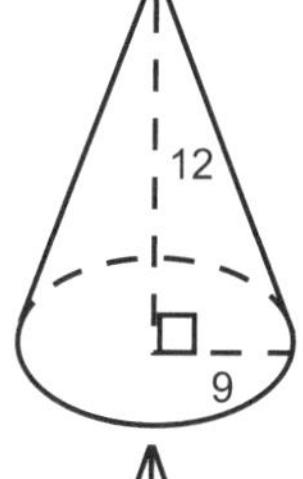

9.
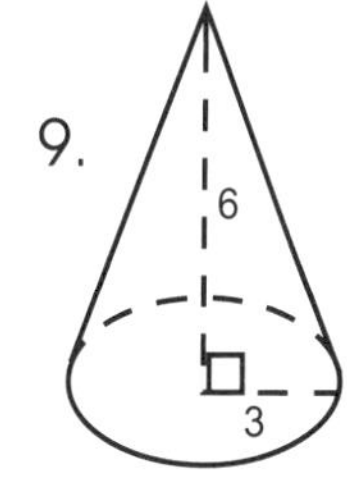

10.
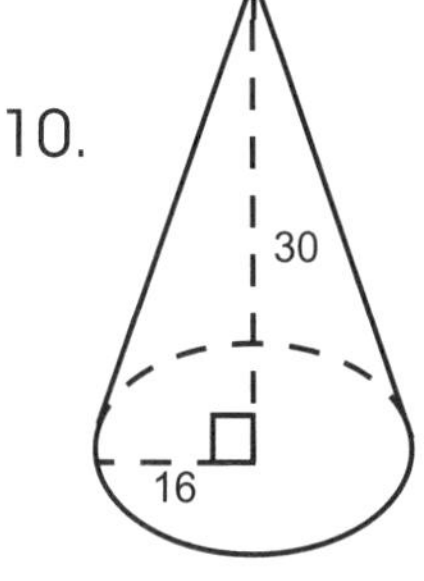

11.
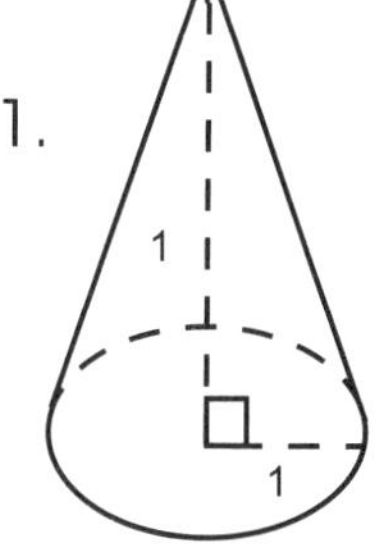

12.
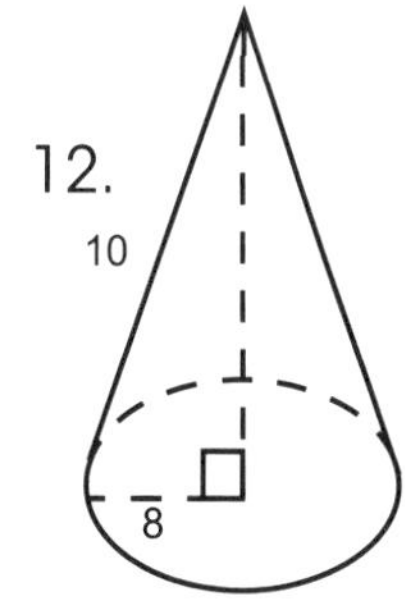

Name________________________________ HSG-CO.D.12

Polygonal Regions

A **polygonal region** is defined to be the union of a finite number of triangular regions in a single plane. The intersection of any two or more triangular regions is either a point or a segment.

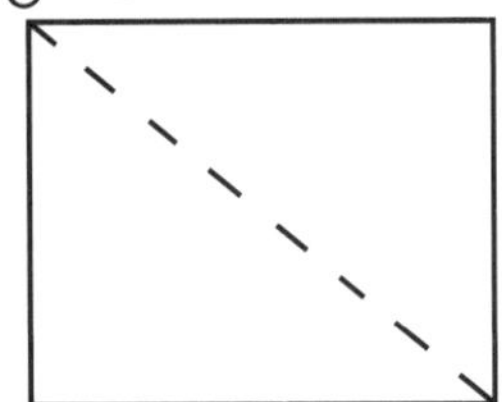

This is polygonal region.

Determine if each region below is polygonal by breaking it into triangular regions.

1. 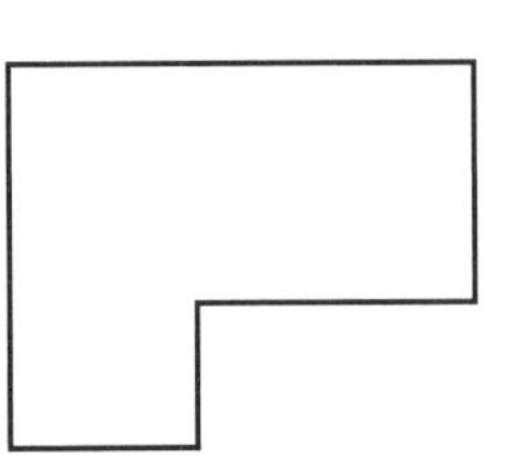

2. 

3. 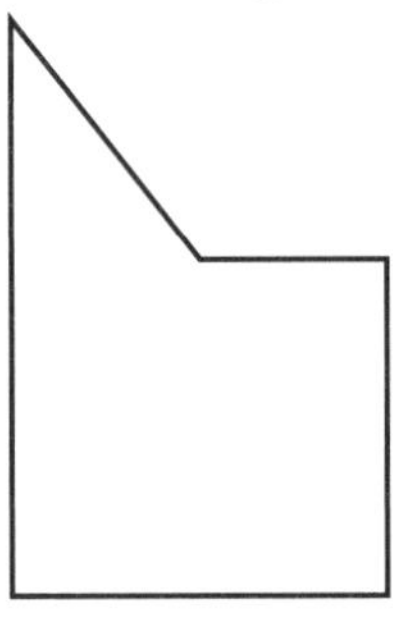

4. 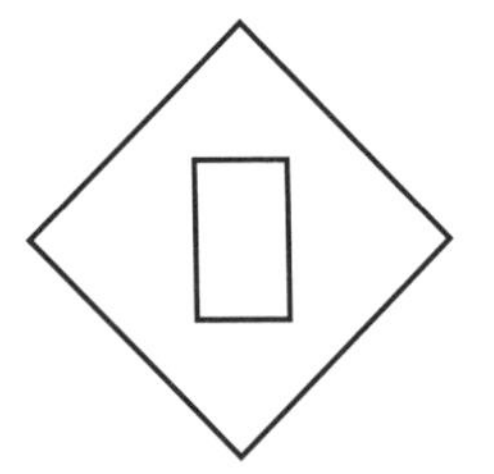

5. 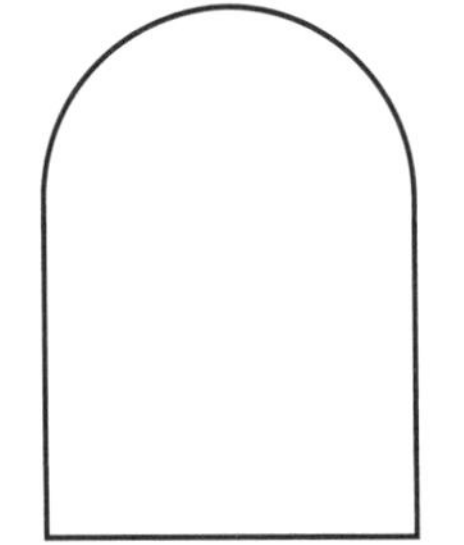

6. 

7. 

8. 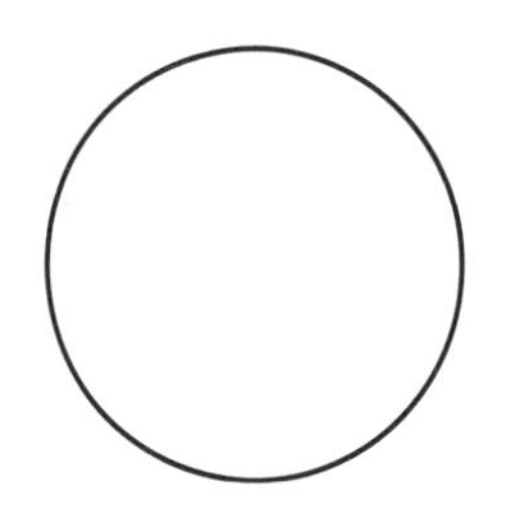

9. 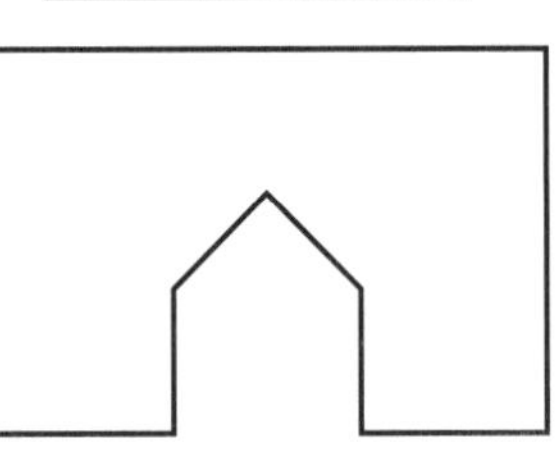

10. 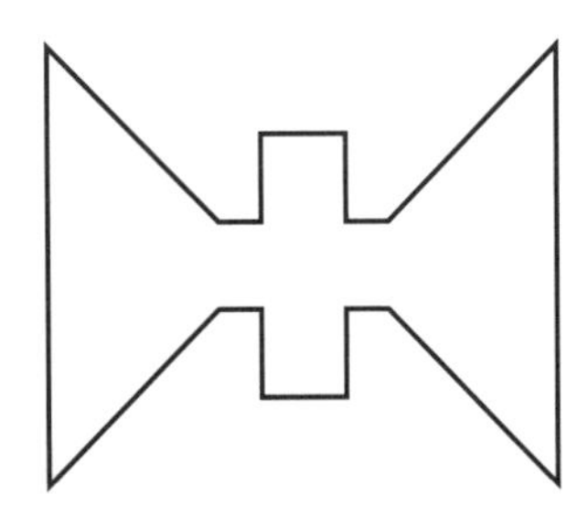

11. 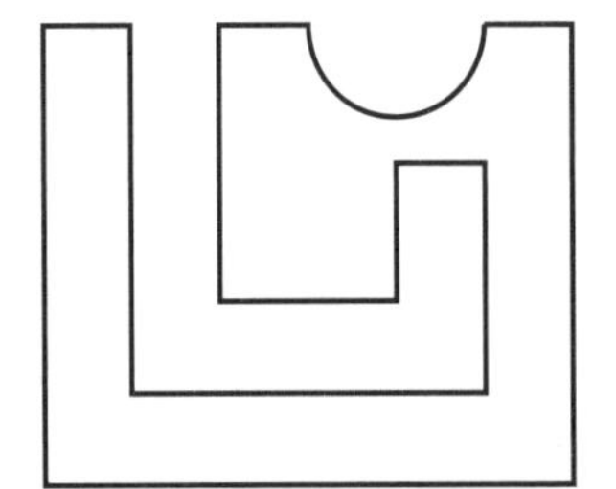

12. 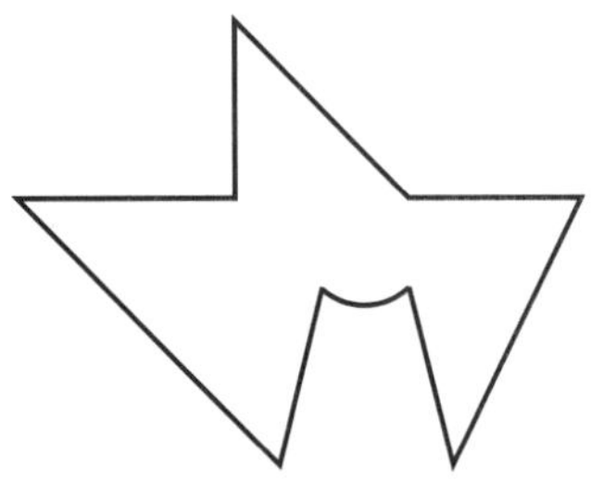

Name____________________________________

8.G.A.5

Polygons

Types of Polygons

Number of Sides	*Name*
3	triangle
4	quadrilateral
5	pentagon
6	hexagon
8	octagon
10	decagon
n	n-gon

- A regular polygon has equal angles and equal sides.
- The sum of the measures of the angles of a convex polygon with n sides is $(n - 2)\,180^\circ$.
- The sum of the measures of the exterior angles of any convex polygon is 360°.

Find the following for each polygon: a) The sum of the measures of the interior angles, b) The sum of the measures of the exterior angles.

1. A 32-sided polygon
2. A 12-sided polygon
3. A 6-sided polygon
4. An 8-sided polygon
5. A 3-sided polygon
6. A 5-sided polygon

Find the following for each regular polygon: a) The measure of each exterior angle, b) the measure of each interior angle.

7. A 6-sided polygon
8. A 5-sided polygon
9. A 3-sided polygon
10. An 8-sided polygon
11. A 4-sided polygon
12. A 10-sided polygon

13. A regular polygon has an exterior angle with a measure of 20°. Find the number of sides.
14. A regular polygon has an interior angle with a measure of 120°. Find the number of sides.
15. A regular polygon has 20 sides. Find the measure of each exterior angle.
16. A regular polygon has 10 sides. Find the measure of each interior angle.

Regular Polygons

A **regular polygon** is a convex polygon with all sides congruent and all angles congruent.

apothem (a) = distance from the center of the polygon to a side

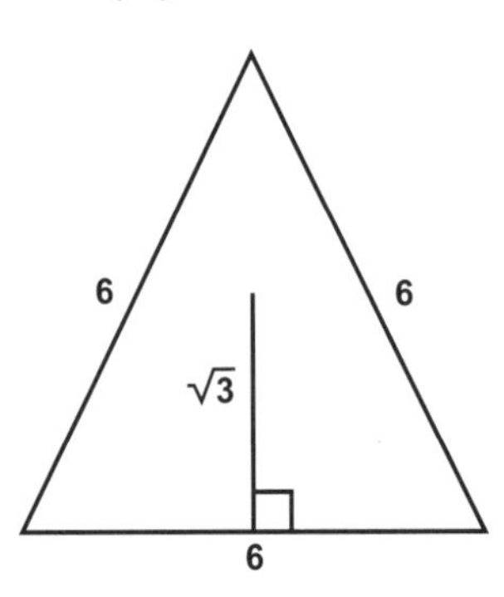

area = $\frac{1}{2}$ ap where p = perimeter

$$A = \frac{1}{2}(\sqrt{3})(6 + 6 + 6)$$

$$= \frac{1}{2}(\sqrt{3})(18)$$

$$A = 9\sqrt{3} \text{ square units}$$

Find the areas of the regular polygonal regions below.

1. 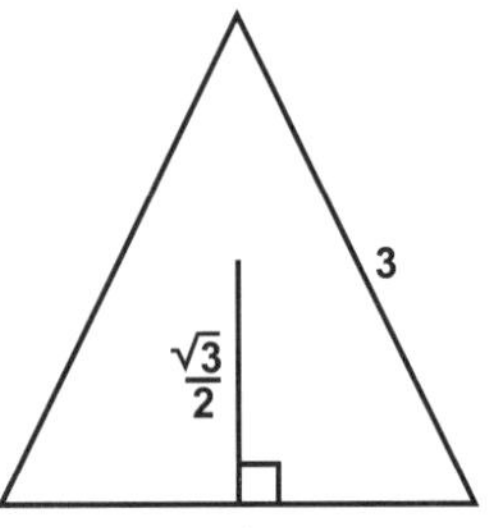

2. 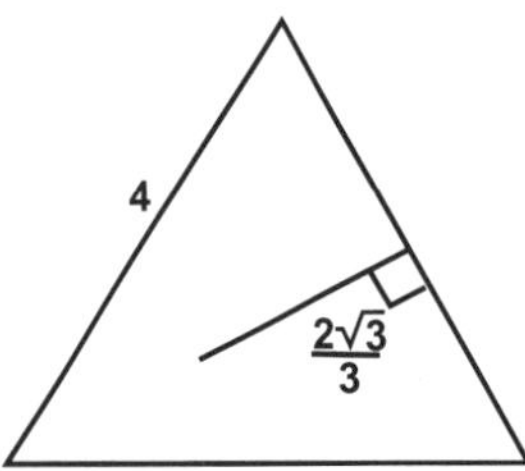

3. 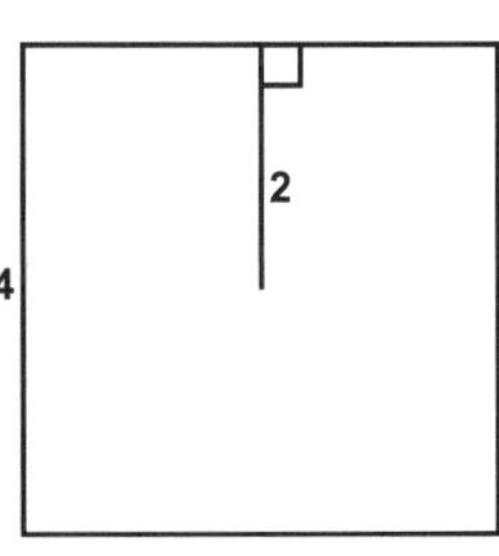

4. 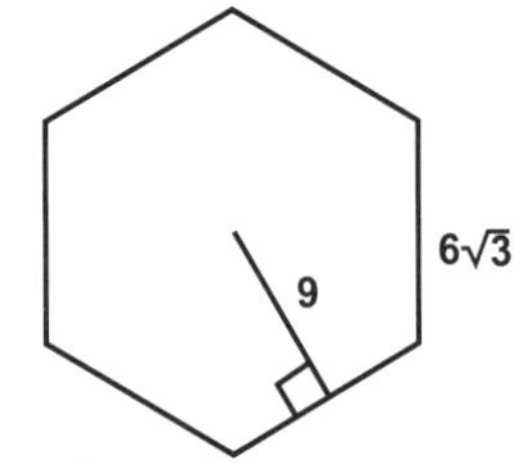

5. 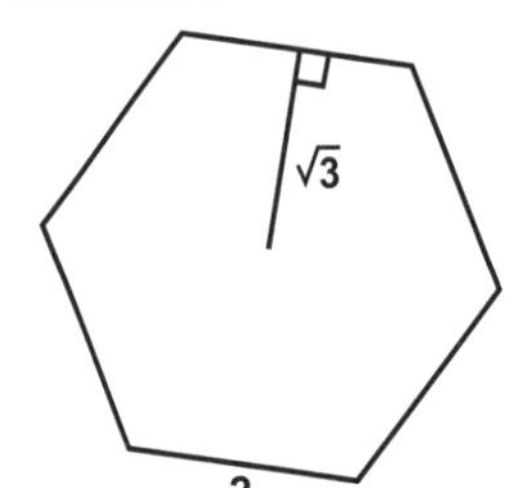

6. 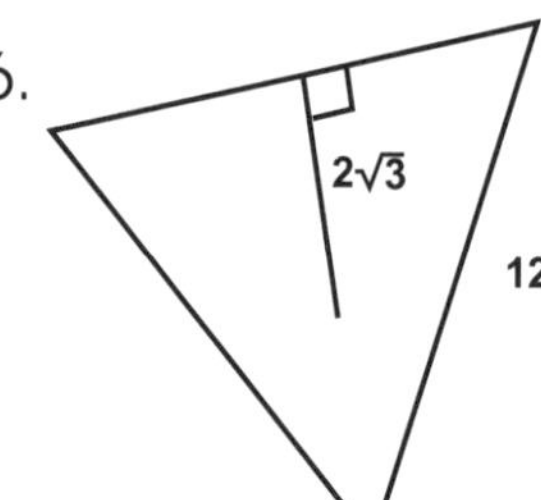

7. 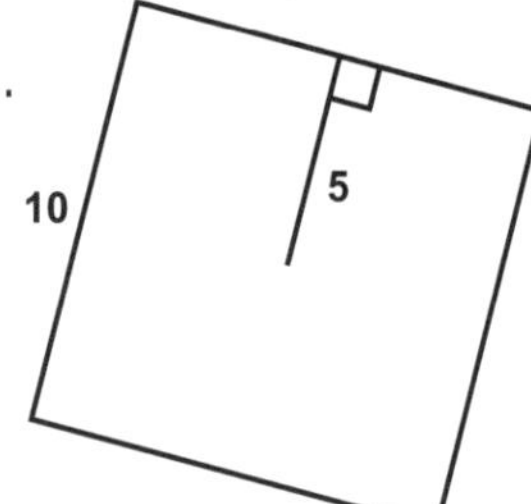

8. 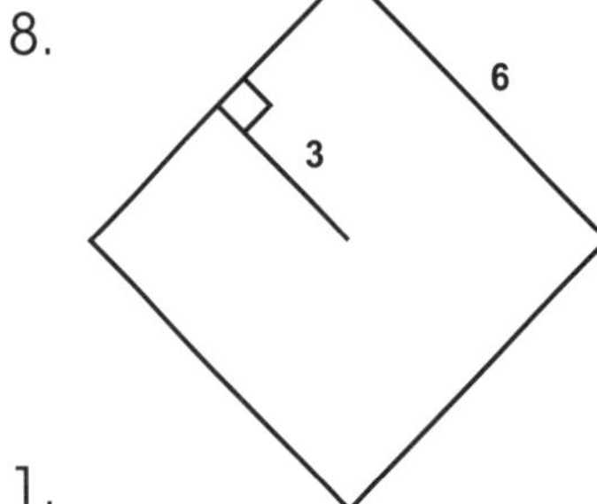

9. 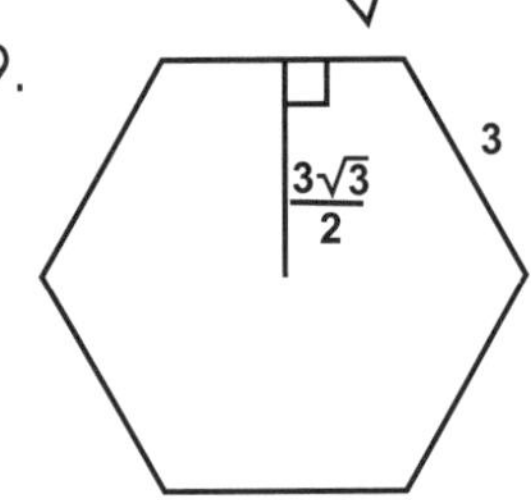

10. 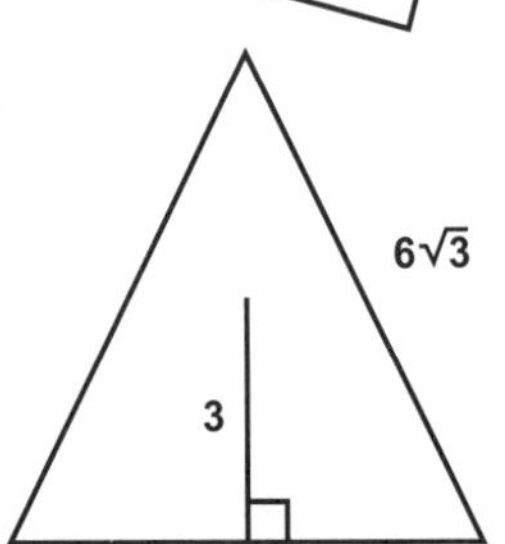

11. 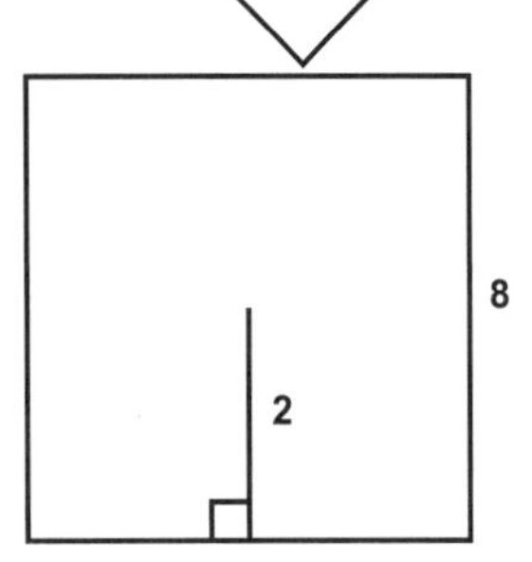

12. 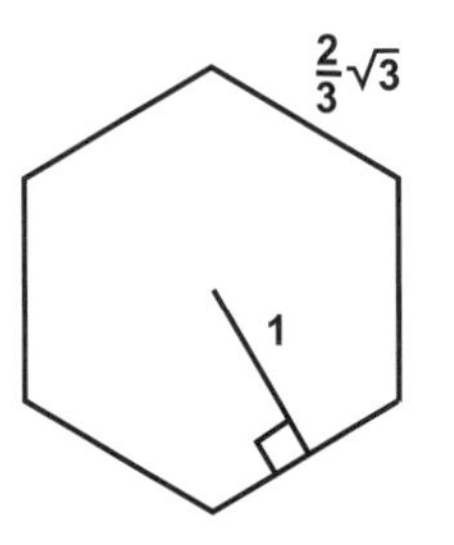

Name________________________________

HSG-GMD.B.4

Platonic Solids

A polyhedron is **regular** if all faces of the solid are congruent regular polygons and the same number of polygons meet at each vertex. There are only 5 regular polyhedra—the Platonic Solids.

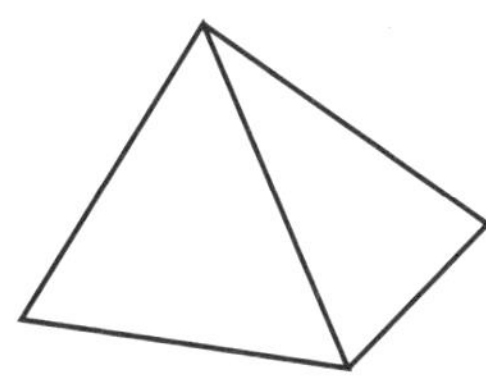
tetrahedron

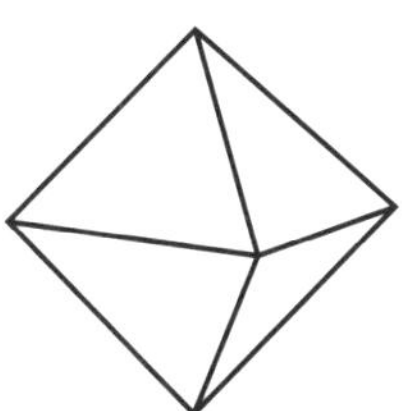
octahedron

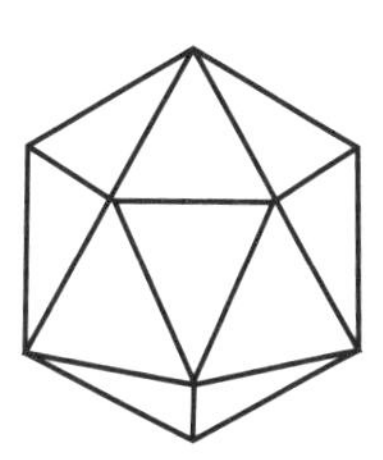
icosahedron

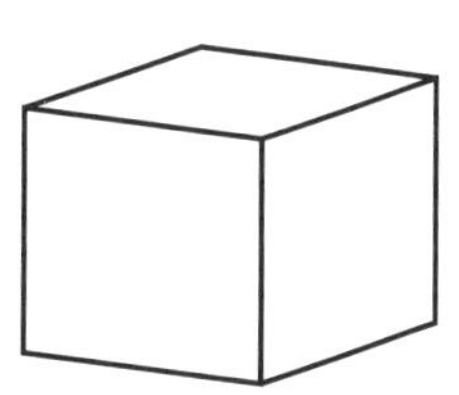
cube

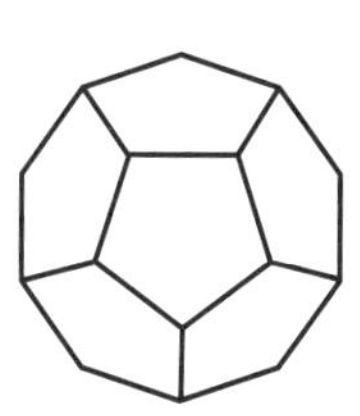
dodecahedron

Use the following steps to make a Platonic Solid.

1. Cut out the circle to use as a pattern.
2. Cut out the number of circles equal to the number of faces of the Platonic Solid you are making: Tetrahedron, 4; octahedron, 8; icosahedron, 20; cube, 6; dodecahedron, 12.
3. Trace the inscribed polygon onto stiff paper to use as a folding template. Use the polygon corresponding to the shape of the face of the Platonic Solid. Triangle-tetrahedron, octahedron, icosahedron; Square-cube; Pentagon-dodecahedron.
4. Place the template on each circle and fold back the flaps.
5. Glue or staple flaps of the faces at each vertex to make the Platonic Solid.

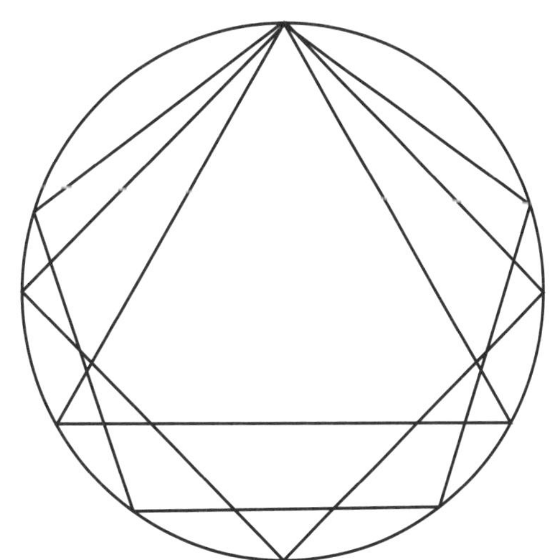

Historical Comment: The Platonic Solids are named after Plato, a Greek philosopher and mathematician (427-347 BC). In ancient Greece, the four basic elements were identified with four of the Platonic Solids: tetrahedron, fire; cube, earth; octahedron, air; and icosahedron, water. The dodecahedron with its 12 faces was related to the universe (12 signs of the zodiac).

Useful Definitions and Theorems

Angle Addition Postulate (AAP) If a point X lies in the interior of $\angle ABC$, then $m\angle ABC = m\angle ABX + m\angle XBC$.

Supplement Postulate If two angles form a linear pair, then these two angles are supplementary.

Side-Angle-Side Postulate (SAS) If two sides and the included angle of one triangle are congruent to the corresponding parts of another triangle, then the two triangles are congruent.

Angle-Side-Angle Postulate (ASA) If two angles and the included side of one triangle are congruent to the corresponding parts of another triangle, then the two triangles are congruent.

Hypotenuse-Leg Postulate (HL) If the hypotenuse and one leg of a right triangle are congruent to the corresponding parts of another right triangle, then the two triangles are congruent.

Parallel Postulate Given a line l and a point p not on l there is exactly one line parallel to l through p.

Congruence Postulate Given two congruent triangles, the triangular regions they determine have the same area.

Angle-Angle Postulate (AA) If two angles of one triangle are congruent to the corresponding angles of another triangle, then the two angles are similar.

Area Addition Postulate If two or more polygonal regions intersect in only points, segments or not at all, then the area of their union is the sum of their individual areas.

Arc Addition Postulate If the arcs $\overset{\frown}{AC}$ and $\overset{\frown}{BC}$ of a circle intersect in the single point B, then $m\overset{\frown}{AB} + m\overset{\frown}{BC} = m\overset{\frown}{ABC}$.

Addition Property of Equality (APOE) If two sides of an equation are equal, then if an equal quantity is added to both sides the equation will still be equal.

Area

triangle = $\frac{1}{2}$ • (base) • (height)

rectangle = (base) • (height)

square = $(\text{side})^2$

parallelogram = (base) • (height)

trapezoid = $\frac{1}{2}$ • (height) • (sum of bases)

circle = $\pi \cdot (\text{radius})^2$

Volume

pyramid = $\frac{1}{3}$ • (area of base) • (height)

cone = $\frac{1}{3} \cdot \pi \cdot (\text{radius})^2 \cdot (\text{height})$

cylinder = $\pi \cdot (\text{radius})^2 \cdot (\text{height})$

prism = (area of base) • (height)

sphere = $\frac{4}{3} \cdot \pi \cdot (\text{radius})^3$

Useful Definitions and Theorems

Between If A is on $\overline{BC}$, then A is between B and C if and only if $\overline{BA} + \overline{AC} = \overline{BC}$.
Bisect A bisects $\overline{BC}$ if A is the midpoint of $\overline{BC}$.
Complementary Angles Two angles are complementary if the sum of their measures is 90°.
CPCTC Corresponding parts of congruent triangles are congruent.
Hinge Theorem: Given that two sides of one triangle are congruent to two sides of a second triangle and the included angle of the first triangle is smaller than the included angle of the second triangle, then the third side of the first triangle is smaller than the third side of the second triangle.
Midpoint A is the midpoint of $\overline{BC}$ if $\overline{BA} \cong \overline{AC}$.
Pythagorean Theorem $a^2 + b^2 = c^2$
Reflexive Property Given a segment, angle, triangle, etc. it is congruent to itself.
Right Angle A right angle is one that has a measure of 90°.
Right Triangle A right triangle is one that contains a right angle.
Substitution Given items a, b and c, if a = b and a = c, then b = c.
Supplementary Angles Two angles are supplementary if the sum of their measures is 180°.
Vertical Angles If the sides of two angles form opposite rays, then the angles are vertical angles.

- Vertical angles are congruent.
- If congruent, then equal.
- If equal, then congruent.
- Perpendicular lines form right angles.
- If parallel lines, then (corresponding, alternate interior, alternate exterior) angles are congruent.
- If parallel lines, then (same side interior, same side exterior) angles are supplementary.
- If (corresponding, alternate interior, alternate exterior) angles are congruent, then lines are parallel.
- If (same side interior, same side exterior) angles are supplementary, then the lines are parallel.
- Two lines parallel to a third are parallel.
- In a triangle, angles opposite congruent sides are congruent.
- In a triangle, sides opposite congruent angles are congruent.
- Complements of congruent angles are congruent.
- Supplements of congruent angles are congruent.
- The two acute angles of a right triangle are complementary.
- The sum of the angles of a triangle is 180°.
- In a triangle, if two angles are not congruent, then the larger side is opposite the larger angle.
- In a triangle, an exterior angle is greater than either remote interior angle.

Answer Key

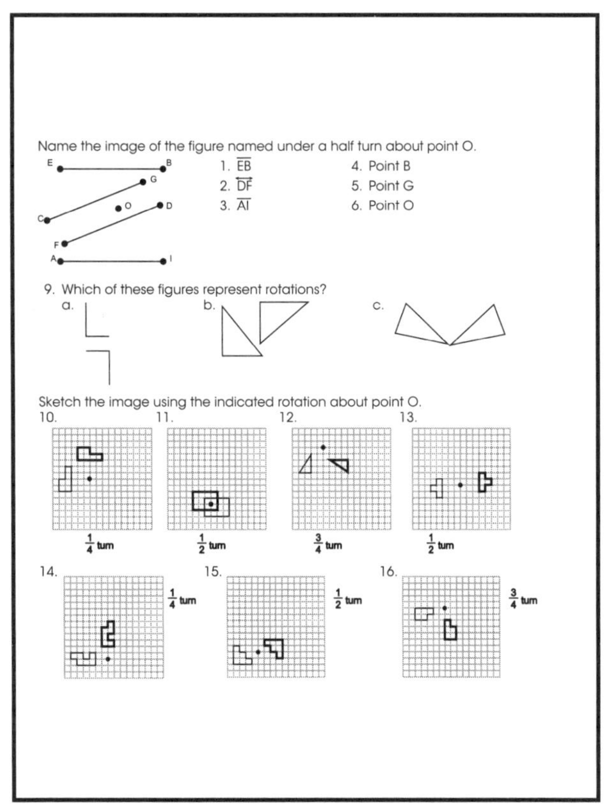
Name the image of the figure named under a half turn about point O.

1. $\overline{EB}$
2. $\overline{DF}$
3. $\overline{AI}$
4. Point B
5. Point G
6. Point O

9. Which of these figures represent rotations?
a. b. c.

Sketch the image using the indicated rotation about point O.

10. $\frac{1}{4}$ turn
11. $\frac{1}{2}$ turn
12. $\frac{3}{4}$ turn
13. $\frac{1}{2}$ turn
14. $\frac{1}{4}$ turn
15. $\frac{1}{2}$ turn
16. $\frac{3}{4}$ turn

7

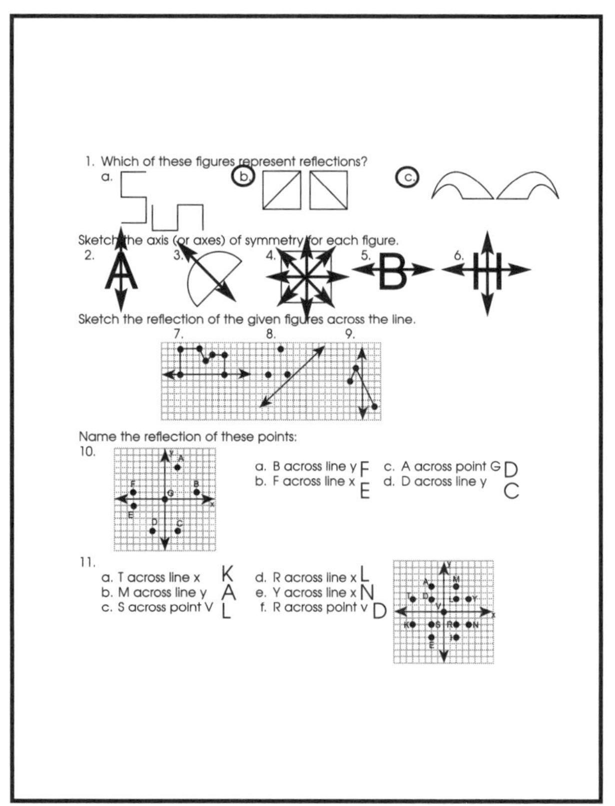
1. Which of these figures represent reflections?
a. (b.) (c.)

Sketch the axis (or axes) of symmetry for each figure.
2. 3. 4. 5. 6.

Sketch the reflection of the given figures across the line.
7. 8. 9.

Name the reflection of these points:

10. a. B across line y F
b. F across line x E
c. A across point G D
d. D across line y C

11. a. T across line x K
b. M across line y A
c. S across point V L
d. R across line x L
e. Y across line x N
f. R across point V D

8

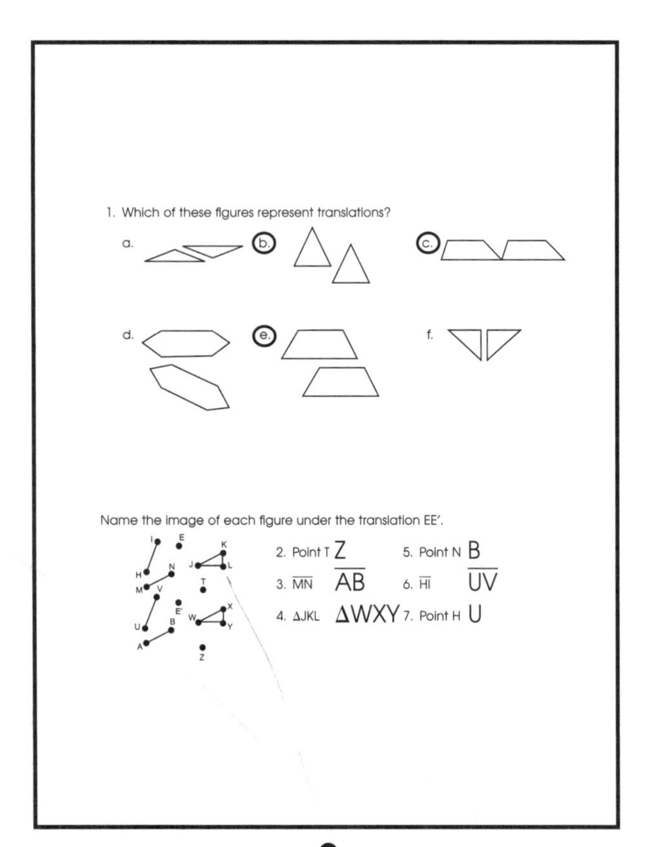
1. Which of these figures represent translations?
a. (b.) (c.) d. (e.) f.

Name the image of each figure under the translation EE′.

2. Point T Z
3. $\overline{MN}$ $\overline{AB}$
4. ΔJKL ΔWXY
5. Point N B
6. $\overline{HI}$ $\overline{UV}$
7. Point H U

9

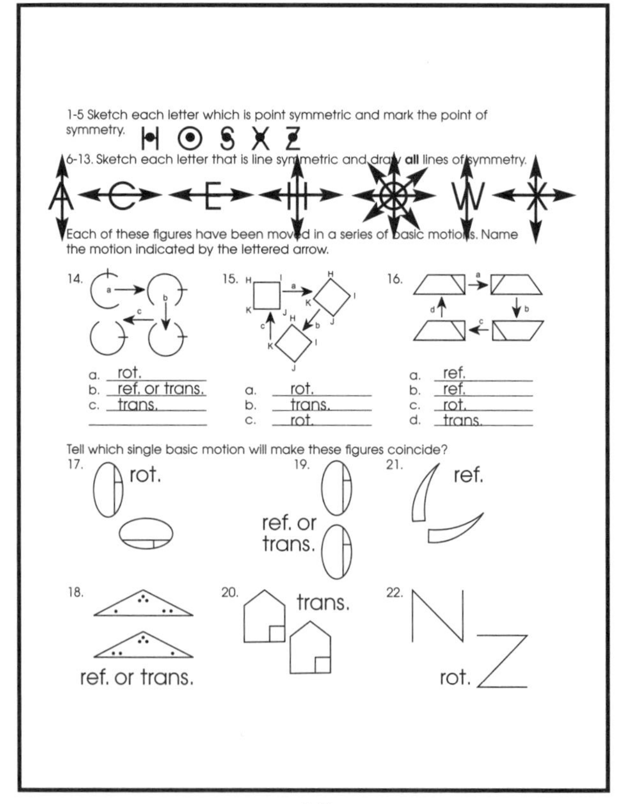
1-5 Sketch each letter which is point symmetric and mark the point of symmetry.
H O S X Z

6-13. Sketch each letter that is line symmetric and draw **all** lines of symmetry.
A C E H W X

Each of these figures have been moved in a series of basic motions. Name the motion indicated by the lettered arrow.

14. a. rot.
b. ref. or trans.
c. trans.

15. a. rot.
b. trans.
c. rot.

16. a. ref.
b. ref.
c. rot.
d. trans.

Tell which single basic motion will make these figures coincide?

17. rot.
18. ref. or trans.
19. ref. or trans.
20. trans.
21. ref.
22. rot.

10

Answer Key

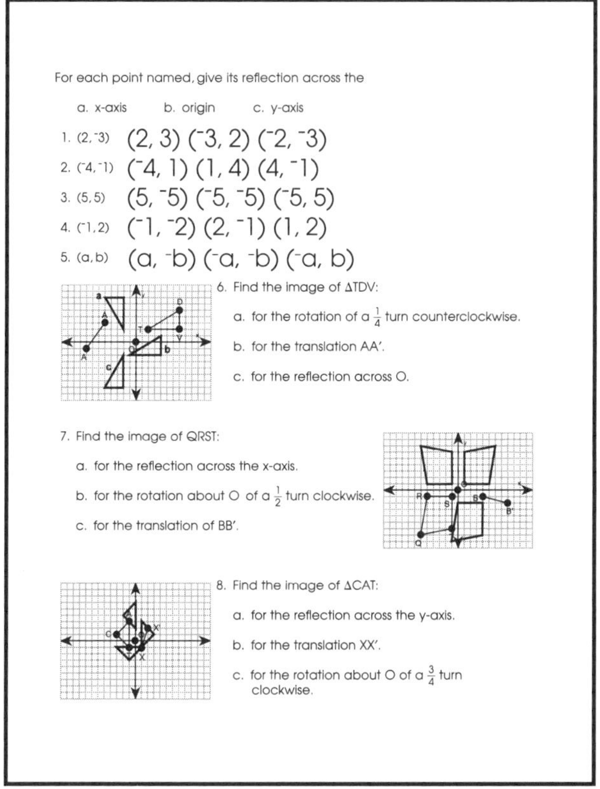

For each point named, give its reflection across the

a. x-axis b. origin c. y-axis

1. $(2, {}^-3)$ $(2, 3)$ $({}^-3, 2)$ $({}^-2, {}^-3)$
2. $({}^-4, {}^-1)$ $({}^-4, 1)$ $(1, 4)$ $(4, {}^-1)$
3. $(5, 5)$ $(5, {}^-5)$ $({}^-5, {}^-5)$ $({}^-5, 5)$
4. $({}^-1, 2)$ $({}^-1, {}^-2)$ $(2, {}^-1)$ $(1, 2)$
5. (a, b) $(a, {}^-b)$ $({}^-a, {}^-b)$ $({}^-a, b)$

6. Find the image of ΔTDV:
   a. for the rotation of a $\frac{1}{4}$ turn counterclockwise.
   b. for the translation AA′.
   c. for the reflection across O.

7. Find the image of QRST:
   a. for the reflection across the x-axis.
   b. for the rotation about O of a $\frac{1}{2}$ turn clockwise.
   c. for the translation of BB′.

8. Find the image of ΔCAT:
   a. for the reflection across the y-axis.
   b. for the translation XX′.
   c. for the rotation about O of a $\frac{3}{4}$ turn clockwise.

11

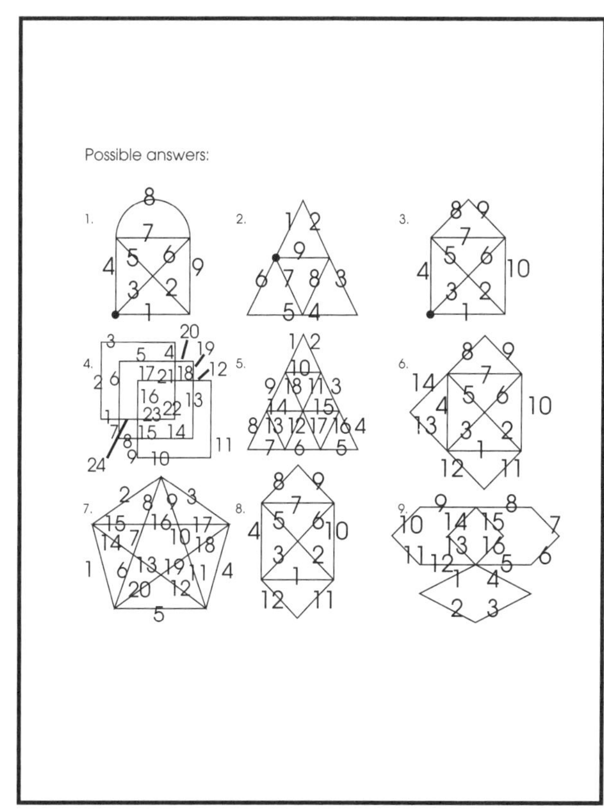

Possible answers:

12

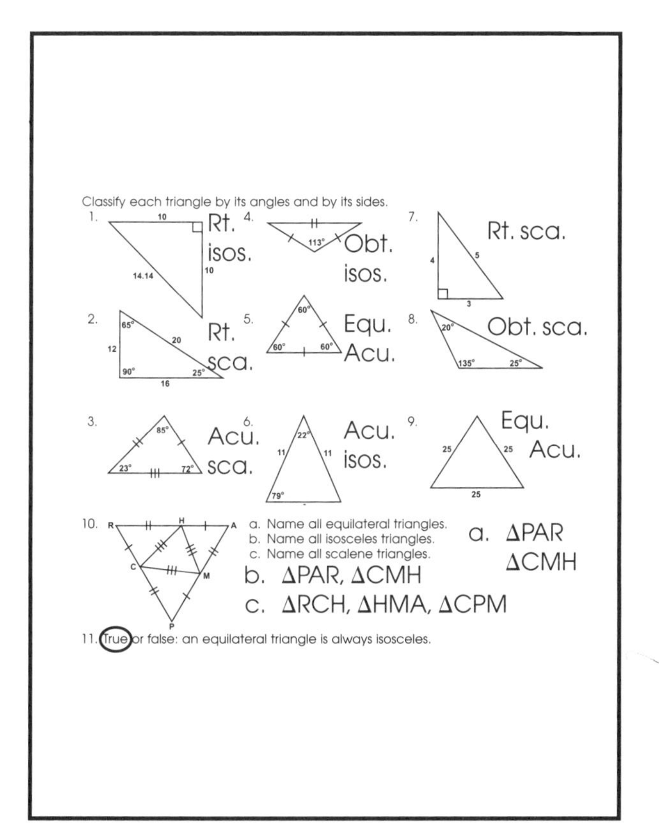

Classify each triangle by its angles and by its sides.

1. Rt. isos.
2. Rt. sca.
3. Acu. sca.
4. Obt. isos.
5. Equ. Acu.
6. Acu. isos.
7. Rt. sca.
8. Obt. sca.
9. Equ. Acu.

10. a. Name all equilateral triangles. b. Name all isosceles triangles. c. Name all scalene triangles.

a. ΔPAR, ΔCMH
b. ΔPAR, ΔCMH
c. ΔRCH, ΔHMA, ΔCPM

11. True or false: an equilateral triangle is always isosceles. (True circled)

13

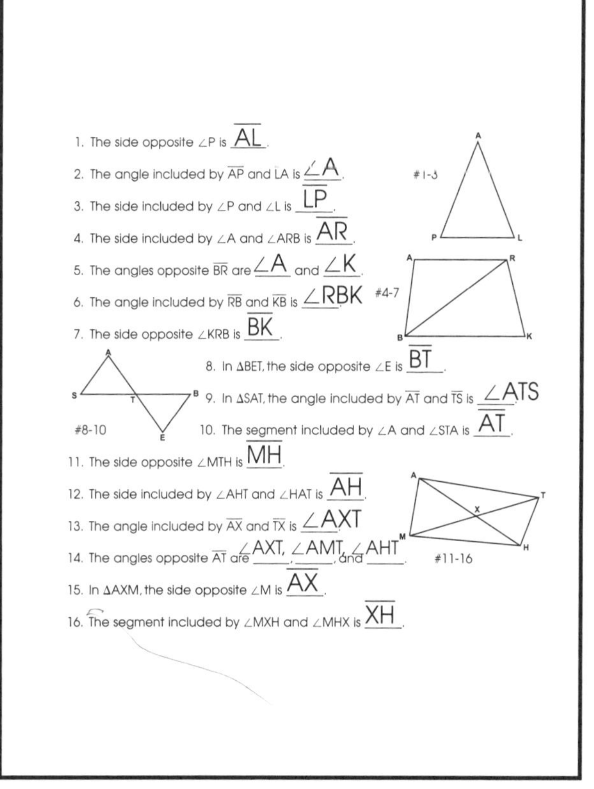

1. The side opposite $\angle P$ is $\overline{AL}$.
2. The angle included by $\overline{AP}$ and $\overline{LA}$ is $\angle A$.
3. The side included by $\angle P$ and $\angle L$ is $\overline{LP}$.
4. The side included by $\angle A$ and $\angle ARB$ is $\overline{AR}$.
5. The angles opposite $\overline{BR}$ are $\angle A$ and $\angle K$.
6. The angle included by $\overline{RB}$ and $\overline{KB}$ is $\angle RBK$.
7. The side opposite $\angle KRB$ is $\overline{BK}$.
8. In ΔBET, the side opposite $\angle E$ is $\overline{BT}$.
9. In ΔSAT, the angle included by $\overline{AT}$ and $\overline{TS}$ is $\angle ATS$.
10. The segment included by $\angle A$ and $\angle STA$ is $\overline{AT}$.
11. The side opposite $\angle MTH$ is $\overline{MH}$.
12. The side included by $\angle AHT$ and $\angle HAT$ is $\overline{AH}$.
13. The angle included by $\overline{AX}$ and $\overline{TX}$ is $\angle AXT$.
14. The angles opposite $\overline{AT}$ are $\angle AXT$, $\angle AMT$, and $\angle AHT$.
15. In ΔAXM, the side opposite $\angle M$ is $\overline{AX}$.
16. The segment included by $\angle MXH$ and $\angle MHX$ is $\overline{XH}$.

#1-3 #4-7 #8-10 #11-16

14

Answer Key

1. $\angle X \cong \angle S, \angle Y \cong \angle T, \angle Z \cong \angle R$
2. $\overline{XY} \cong \overline{ST}, \overline{XZ} \cong \overline{SR}, \overline{YZ} \cong \overline{TR}$

3, 4, 5, 7, 8 are true. 6 is false.

9. $\overline{ML} \cong \overline{BV}, \overline{LB} \cong \overline{VM}, \overline{MB} \cong \overline{BM}, \angle LMB \cong \angle VBM, \angle BLM \cong \angle MVB, \angle LBM \cong \angle VMB$
10. $\overline{LM} \cong \overline{VB}, \overline{MP} \cong \overline{BP}, \overline{LP} \cong \overline{VP}, \angle LMP \cong \angle VBP, \angle MPL \cong \angle VPB, \angle PLM \cong \angle PVB$
11. $\overline{LP} \cong \overline{VP}, \overline{PB} \cong \overline{PM}, \overline{LB} \cong \overline{VM}, \angle LPB \cong \angle VPM, \angle PBL \cong \angle PMV, \angle BLP \cong \angle MV$
12. $\overline{MJ} \cong \overline{TJ}, \overline{MB} \cong \overline{TC}, \overline{JB} \cong \overline{JC}, \angle MJB \cong \angle TJC, \angle JBM \cong \angle JCT, \angle BMJ \cong \angle CTJ$
13. $\overline{TK} \cong \overline{MA}, \overline{KJ} \cong \overline{AJ}, \overline{TJ} \cong \overline{MJ}, \angle KTJ \cong \angle AMJ, \angle TJK \cong \angle MJA, \angle JKT \cong \angle JAM, \angle KTJ \cong \angle JMA$
14. $\overline{CB} \cong \overline{DF}, \overline{CD} \cong \overline{DE}, \overline{BD} \cong \overline{EF}, \angle CBD \cong \angle DFE, \angle BDC \cong \angle FED, \angle DCE \cong \angle EDF$
15. $\overline{BA} \cong \overline{BD}, \overline{BF} \cong \overline{BF}, \overline{AF} \cong \overline{DF}, \angle BAF \cong \angle BDF, \angle AFB \cong \angle DFB, \angle FBA \cong \angle FBD$
16. $\overline{CB} \cong \overline{FB}, \overline{FD} \cong \overline{CD}, \overline{BD} \cong \overline{BD}, \angle CBD \cong \angle DFE, \angle BCD \cong \angle BFD, \angle CBD \cong \angle DBF, \angle BDC \cong \angle BDF$

15

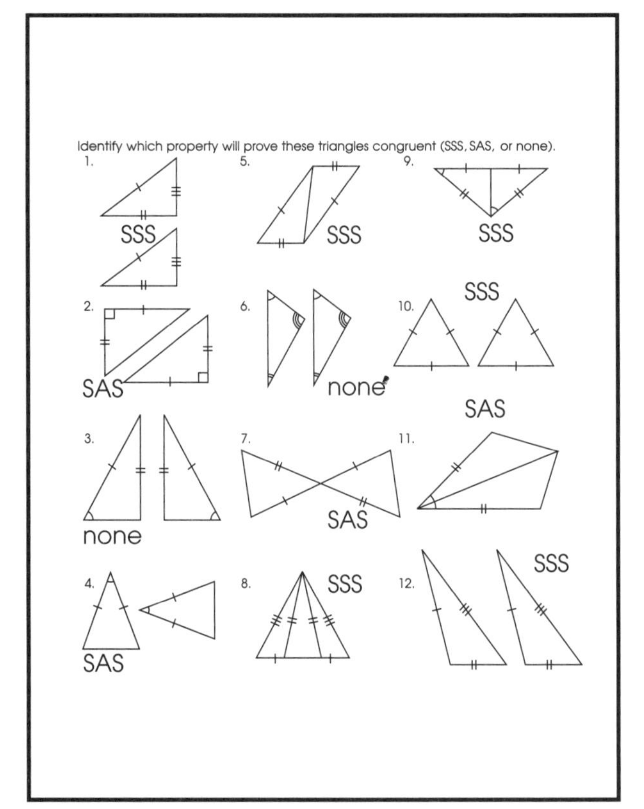

16

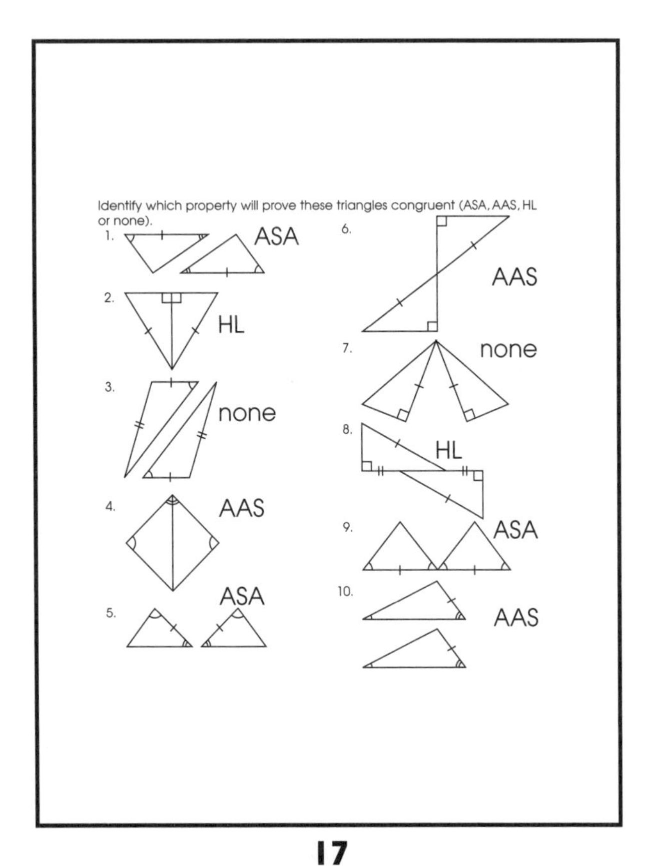

17

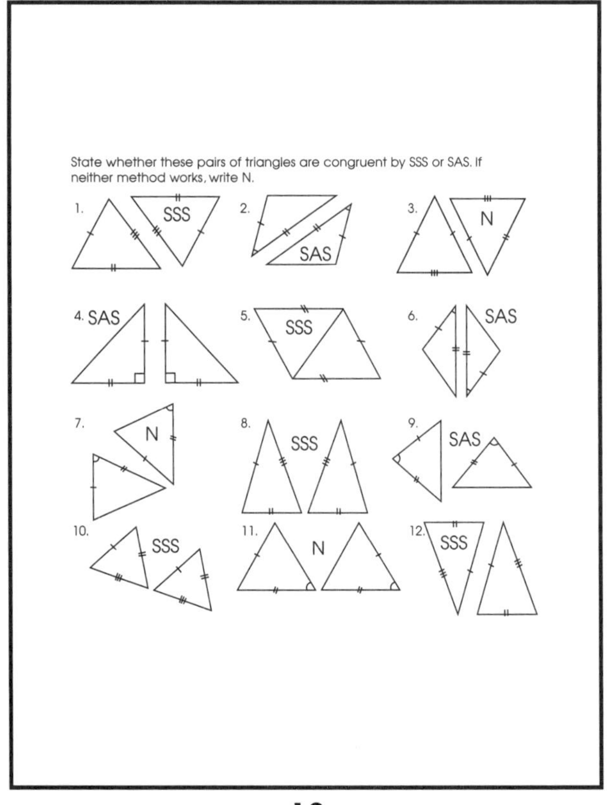

18

Answer Key

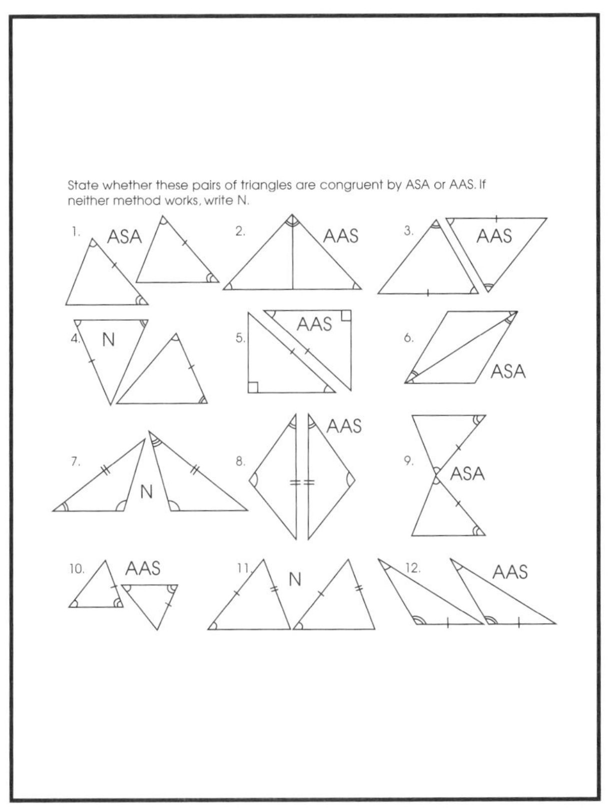

State whether these pairs of triangles are congruent by ASA or AAS. If neither method works, write N.

1. ASA
2. AAS
3. AAS
4. N
5. AAS
6. ASA
7. N
8. AAS
9. ASA
10. AAS
11. N
12. AAS

19

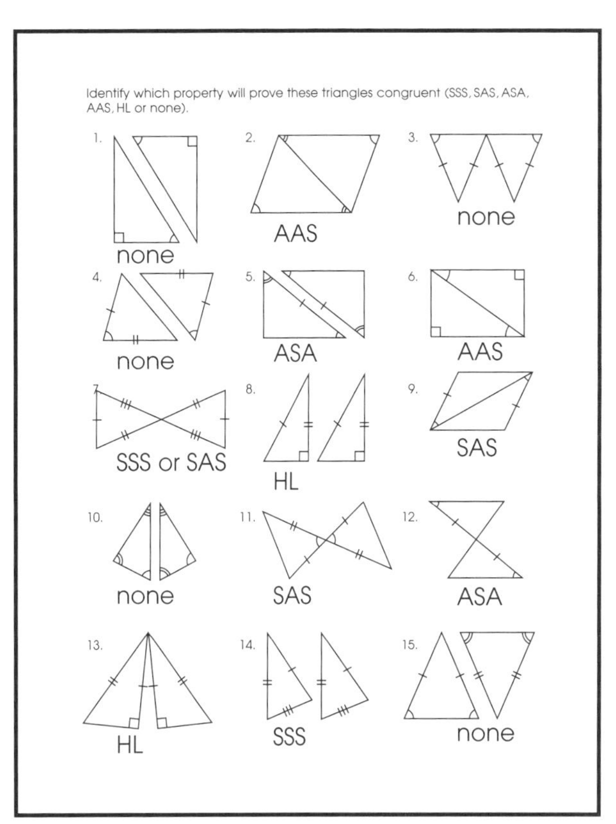

Identify which property will prove these triangles congruent (SSS, SAS, ASA, AAS, HL or none).

1. none
2. AAS
3. none
4. none
5. ASA
6. AAS
7. SSS or SAS
8. HL
9. SAS
10. none
11. SAS
12. ASA
13. HL
14. SSS
15. none

20

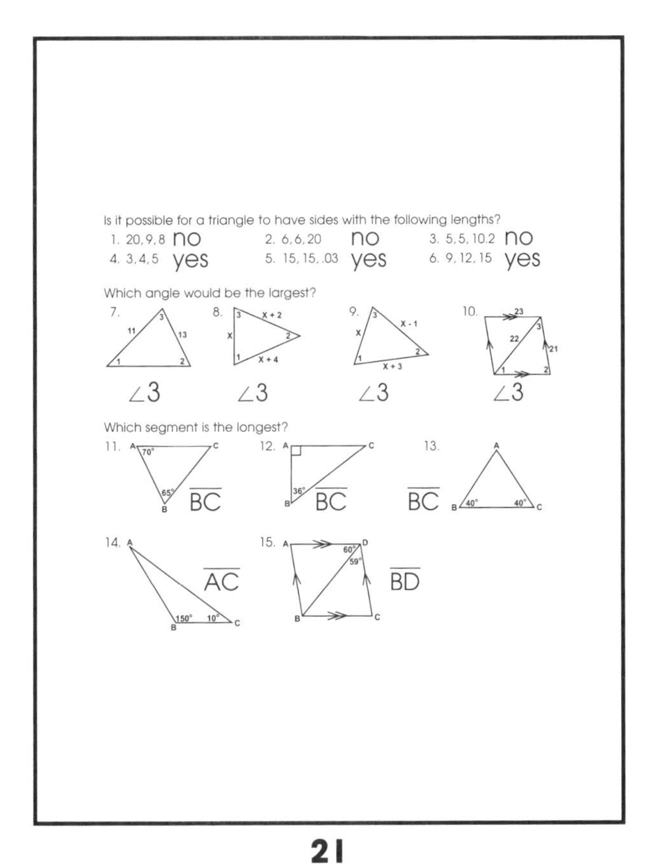

Is it possible for a triangle to have sides with the following lengths?

1. 20, 9, 8 no
2. 6, 6, 20 no
3. 5, 5, 10.2 no
4. 3, 4, 5 yes
5. 15, 15, .03 yes
6. 9, 12, 15 yes

Which angle would be the largest?

7. ∠3
8. ∠3
9. ∠3
10. ∠3

Which segment is the longest?

11. $\overline{BC}$
12. $\overline{BC}$
13. $\overline{BC}$
14. $\overline{AC}$
15. $\overline{BD}$

21

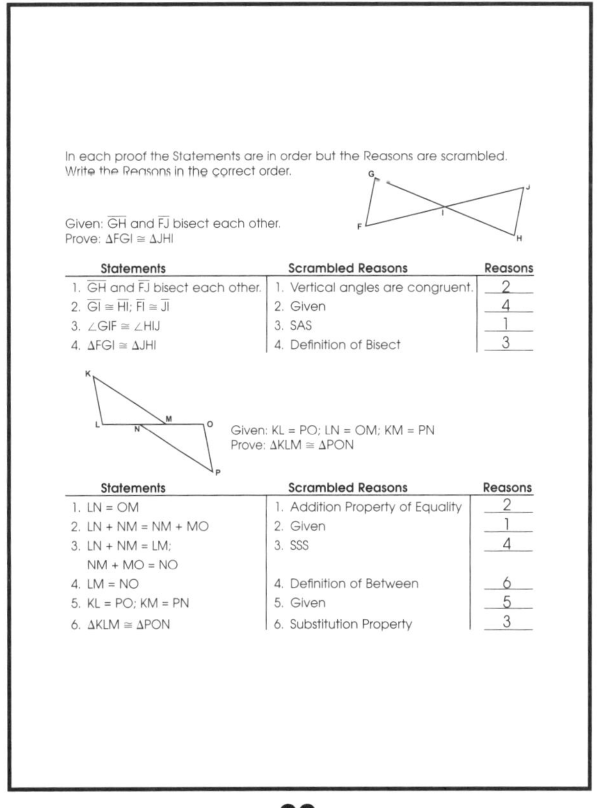

In each proof the Statements are in order but the Reasons are scrambled. Write the Reasons in the correct order.

Given: $\overline{GH}$ and $\overline{FJ}$ bisect each other.
Prove: ΔFGI ≅ ΔJHI

Statements	Scrambled Reasons	Reasons
1. $\overline{GH}$ and $\overline{FJ}$ bisect each other.	1. Vertical angles are congruent.	2
2. $\overline{GI} \cong \overline{HI}$; $\overline{FI} \cong \overline{JI}$	2. Given	4
3. ∠GIF ≅ ∠HIJ	3. SAS	1
4. ΔFGI ≅ ΔJHI	4. Definition of Bisect	3

Given: KL = PO; LN = OM; KM = PN
Prove: ΔKLM ≅ ΔPON

Statements	Scrambled Reasons	Reasons
1. LN = OM	1. Addition Property of Equality	2
2. LN + NM = NM + MO	2. Given	1
3. LN + NM = LM; NM + MO = NO	3. SSS	4
4. LM = NO	4. Definition of Between	6
5. KL = PO; KM = PN	5. Given	5
6. ΔKLM ≅ ΔPON	6. Substitution Property	3

22

Answer Key

Complete the following proofs.
Given: m∠1 = 40°; m∠3 = 40°, ∠2 ≅ ∠4
Prove: ΔRTQ ≅ ΔTRS

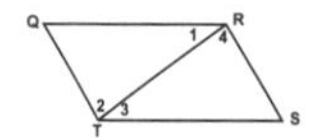

Statements	Reasons
1. m∠1 = 40°; m∠3 = 40°, ∠2 ≅ ∠4	1. Given
2. ∠1 ≅ ∠3	2. Definition of ≅ ∠'s.
3. $\overline{RT}$ ≅ $\overline{TR}$	3. Reflexive
4. ΔRTQ ≅ ΔTRS	4. ASA

Given: $\overline{WY}$ ≅ $\overline{XV}$; $\overline{VW}$ ⊥ $\overline{WX}$; $\overline{YX}$ ⊥ $\overline{WX}$
Prove: ΔXWV ≅ ΔWXY

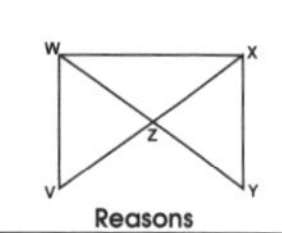

Statements	Reasons
1. $\overline{VW}$ ⊥ $\overline{WX}$ and $\overline{YX}$ ⊥ $\overline{WX}$	1. Given
2. ∠VWX, ∠YXW are rt. ∠'s.	2. Definition Perpendicular Lines
3. ΔXWV, ΔWXY are right Δs	3. Definition of rt. Δ's.
4. WY ≅ XV	4. Given
5. $\overline{WX}$ ≅ $\overline{WX}$	5. Reflexive
6. ΔXWV ≅ ΔWXY	6. HL

Given: ∠1 ≅ ∠6; ∠3 ≅ ∠4; B is the midpoint of $\overline{AC}$
Prove: ΔABE ≅ ΔCBD

Statements	Reasons
1. ∠1 ≅ ∠6, ∠3 ≅ ∠4	1. Given
2. $\overline{AB}$ ≅ $\overline{BC}$ B is midpoint to AC.	2. Definition of midpoint.
3. ∠1 is supplement to ∠2	3. Definition of Supplementary
4. ∠5 is supplementary to ∠6	4. Definition of supplementary.
5. ∠2 ≅ ∠5	5. Supplements of ≅ ∠'s are ≅.
6. ΔABE ≅ ΔCBD	6. ASA

23

Example: Draw a rectangle. Perform the iteration of connecting the midpoints of the adjacent sides. Every interior rectangle looks like the original—self-similar.

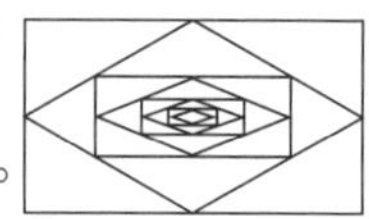

On another sheet of paper, complete Steps 0-3 to begin the Koch Curve.

Step 0 Draw a line segment 6 inches long. Consider its length to be one unit. Ex. Step 0

Step 1 Draw an equilateral triangle whose base is the middle third of the line segment. Do not draw the base. Step 1

Step 2 Draw an equilateral triangle on each segment so the base (not drawn) of each triangle is the middle third of the corresponding segment.

Step 3 Repeat Step 2.

Complete the table

Step	Number of Segments	Length of 1 Segment	Total Length
0	1	1	1
1	4	$\frac{1}{3}$	4/3
2	16	1/9	16/9
3	64	1/27	64/27

1. Describe the pattern in each column.

A. Number of segments is 4 times greater or 4^S (s = step number).

B. Length of 1 segment is 1/3 as much or $(1/3)^S$ (s = step number).

C. Total length is the product of answers A and B or $4^S(1/3)^S$ or $(4/3)^S$ (s = step number).

2. What would the values be for Step 5? 1,024 segments, each 1/243 units long for a total length of 1,024/243.

Historical Comment: The curve is the basis for the Koch Snowflake designed by Helge von Koch in 1904. Step 0 starts with an equilateral triangle. Steps 1, 2, 3, etc., are the same.

24

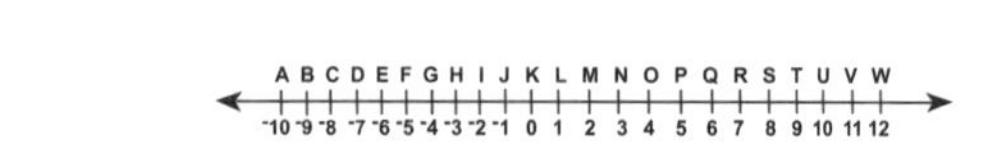

Find the length of each segment and link the segments in Columns A and B that have equal lengths.

Column A	Column B
16 1. Segment $\overline{GW}$	A. Segment $\overline{EL}$
12 2. Segment $\overline{BN}$	B. Segment $\overline{AF}$
6 3. Segment $\overline{PV}$	C. Segment $\overline{CS}$
7 4. Segment $\overline{KR}$	D. Segment $\overline{MS}$
5 5. Segment with endpoints $\frac{3}{4}$ and $5\frac{3}{4}$	E. Segment with endpoints ⁻2 and 10
4 1/2 6. Segment with endpoints ⁻3 and ⁻$7\frac{1}{2}$	F. Segment with endpoints ⁻$1\frac{1}{2}$ and 3
3 1/2 7. Segment with endpoints - $\frac{1}{4}$ and $3\frac{1}{4}$	G. Segment with endpoints ⁻$1\frac{3}{4}$ and $1\frac{3}{4}$
3 8. Segment with endpoints $\frac{1}{4}$ and $3\frac{1}{4}$	H. Segment with endpoints ⁻5 and ⁻2

In a Magic Square, each row, column, and diagonal has the same: Magic Sum. Find the length of the segments and determine the Magic Sum: 17.

Endpoints $1\frac{1}{2}$ and 2 1/2	Endpoints ⁻3 and $4\frac{1}{2}$ 7 1/2	Segment $\overline{DK}$ 7	Segment $\overline{HJ}$ 2
Segment $\overline{HN}$ 6	Segment $\overline{QT}$ 3	Endpoints $6\frac{1}{2}$ and 3 3 1/2	Endpoints $5\frac{3}{4}$ and $1\frac{1}{4}$ 4 1/2
Segment $\overline{DH}$ 4	Segment $\overline{JO}$ 5	Endpoints $1\frac{3}{4}$ and ⁻$3\frac{3}{4}$ 5 1/2	Endpoints $1\frac{1}{2}$ and ⁻1 2 1/2
Endpoints ⁻$2\frac{1}{4}$ and $4\frac{1}{4}$ 6 1/2	Endpoints ⁻$5\frac{1}{2}$ and ⁻4 1 1/2	Segment $\overline{JK}$ 1	Segment $\overline{AI}$ 8

25

True or False.

1. $\overline{TV}$ ≅ $\overline{ML}$ T
2. $\overline{KJ}$ ≅ $\overline{TV}$ T
3. $\overline{LB}$ ≅ $\overline{JV}$ T
4. $\overline{TV}$ ≅ $\overline{BV}$ F
5. $\overline{VB}$ ≅ $\overline{LB}$ T
6. $\overline{KJ}$ ≅ $\overline{VB}$ F

Complete.

7. QR + RS = 7. QS
8. RU – SU = 8. RS
9. RS + SU = 9. RU
10. QS + SU = 10. QU
11. QU – QR – TU = 11. RT
12. QR + RS + ST = 12. QT

Find the length of the indicated segments.

13. JD = 8.4 GB = 11 JB = 12.4
14. SK = 17 DT = 10 MT = 39.5
15. BC = 15 DE = 24 AE = 95
16. MN = 14 NO = 14 OP = 15
17. Which segments are congruent in #15? $\overline{AB}$ ≅ $\overline{CD}$

26

Answer Key

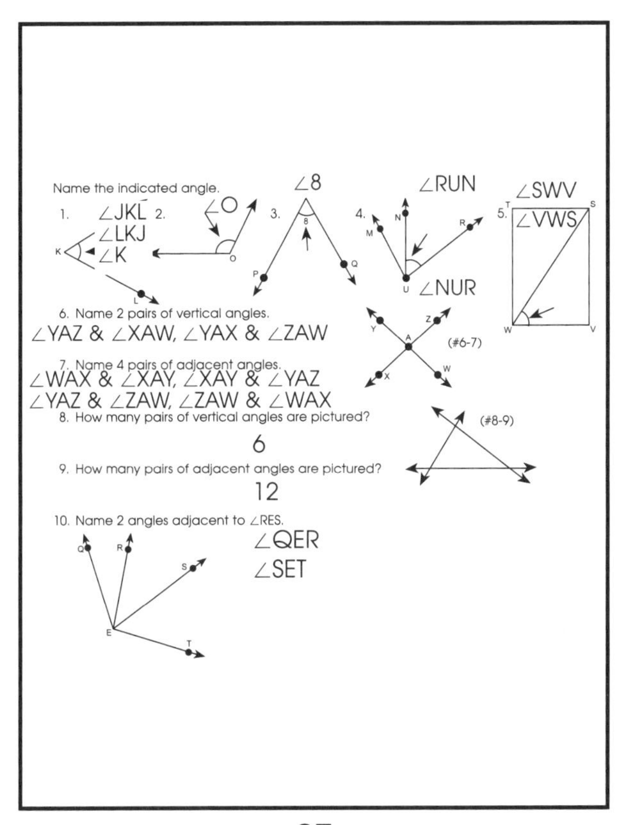

Name the indicated angle.
1. ∠JKL ∠LKJ ∠K
2. ∠O
3. ∠8
4. ∠RUN ∠NUR
5. ∠SWV ∠VWS
6. Name 2 pairs of vertical angles.
∠YAZ & ∠XAW, ∠YAX & ∠ZAW
7. Name 4 pairs of adjacent angles.
∠WAX & ∠XAY, ∠XAY & ∠YAZ
∠YAZ & ∠ZAW, ∠ZAW & ∠WAX
8. How many pairs of vertical angles are pictured?
6
9. How many pairs of adjacent angles are pictured?
12
10. Name 2 angles adjacent to ∠RES.
∠QER
∠SET

27

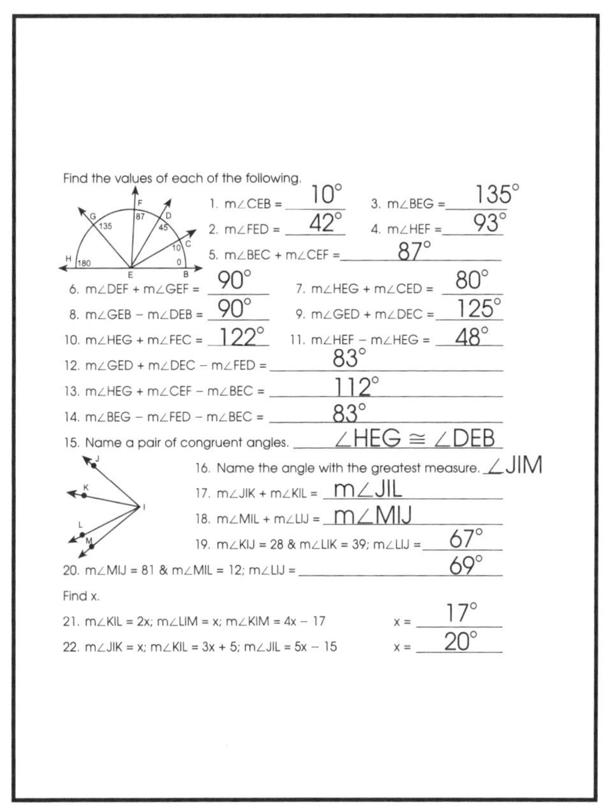

Find the values of each of the following.
1. m∠CEB = 10°
2. m∠FED = 42°
3. m∠BEG = 135°
4. m∠HEF = 93°
5. m∠BEC + m∠CEF = 87°
6. m∠DEF + m∠GEF = 90°
7. m∠HEG + m∠CED = 80°
8. m∠GEB − m∠DEB = 90°
9. m∠GED + m∠DEC = 125°
10. m∠HEG + m∠FEC = 122°
11. m∠HEF − m∠HEG = 48°
12. m∠GED + m∠DEC − m∠FED = 83°
13. m∠HEG + m∠CEF − m∠BEC = 112°
14. m∠BEG − m∠FED − m∠BEC = 83°
15. Name a pair of congruent angles. ∠HEG ≅ ∠DEB
16. Name the angle with the greatest measure. ∠JIM
17. m∠JIK + m∠KIL = m∠JIL
18. m∠MIL + m∠LIJ = m∠MIJ
19. m∠KIJ = 28 & m∠LIK = 39; m∠LIJ = 67°
20. m∠MIJ = 81 & m∠MIL = 12; m∠LIJ = 69°
Find x.
21. m∠KIL = 2x; m∠LIM = x; m∠KIM = 4x − 17 x = 17°
22. m∠JIK = x; m∠KIL = 3x + 5; m∠JIL = 5x − 15 x = 20°

28

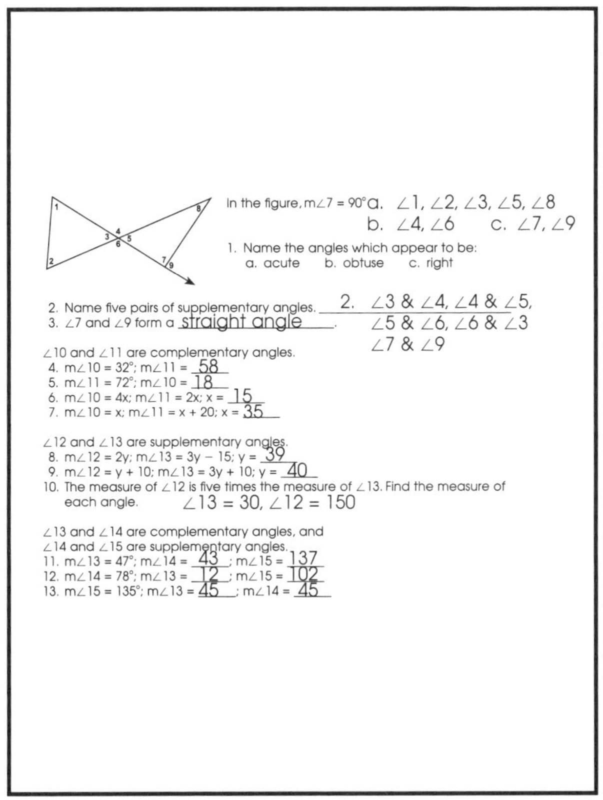

In the figure, m∠7 = 90°
1. Name the angles which appear to be: a. acute b. obtuse c. right
a. ∠1, ∠2, ∠3, ∠5, ∠8
b. ∠4, ∠6
c. ∠7, ∠9
2. Name five pairs of supplementary angles.
2. ∠3 & ∠4, ∠4 & ∠5, ∠5 & ∠6, ∠6 & ∠3, ∠7 & ∠9
3. ∠7 and ∠9 form a straight angle.
∠10 and ∠11 are complementary angles.
4. m∠10 = 32°; m∠11 = 58
5. m∠11 = 72°; m∠10 = 18
6. m∠10 = 4x; m∠11 = 2x; x = 15
7. m∠10 = x; m∠11 = x + 20; x = 35
∠12 and ∠13 are supplementary angles.
8. m∠12 = 2y; m∠13 = 3y − 15; y = 39
9. m∠12 = y + 10; m∠13 = 3y + 10; y = 40
10. The measure of ∠12 is five times the measure of ∠13. Find the measure of each angle. ∠13 = 30, ∠12 = 150
∠13 and ∠14 are complementary angles, and ∠14 and ∠15 are supplementary angles.
11. m∠13 = 47°; m∠14 = 43; m∠15 = 137
12. m∠14 = 78°; m∠13 = 12; m∠15 = 102
13. m∠15 = 135°; m∠13 = 45; m∠14 = 45

29

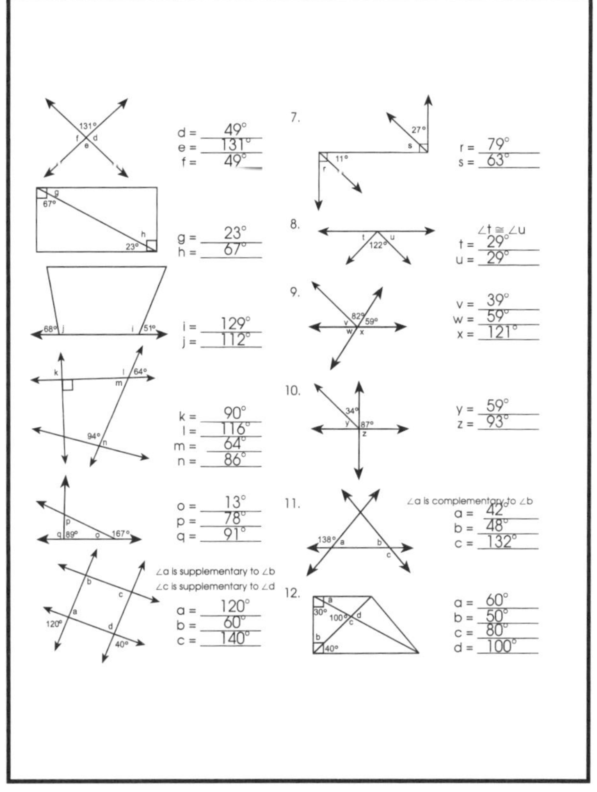

30

Answer Key

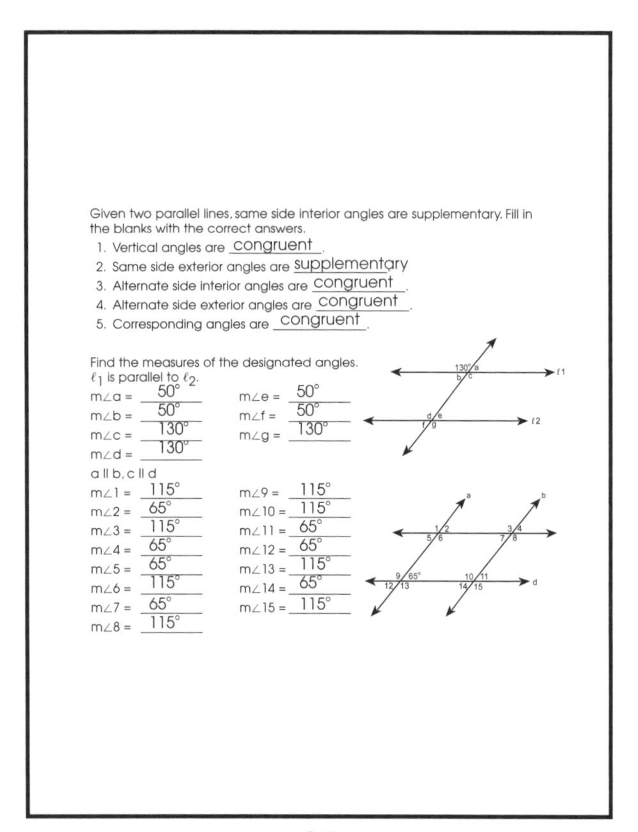

Given two parallel lines, same side interior angles are supplementary. Fill in the blanks with the correct answers.

1. Vertical angles are congruent.
2. Same side exterior angles are supplementary
3. Alternate side interior angles are congruent.
4. Alternate side exterior angles are congruent.
5. Corresponding angles are congruent.

Find the measures of the designated angles. ℓ_1 is parallel to ℓ_2.

m∠a =	50°	m∠e =	50°
m∠b =	50°	m∠f =	50°
m∠c =	130°	m∠g =	130°
m∠d =	130°		

a ll b, c ll d

m∠1 =	115°	m∠9 =	115°
m∠2 =	65°	m∠10 =	115°
m∠3 =	115°	m∠11 =	65°
m∠4 =	65°	m∠12 =	65°
m∠5 =	65°	m∠13 =	115°
m∠6 =	115°	m∠14 =	65°
m∠7 =	65°	m∠15 =	115°
m∠8 =	115°		

31

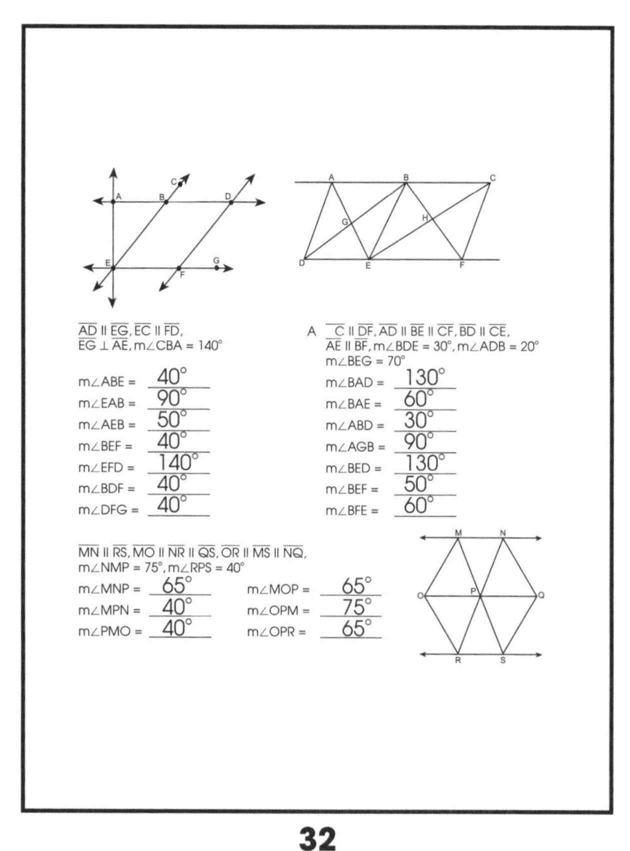

$\overline{AD}$ ll $\overline{EG}$, EC ll FD, $\overline{EG}$ ⊥ AE, m∠CBA = 140°

m∠ABE =	40°
m∠EAB =	90°
m∠AEB =	50°
m∠BEF =	40°
m∠EFD =	140°
m∠BDF =	40°
m∠DFG =	40°

A $\overline{C}$ ll $\overline{DF}$, $\overline{AD}$ ll $\overline{BE}$ ll $\overline{CF}$, $\overline{BD}$ ll $\overline{CE}$, $\overline{AE}$ ll $\overline{BF}$, m∠BDE = 30°, m∠ADB = 20°, m∠BEG = 70°

m∠BAD =	130°
m∠BAE =	60°
m∠ABD =	30°
m∠AGB =	90°
m∠BED =	130°
m∠BEF =	50°
m∠BFE =	60°

$\overline{MN}$ ll $\overline{RS}$, $\overline{MO}$ ll $\overline{NR}$ ll $\overline{QS}$, $\overline{OR}$ ll $\overline{MS}$ ll $\overline{NQ}$, m∠NMP = 75°, m∠RPS = 40°

m∠MNP =	65°	m∠MOP =	65°
m∠MPN =	40°	m∠OPM =	75°
m∠PMO =	40°	m∠OPR =	65°

32

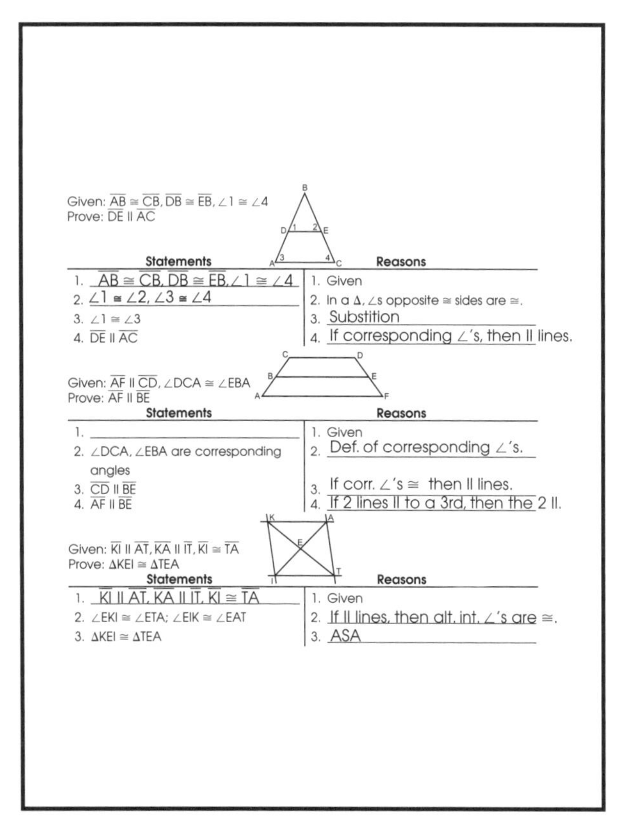

Given: $\overline{AB}$ ≅ $\overline{CB}$, $\overline{DB}$ ≅ $\overline{EB}$, ∠1 ≅ ∠4
Prove: $\overline{DE}$ ll $\overline{AC}$

Statements	Reasons
1. AB ≅ CB, DB ≅ EB, ∠1 ≅ ∠4	1. Given
2. ∠1 ≅ ∠2, ∠3 ≅ ∠4	2. In a Δ, ∠s opposite ≅ sides are ≅.
3. ∠1 ≅ ∠3	3. Substition
4. $\overline{DE}$ ll $\overline{AC}$	4. If corresponding ∠'s, then ll lines.

Given: $\overline{AF}$ ll $\overline{CD}$, ∠DCA ≅ ∠EBA
Prove: AF ll BE

Statements	Reasons
1.	1. Given
2. ∠DCA, ∠EBA are corresponding angles	2. Def. of corresponding ∠'s.
3. $\overline{CD}$ ll $\overline{BE}$	3. If corr. ∠'s ≅ then ll lines.
4. $\overline{AF}$ ll $\overline{BE}$	4. If 2 lines ll to a 3rd, then the 2 ll.

Given: $\overline{KI}$ ll $\overline{AT}$, $\overline{KA}$ ll $\overline{IT}$, $\overline{KI}$ ≅ $\overline{TA}$
Prove: ΔKEI ≅ ΔTEA

Statements	Reasons
1. KI ll AT, KA ll IT, KI ≅ TA	1. Given
2. ∠EKI ≅ ∠ETA; ∠EIK ≅ ∠EAT	2. If ll lines, then alt. int. ∠'s are ≅.
3. ΔKEI ≅ ΔTEA	3. ASA

33

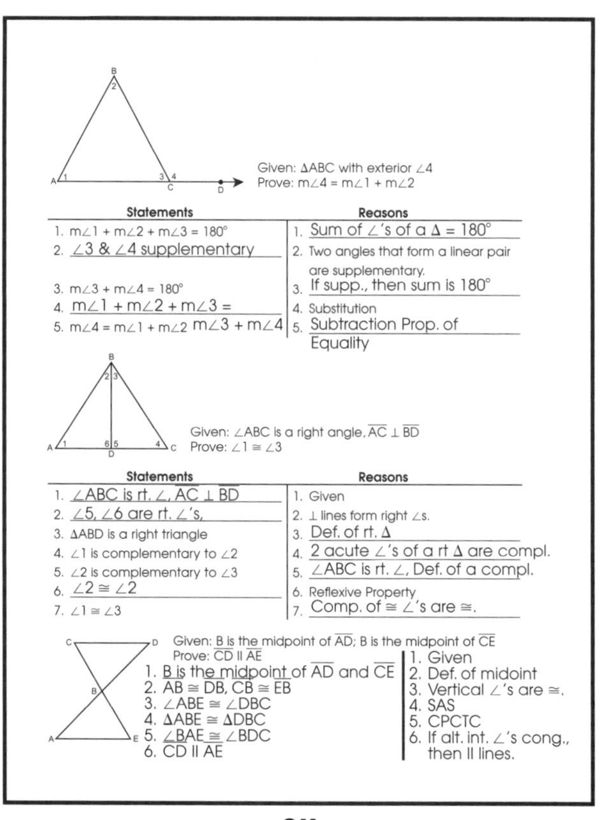

Given: ΔABC with exterior ∠4
Prove: m∠4 = m∠1 + m∠2

Statements	Reasons
1. m∠1 + m∠2 + m∠3 = 180°	1. Sum of ∠'s of a Δ = 180°
2. ∠3 & ∠4 supplementary	2. Two angles that form a linear pair are supplementary.
3. m∠3 + m∠4 = 180°	3. If supp., then sum is 180°
4. m∠1 + m∠2 + m∠3 =	4. Substitution
5. m∠4 = m∠1 + m∠2 m∠3 + m∠4	5. Subtraction Prop. of Equality

Given: ∠ABC is a right angle, $\overline{AC}$ ⊥ $\overline{BD}$
Prove: ∠1 ≅ ∠3

Statements	Reasons
1. ∠ABC is rt. ∠, AC ⊥ BD	1. Given
2. ∠5, ∠6 are rt. ∠'s,	2. ⊥ lines form right ∠s.
3. ΔABD is a right triangle	3. Def. of rt. Δ
4. ∠1 is complementary to ∠2	4. 2 acute ∠'s of a rt Δ are compl.
5. ∠2 is complementary to ∠3	5. ∠ABC is rt. ∠, Def. of a compl.
6. ∠2 ≅ ∠2	6. Reflexive Property
7. ∠1 ≅ ∠3	7. Comp. of ≅ ∠'s are ≅.

Given: B is the midpoint of $\overline{AD}$; B is the midpoint of $\overline{CE}$
Prove: $\overline{CD}$ ll $\overline{AE}$

Statements	Reasons
1. B is the midpoint of $\overline{AD}$ and $\overline{CE}$	1. Given
2. AB ≅ DB, CB ≅ EB	2. Def. of midoint
3. ∠ABE ≅ ∠DBC	3. Vertical ∠'s are ≅.
4. ΔABE ≅ ΔDBC	4. SAS
5. ∠BAE ≅ ∠BDC	5. CPCTC
6. CD ll AE	6. If alt. int. ∠'s cong., then ll lines.

34

Answer Key

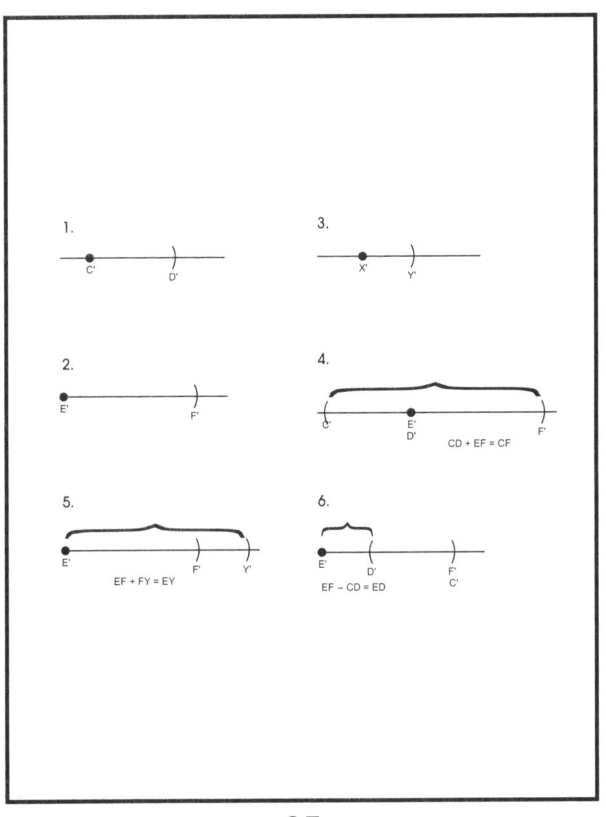

35

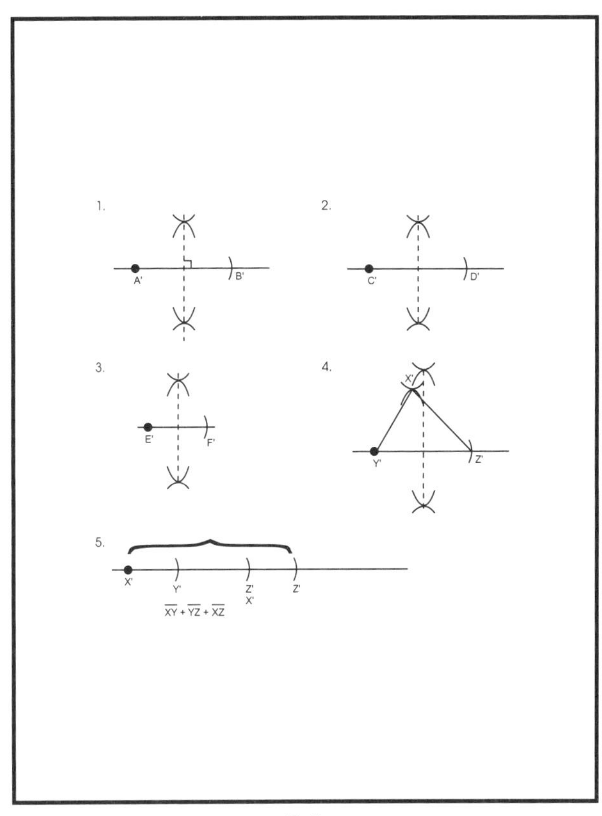

36

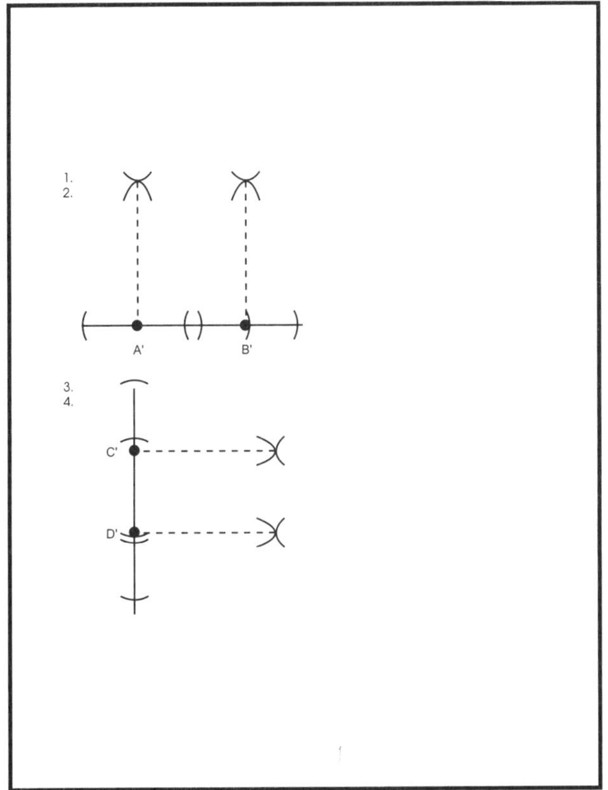

37

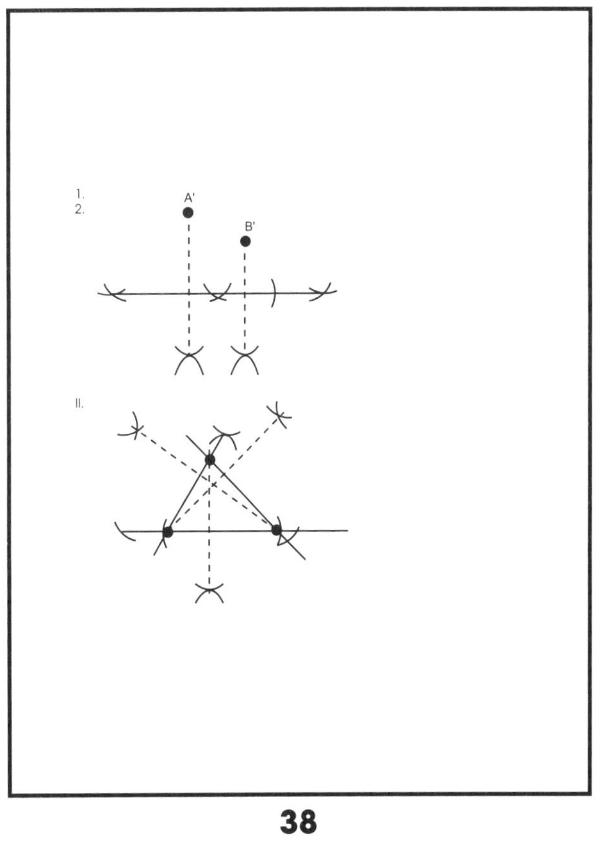

38

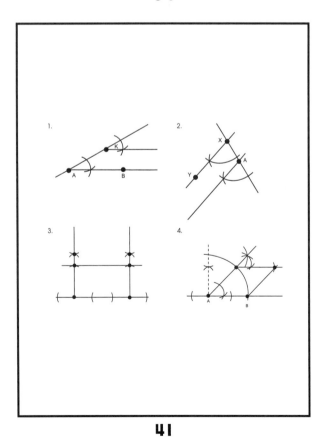

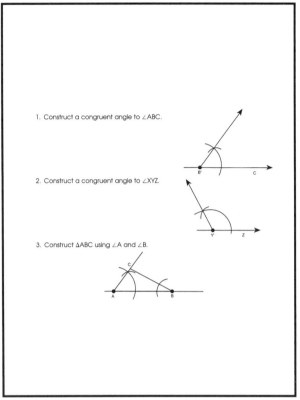

Answer Key

1. Construct a congruent angle to ∠ABC.

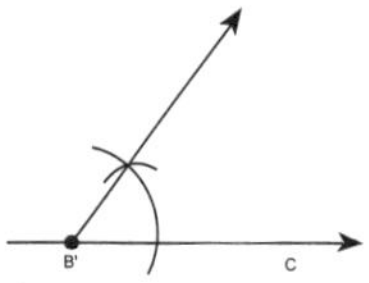

B'
C

2. Construct a congruent angle to ∠XYZ.

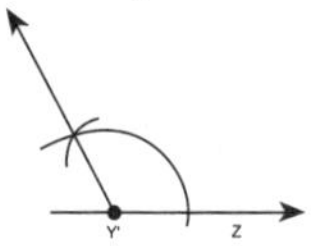

Y'
Z

3. Construct ΔABC using ∠A and ∠B.

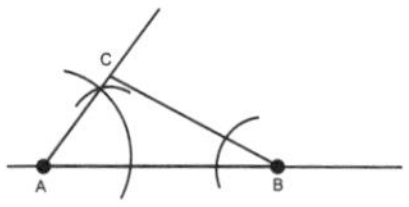

C
A
B

39

1.

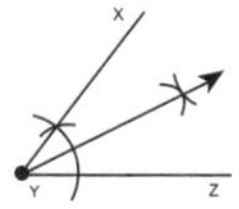

X
Y
Z

2.

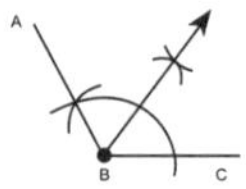

A
B
C

3.

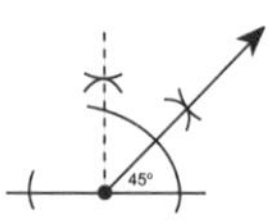

45°

4.

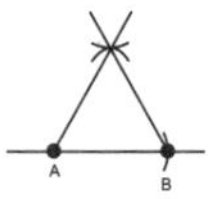

A
B

5. 60°

6.

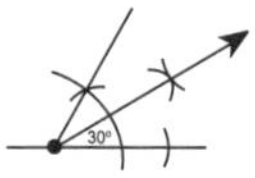

30°

40

1.

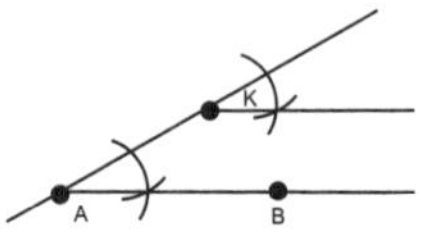

K
A
B

2.

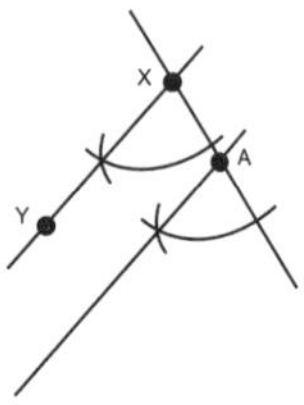

X
A
Y

3.

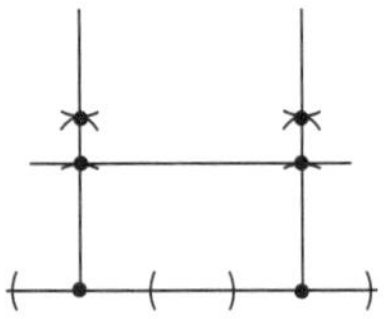

4.

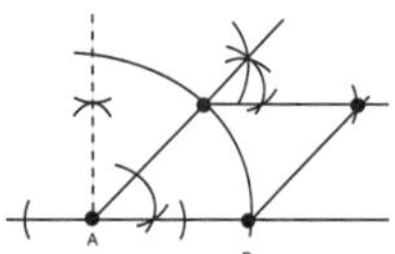

A
B

41

Complete the following ▱ABCD.

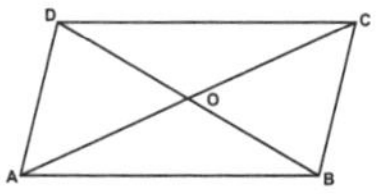

D
C
O
A
B

1. $\overline{AB} \parallel \overline{DC}$
2. $\overline{AB} \cong \overline{DC}$
3. $\angle A \cong \angle C$
4. $\overline{OA} \cong \overline{OC}$
5. $\overline{OB} \cong \overline{OD}$
6. $\overline{AD} \cong \overline{BC}$

Find the missing values for each parallelogram.

7.

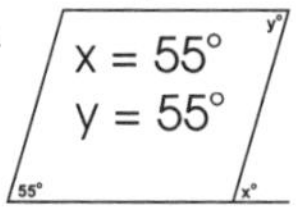

y°
55°
x°

x = 55°
y = 55°

8.

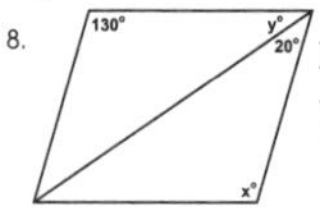

130°
y°
20°
x°

x = 130°
y = 30°

9.

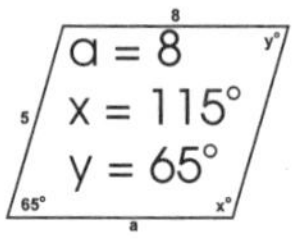

8
y°
5
65°
x°
a

a = 8
x = 115°
y = 65°

10.

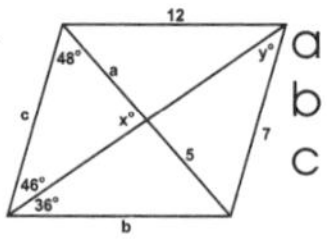

12
48°
y°
a
c
x°
7
5
46°
36°
b

a = 5
b = 12
c = 7
x = 86
y = 46

11.

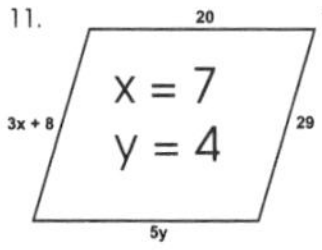

20
3x + 8
29
5y

x = 7
y = 4

12.

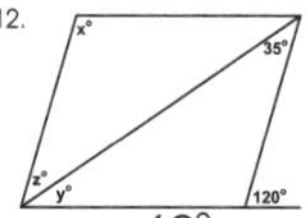

x°
35°
z°
y°
120°

x = 60°
y = 85°
z = 35°

13.

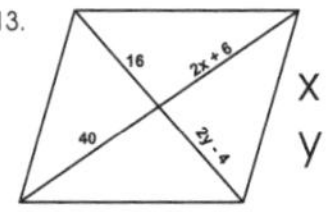

16
2x + 6
40
2y - 4

x = 17
y = 10

42

Answer Key

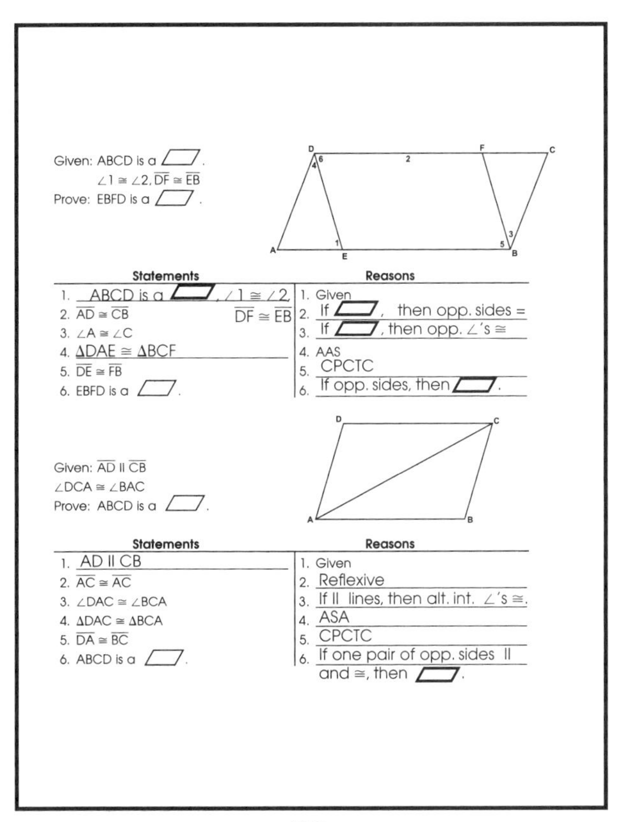

Given: ABCD is a ▱.
$\angle 1 \cong \angle 2$, $\overline{DF} \cong \overline{EB}$
Prove: EBFD is a ▱.

Statements	Reasons
1. ABCD is a ▱, $\angle 1 \cong \angle 2$, $\overline{DF} \cong \overline{EB}$	1. Given
2. $\overline{AD} \cong \overline{CB}$	2. If ▱, then opp. sides =
3. $\angle A \cong \angle C$	3. If ▱, then opp. ∠'s ≅
4. ΔDAE ≅ ΔBCF	4. AAS
5. $\overline{DE} \cong \overline{FB}$	5. CPCTC
6. EBFD is a ▱.	6. If opp. sides, then ▱.

Given: $\overline{AD} \parallel \overline{CB}$
$\angle DCA \cong \angle BAC$
Prove: ABCD is a ▱.

Statements	Reasons
1. AD ∥ CB	1. Given
2. $\overline{AC} \cong \overline{AC}$	2. Reflexive
3. $\angle DAC \cong \angle BCA$	3. If ∥ lines, then alt. int. ∠'s ≅.
4. ΔDAC ≅ ΔBCA	4. ASA
5. $\overline{DA} \cong \overline{BC}$	5. CPCTC
6. ABCD is a ▱.	6. If one pair of opp. sides ∥ and ≅, then ▱.

43

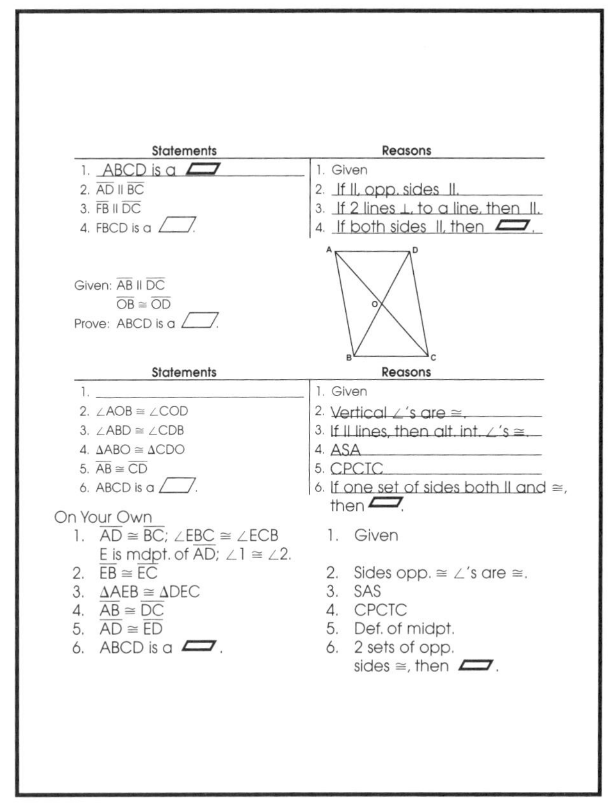

Statements	Reasons
1. ABCD is a ▱	1. Given
2. $\overline{AD} \parallel \overline{BC}$	2. If ∥, opp. sides ∥.
3. $\overline{FB} \parallel \overline{DC}$	3. If 2 lines ⊥ to a line, then ∥.
4. FBCD is a ▱.	4. If both sides ∥, then ▱.

Given: $\overline{AB} \parallel \overline{DC}$
$\overline{OB} \cong \overline{OD}$
Prove: ABCD is a ▱.

Statements	Reasons
1.	1. Given
2. $\angle AOB \cong \angle COD$	2. Vertical ∠'s are ≅.
3. $\angle ABD \cong \angle CDB$	3. If ∥ lines, then alt. int. ∠'s ≅.
4. ΔABO ≅ ΔCDO	4. ASA
5. $\overline{AB} \cong \overline{CD}$	5. CPCTC
6. ABCD is a ▱.	6. If one set of sides both ∥ and ≅, then ▱.

On Your Own

1. $\overline{AD} \cong \overline{BC}$; $\angle EBC \cong \angle ECB$; E is mdpt. of $\overline{AD}$; $\angle 1 \cong \angle 2$.	1. Given
2. $\overline{EB} \cong \overline{EC}$	2. Sides opp. ≅ ∠'s are ≅.
3. ΔAEB ≅ ΔDEC	3. SAS
4. $\overline{AB} \cong \overline{DC}$	4. CPCTC
5. $\overline{AD} \cong \overline{ED}$	5. Def. of midpt.
6. ABCD is a ▱.	6. 2 sets of opp. sides ≅, then ▱.

44

a, b, c, d 1. Diagonals bisect each other.
c, d 3. All sides are congruent.
a, b, c, d 5. Opposite angles are congruent.
c, d 7. Diagonals are perpendicular.

2. All ∠'s are right ∠'s. b, d
4. Opposite sides are congruent. a, c, b, d
6. Diagonals are congruent. b, d
8. Opposite sides are parallel. a, b, c, d

9. ABCD is a rhombus. If $m\angle 8 = 35$, find the measures of $\angle 1, \angle 2, \angle 3, \angle 4, \angle 5, \angle 6, \angle 7$.

∠'s 1-4 = 90°
∠'s 5-6 = 55°
∠7 = 35°

10. ABCD is a rectangle. If $m\angle 1 = 20$, find the measures of $\angle 2, \angle 3, \angle 4, \angle 5, \angle 6$.

∠2 = 20°
∠'s 3-4 = 40°
∠'s 5-6 = 70°

11. ABCD is a square. If $\overline{AC} = 16$ and $\overline{BD} = 2x + 4$, find x.

x = 6

12. ABCD is a parallelogram. $\overline{AR} = 2x + 3$, $\overline{RC} = 35$, $\overline{BR} = 4y - 10$, $\overline{DR} = 90$. Find x and y.

x = 16, y = 25

45

Find the missing values.

1. x - 3; 5; 4; 10; 5; 4; 2x - 4 — 9
2. 15; x; 35 — 25
3. x; 30; 5x — 10
4. 2x; 5x; 6x + 12 — 6
5. y°; 4; x°; 4; 4; 4; 40° — x = 40°, y = 140°
6. 8; y°; 8; 60°; x — x = 16, y = 60°

7. If BG = 8, then CF = 16 and DE = 24.
8. If CF = 10, then BG = 5 and DE = 15.
9. If DE = 15 and BG = 7, then CF = 11.
10. If CF = 2x + 4, BG = 2x + 1, and DE = 3x + 2, then x = 5.

46

Answer Key

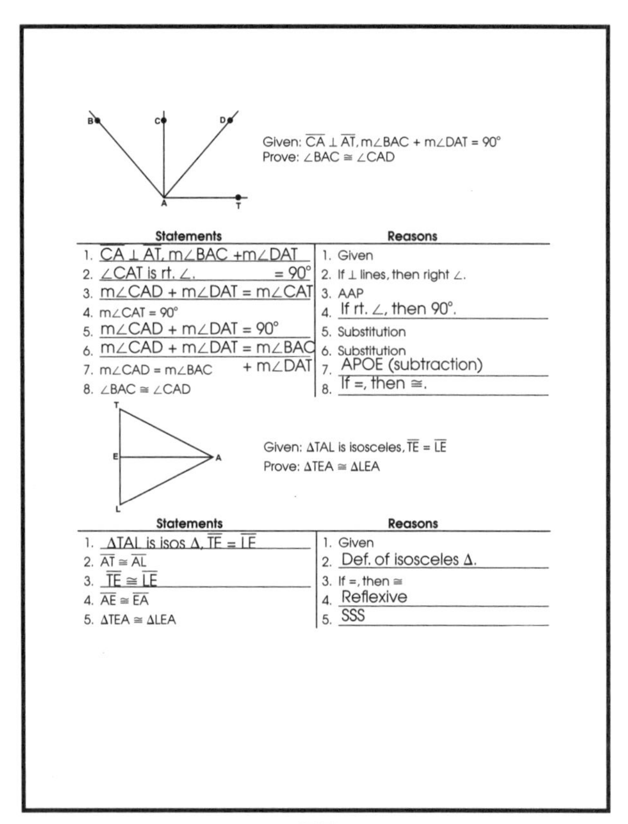

Given: $\overline{CA} \perp \overline{AT}$, $m\angle BAC + m\angle DAT = 90°$
Prove: $\angle BAC \cong \angle CAD$

Statements	Reasons
1. $\overline{CA} \perp \overline{AT}$, $m\angle BAC + m\angle DAT = 90°$	1. Given
2. $\angle CAT$ is rt. $\angle$.	2. If $\perp$ lines, then right $\angle$.
3. $m\angle CAD + m\angle DAT = m\angle CAT$	3. AAP
4. $m\angle CAT = 90°$	4. If rt. $\angle$, then 90°.
5. $m\angle CAD + m\angle DAT = 90°$	5. Substitution
6. $m\angle CAD + m\angle DAT = m\angle BAC + m\angle DAT$	6. Substitution
7. $m\angle CAD = m\angle BAC$	7. APOE (subtraction)
8. $\angle BAC \cong \angle CAD$	8. If =, then $\cong$.

Given: ΔTAL is isosceles, $\overline{TE} = \overline{LE}$
Prove: $\Delta TEA \cong \Delta LEA$

Statements	Reasons
1. ΔTAL is isos Δ, $\overline{TE} = \overline{LE}$	1. Given
2. $\overline{AT} \cong \overline{AL}$	2. Def. of isosceles Δ.
3. $\overline{TE} \cong \overline{LE}$	3. If =, then $\cong$
4. $\overline{AE} \cong \overline{EA}$	4. Reflexive
5. $\Delta TEA \cong \Delta LEA$	5. SSS

47

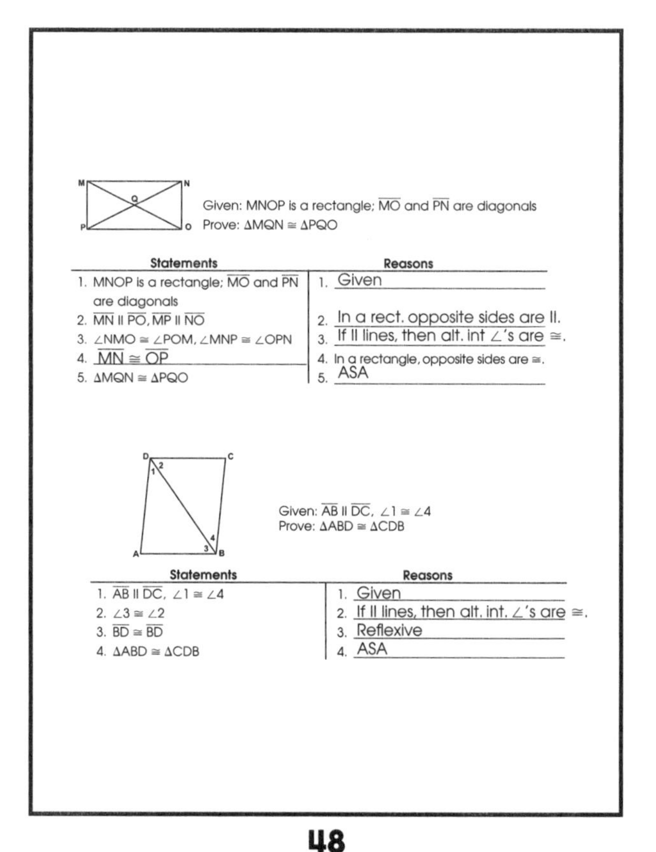

Given: MNOP is a rectangle; $\overline{MO}$ and $\overline{PN}$ are diagonals
Prove: $\Delta MQN \cong \Delta PQO$

Statements	Reasons
1. MNOP is a rectangle; $\overline{MO}$ and $\overline{PN}$ are diagonals	1. Given
2. $\overline{MN} \parallel \overline{PO}$, $\overline{MP} \parallel \overline{NO}$	2. In a rect. opposite sides are $\parallel$.
3. $\angle NMO \cong \angle POM$, $\angle MNP \cong \angle OPN$	3. If $\parallel$ lines, then alt. int $\angle$'s are $\cong$.
4. $\overline{MN} \cong \overline{OP}$	4. In a rectangle, opposite sides are $\cong$.
5. $\Delta MQN \cong \Delta PQO$	5. ASA

Given: $\overline{AB} \parallel \overline{DC}$, $\angle 1 \cong \angle 4$
Prove: $\Delta ABD \cong \Delta CDB$

Statements	Reasons
1. $\overline{AB} \parallel \overline{DC}$, $\angle 1 \cong \angle 4$	1. Given
2. $\angle 3 \cong \angle 2$	2. If $\parallel$ lines, then alt. int. $\angle$'s are $\cong$.
3. $\overline{BD} \cong \overline{BD}$	3. Reflexive
4. $\Delta ABD \cong \Delta CDB$	4. ASA

48

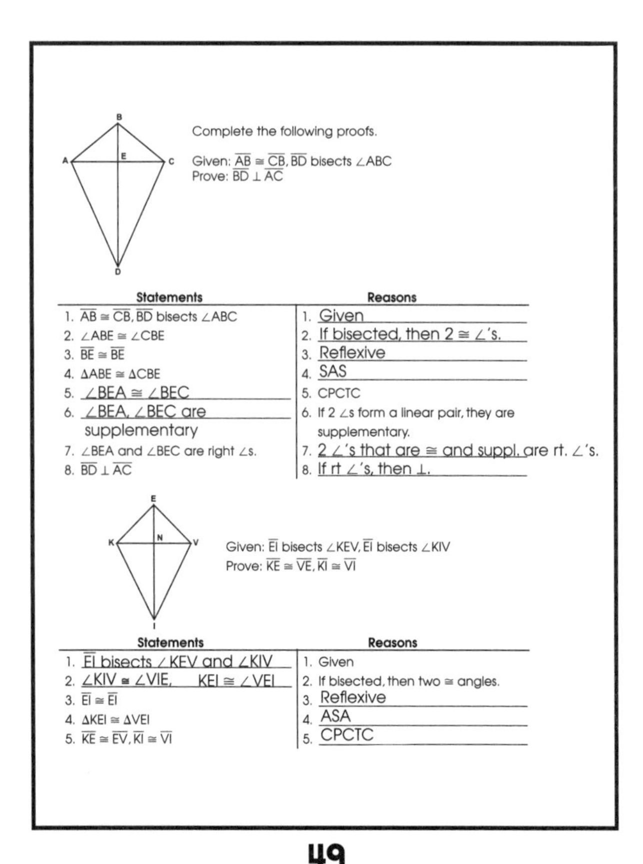

Complete the following proofs.

Given: $\overline{AB} \cong \overline{CB}$, $\overline{BD}$ bisects $\angle ABC$
Prove: $\overline{BD} \perp \overline{AC}$

Statements	Reasons
1. $\overline{AB} \cong \overline{CB}$, $\overline{BD}$ bisects $\angle ABC$	1. Given
2. $\angle ABE \cong \angle CBE$	2. If bisected, then 2 $\cong \angle$'s.
3. $\overline{BE} \cong \overline{BE}$	3. Reflexive
4. $\Delta ABE \cong \Delta CBE$	4. SAS
5. $\angle BEA \cong \angle BEC$	5. CPCTC
6. $\angle BEA$, $\angle BEC$ are supplementary	6. If 2 $\angle$s form a linear pair, they are supplementary.
7. $\angle BEA$ and $\angle BEC$ are right $\angle$s.	7. 2 $\angle$'s that are $\cong$ and suppl. are rt. $\angle$'s.
8. $\overline{BD} \perp \overline{AC}$	8. If rt $\angle$'s, then $\perp$.

Given: $\overline{EI}$ bisects $\angle KEV$, $\overline{EI}$ bisects $\angle KIV$
Prove: $\overline{KE} \cong \overline{VE}$, $\overline{KI} \cong \overline{VI}$

Statements	Reasons
1. $\overline{EI}$ bisects $\angle KEV$ and $\angle KIV$	1. Given
2. $\angle KIV \cong \angle VIE$, $\angle KEI \cong \angle VEI$	2. If bisected, then two $\cong$ angles.
3. $\overline{EI} \cong \overline{EI}$	3. Reflexive
4. $\Delta KEI \cong \Delta VEI$	4. ASA
5. $\overline{KE} \cong \overline{EV}$, $\overline{KI} \cong \overline{VI}$	5. CPCTC

49

Given: B is the midpoint of $\overline{AC}$, D is the midpoint of $\overline{CE}$, F is the midpoint of $\overline{AE}$
Prove: $\Delta CBD \cong \Delta BAF \cong \Delta DFE \cong \Delta FDB$

Statements	Reasons
1. B, D, F are midpoints	1. Given
2. $BF = \frac{1}{2}CE$, $BD = \frac{1}{2}AE$, $FD = \frac{1}{2}AC$	2. Seg. joining midpts. = 1/2 3rd side.
3. $\overline{AB} \cong \overline{BC}$, $\overline{AE} \cong \overline{FE}$, $\overline{CD} \cong \overline{DE}$	3. If midpoint, then two $\cong$ segments.
4. $AB = BC$, $AF = FE$, $CD = DE$	4. If $\cong$, then =.
5. $AB + BC = AC$, $CD + DE = CE$, $AF + FE = AE$	5. Definition of Between
6. $AB + AB = AC$, $CD + CD = CE$, $AF + AF = AE$	6. Substitution
7. $2AB = AC$, $2CD = CE$, $2AF = AE$	7. Combining Similar Terms
8. $AB = \frac{1}{2}AC$, $CD = \frac{1}{2}CE$, $AF = \frac{1}{2}AE$	8. MPOE
9. $AB = FD$, $CD = BF$, $AF = BD$	9. Substitution
10. $\overline{AB} \cong \overline{FD}$, $\overline{CD} \cong \overline{BF}$, $\overline{AF} \cong \overline{BD}$	10. If =, then $\cong$.
11. $\overline{BC} \cong \overline{FD}$, $\overline{FE} \cong \overline{BD}$, $\overline{DE} \cong \overline{BF}$	11. Substitution
12. $\Delta CBD \cong \Delta BAF \cong \Delta DFE \cong \Delta FDB$	12. SSS

How are the four small triangles and the one large triangle related?

The four small triangles are similar to the large triangle.

50

Answer Key

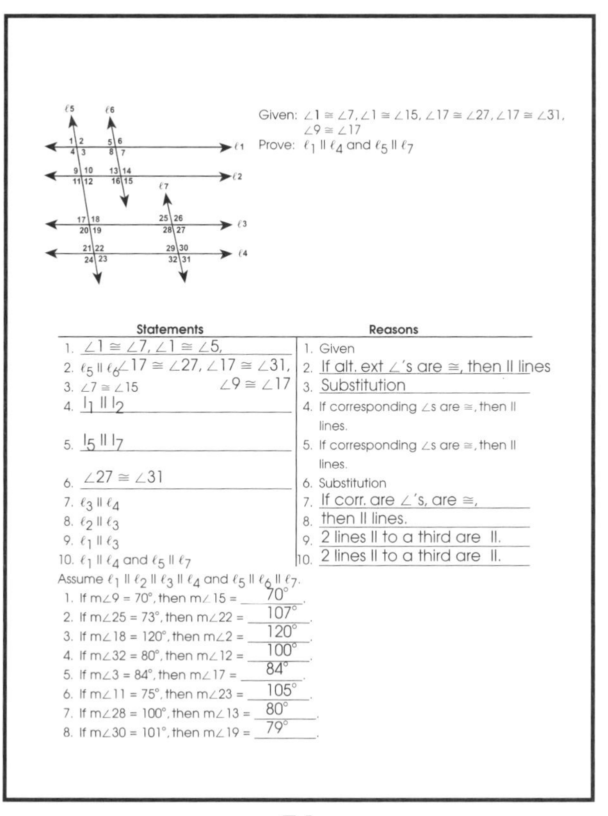

Given: $\angle 1 \cong \angle 7$, $\angle 1 \cong \angle 15$, $\angle 17 \cong \angle 27$, $\angle 17 \cong \angle 31$, $\angle 9 \cong \angle 17$
Prove: $\ell_1 \parallel \ell_4$ and $\ell_5 \parallel \ell_7$

Statements	Reasons
1. $\angle 1 \cong \angle 7$, $\angle 1 \cong \angle 5$,	1. Given
2. $\ell_5 \parallel \ell_6$ $\angle 17 \cong \angle 27$, $\angle 17 \cong \angle 31$, $\angle 9 \cong \angle 17$	2. If alt. ext $\angle$'s are $\cong$, then $\parallel$ lines
3. $\angle 7 \cong \angle 15$	3. Substitution
4. $l_1 \parallel l_2$	4. If corresponding $\angle$s are $\cong$, then $\parallel$ lines.
5. $l_5 \parallel l_7$	5. If corresponding $\angle$s are $\cong$, then $\parallel$ lines.
6. $\angle 27 \cong \angle 31$	6. Substitution
7. $\ell_3 \parallel \ell_4$	7. If corr. are $\angle$'s, are $\cong$,
8. $\ell_2 \parallel \ell_3$	8. then $\parallel$ lines.
9. $\ell_1 \parallel \ell_3$	9. 2 lines $\parallel$ to a third are $\parallel$.
10. $\ell_1 \parallel \ell_4$ and $\ell_5 \parallel \ell_7$	10. 2 lines $\parallel$ to a third are $\parallel$.

Assume $\ell_1 \parallel \ell_2 \parallel \ell_3 \parallel \ell_4$ and $\ell_5 \parallel \ell_6 \parallel \ell_7$.

1. If $m\angle 9 = 70°$, then $m\angle 15$ = 70°.
2. If $m\angle 25 = 73°$, then $m\angle 22$ = 107°.
3. If $m\angle 18 = 120°$, then $m\angle 2$ = 120°.
4. If $m\angle 32 = 80°$, then $m\angle 12$ = 100°.
5. If $m\angle 3 = 84°$, then $m\angle 17$ = 84°.
6. If $m\angle 11 = 75°$, then $m\angle 23$ = 105°.
7. If $m\angle 28 = 100°$, then $m\angle 13$ = 80°.
8. If $m\angle 30 = 101°$, then $m\angle 19$ = 79°.

51

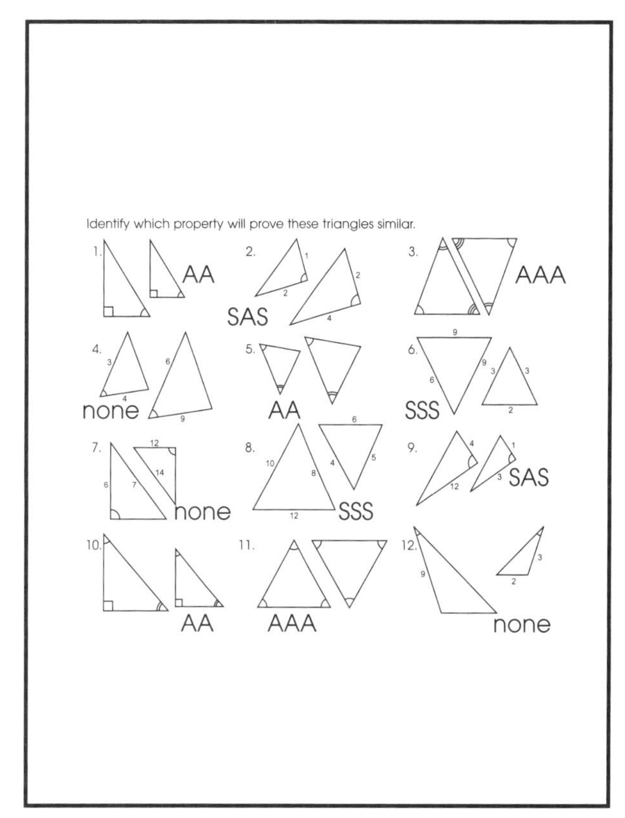

Identify which property will prove these triangles similar.

1. AA
2. SAS
3. AAA
4. none
5. AA
6. SSS
7. none
8. SSS
9. SAS
10. AA
11. AAA
12. none

52

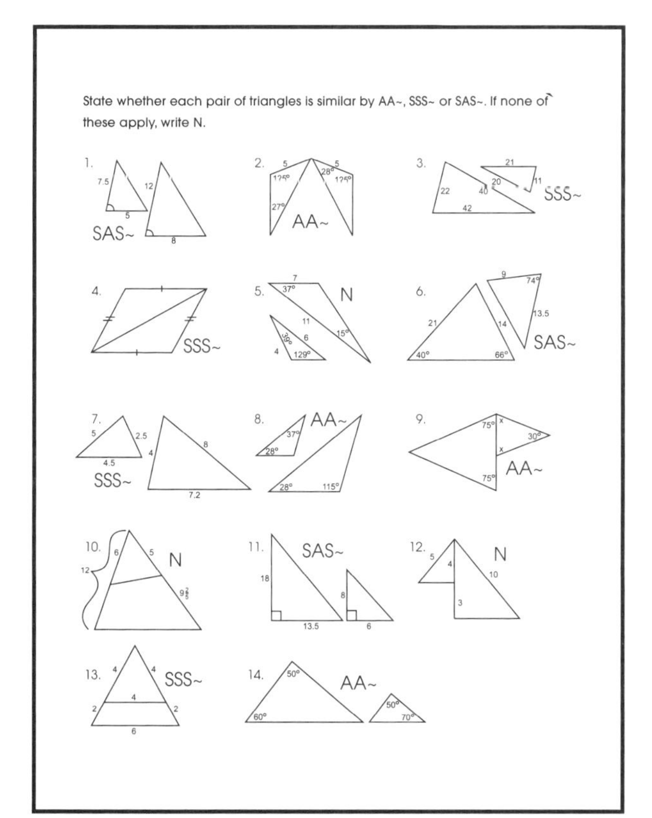

State whether each pair of triangles is similar by AA~, SSS~ or SAS~. If none of these apply, write N.

1. SAS~
2. AA~
3. SSS~
4. SSS~
5. N
6. SAS~
7. SSS~
8. AA~
9. AA~
10. N
11. SAS~
12. N
13. SSS~
14. AA~

53

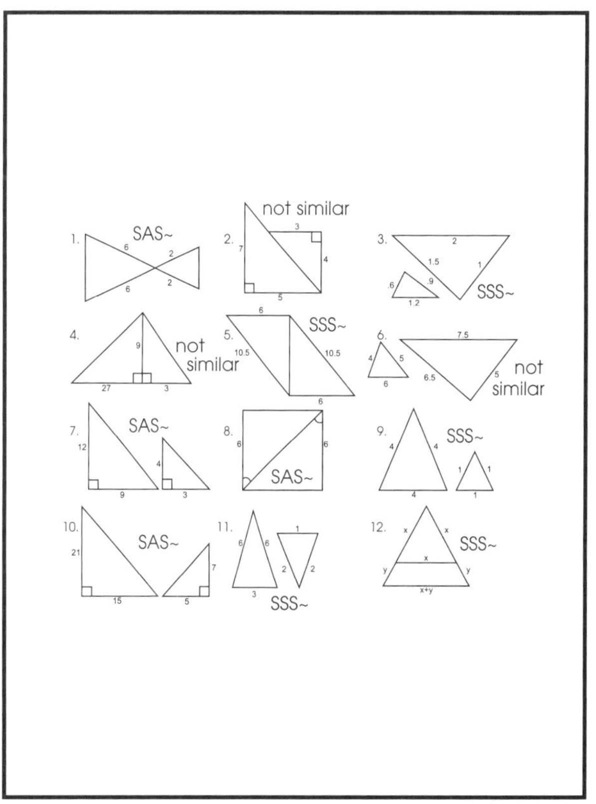

1. SAS~
2. not similar
3. SSS~
4. not similar
5. SSS~
6. not similar
7. SAS~
8. SAS~
9. SSS~
10. SAS~
11. SSS~
12. SSS~

54

Answer Key

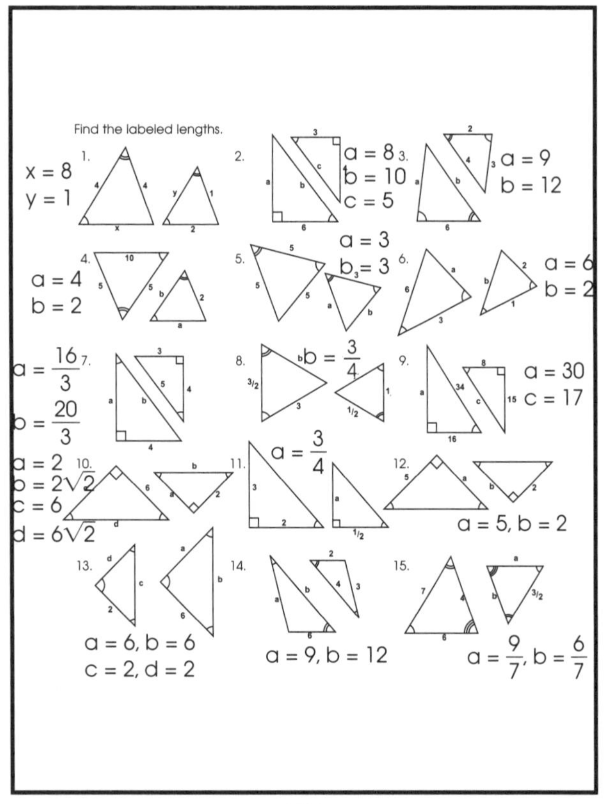

Find the labeled lengths.

1. x = 8, y = 1
2. a = 8, b = 10, c = 5
3. a = 9, b = 12
4. a = 4, b = 2
5. a = 3, b = 3
6. a = 6, b = 2
7. $a = \frac{16}{3}$, $b = \frac{20}{3}$
8. $b = \frac{3}{4}$
9. a = 30, c = 17
10. a = 2, $b = 2\sqrt{2}$, c = 6, $d = 6\sqrt{2}$
11. $a = \frac{3}{4}$
12. a = 5, b = 2
13. a = 6, b = 6, c = 2, d = 2
14. a = 9, b = 12
15. $a = \frac{9}{7}$, $b = \frac{6}{7}$

55

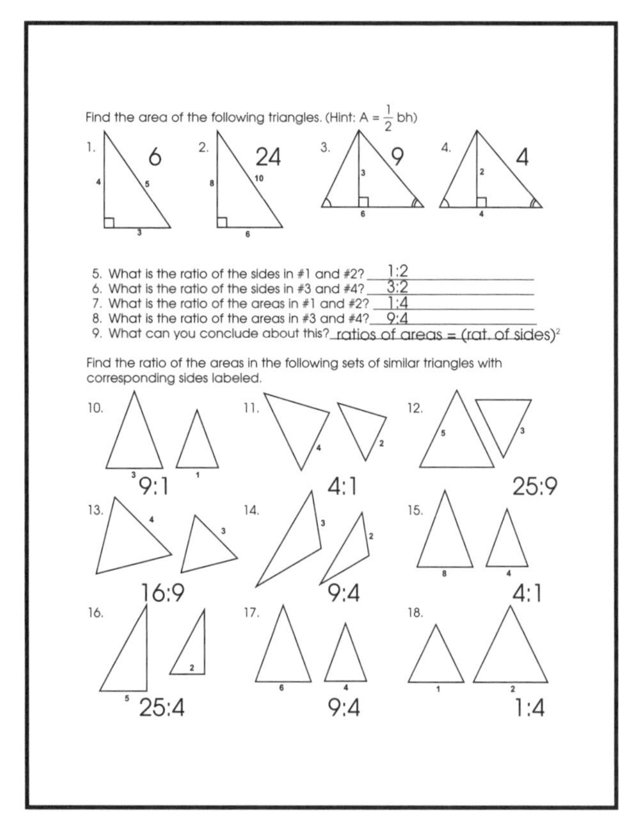

Find the area of the following triangles. (Hint: $A = \frac{1}{2}bh$)

1. 6
2. 24
3. 9
4. 4

5. What is the ratio of the sides in #1 and #2? 1:2
6. What is the ratio of the sides in #3 and #4? 3:2
7. What is the ratio of the areas in #1 and #2? 1:4
8. What is the ratio of the areas in #3 and #4? 9:4
9. What can you conclude about this? ratios of areas = (rat. of sides)^2

Find the ratio of the areas in the following sets of similar triangles with corresponding sides labeled.

10. 9:1
11. 4:1
12. 25:9
13. 16:9
14. 9:4
15. 4:1
16. 25:4
17. 9:4
18. 1:4

56

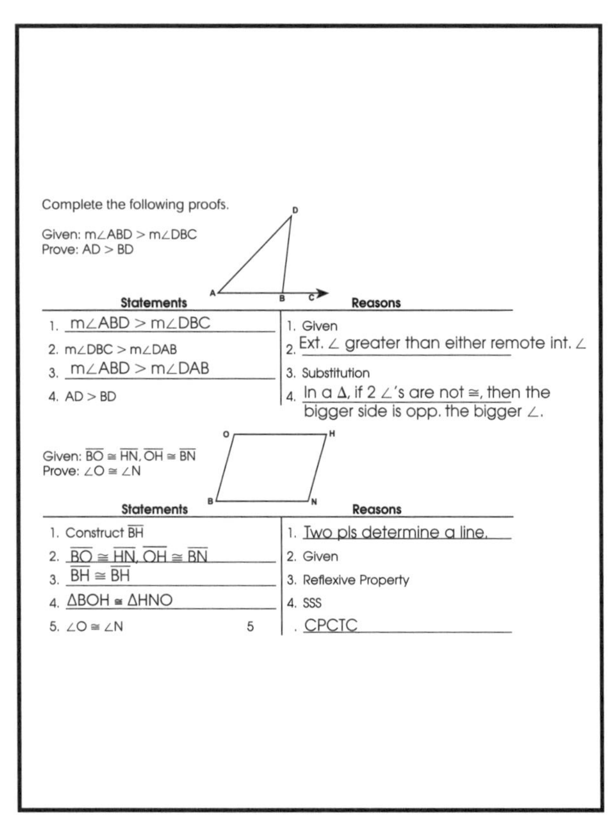

Complete the following proofs.

Given: m∠ABD > m∠DBC
Prove: AD > BD

Statements	Reasons
1. m∠ABD > m∠DBC	1. Given
2. m∠DBC > m∠DAB	2. Ext. ∠ greater than either remote int. ∠
3. m∠ABD > m∠DAB	3. Substitution
4. AD > BD	4. In a Δ, if 2 ∠'s are not ≅, then the bigger side is opp. the bigger ∠.

Given: $\overline{BO} \cong \overline{HN}$, $\overline{OH} \cong \overline{BN}$
Prove: ∠O ≅ ∠N

Statements	Reasons
1. Construct $\overline{BH}$	1. Two pts determine a line.
2. $\overline{BO} \cong \overline{HN}$, $\overline{OH} \cong \overline{BN}$	2. Given
3. $\overline{BH} \cong \overline{BH}$	3. Reflexive Property
4. ΔBOH ≅ ΔHNO	4. SSS
5. ∠O ≅ ∠N	5. CPCTC

57

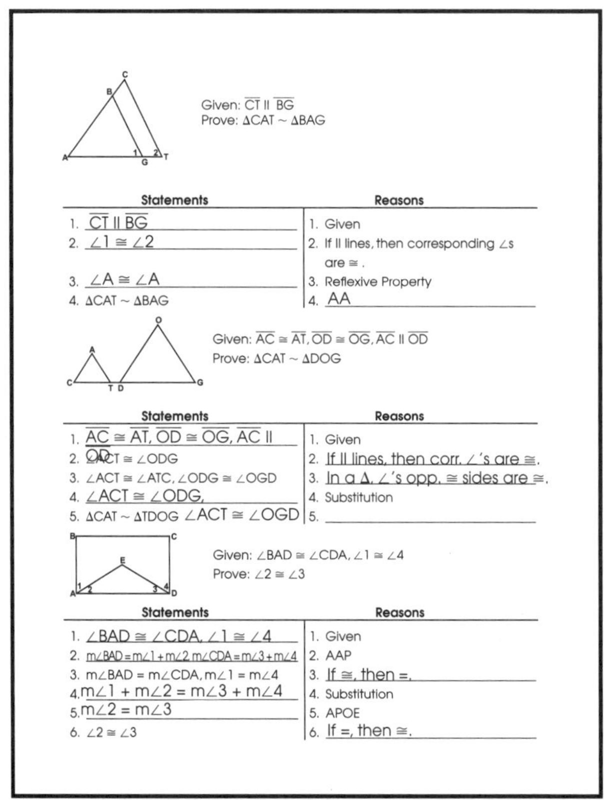

Given: $\overline{CT} \parallel \overline{BG}$
Prove: ΔCAT ~ ΔBAG

Statements	Reasons
1. $\overline{CT} \parallel \overline{BG}$	1. Given
2. ∠1 ≅ ∠2	2. If ∥ lines, then corresponding ∠s are ≅.
3. ∠A ≅ ∠A	3. Reflexive Property
4. ΔCAT ~ ΔBAG	4. AA

Given: $\overline{AC} \cong \overline{AT}$, $\overline{OD} \cong \overline{OG}$, $\overline{AC} \parallel \overline{OD}$
Prove: ΔCAT ~ ΔDOG

Statements	Reasons
1. $\overline{AC} \cong \overline{AT}$, $\overline{OD} \cong \overline{OG}$, $\overline{AC} \parallel \overline{OD}$	1. Given
2. ∠ACT ≅ ∠ODG	2. If ∥ lines, then corr. ∠'s are ≅.
3. ∠ACT ≅ ∠ATC, ∠ODG ≅ ∠OGD	3. In a Δ, ∠'s opp. ≅ sides are ≅.
4. ∠ACT ≅ ∠ODG,	4. Substitution
5. ΔCAT ~ ΔTDOG ∠ACT ≅ ∠OGD	5.

Given: ∠BAD ≅ ∠CDA, ∠1 ≅ ∠4
Prove: ∠2 ≅ ∠3

Statements	Reasons
1. ∠BAD ≅ ∠CDA, ∠1 ≅ ∠4	1. Given
2. m∠BAD = m∠1 + m∠2, m∠CDA = m∠3 + m∠4	2. AAP
3. m∠BAD = m∠CDA, m∠1 = m∠4	3. If ≅, then =.
4. m∠1 + m∠2 = m∠3 + m∠4	4. Substitution
5. m∠2 = m∠3	5. APOE
6. ∠2 ≅ ∠3	6. If =, then ≅.

58

Answer Key

	sin	cos	tan	csc	sec	cot
1.	3/5	4/5	3/4	5/3	5/4	4/3
2.	15/17	8/17	15/8	17/15	17/8	8/15
3.	7/25	24/25	7/24	25/7	25/24	24/7
4.	4/5	3/5	4/3	5/4	5/3	3/4
5.	4/5	3/5	4/3	5/4	5/3	3/4
6.	3/5	4/5	3/4	5/3	5/4	4/3
7.	$\sqrt{2}/2$	$\sqrt{2}/2$	1	$\sqrt{2}$	$\sqrt{2}$	1
8.	8/17	15/17	8/15	17/8	17/15	15/8
9.	8/17	15/17	8/15	17/8	17/15	15/8
10.	24/25	7/25	24/7	25/24	25/7	7/24
11.	$\sqrt{2}/2$	$\sqrt{2}/2$	1	$\sqrt{2}$	$\sqrt{2}$	1
12.	$3\sqrt{13}/13$	$2\sqrt{13}/13$	3/2	$\sqrt{13}/3$	$\sqrt{13}/2$	2/3

59

1. $\sin\theta\cos\theta - \cos\theta\sin\beta$
2. $\dfrac{\tan\theta + \tan\beta}{1 - \tan\theta\tan\beta}$
3. $\dfrac{\tan\theta - \tan\beta}{1 + \tan\theta\tan\beta}$
4. $-\sin\theta$
5. $-\cos\theta$
6. $\cos\theta$
7. $-\sin\theta$
8. $2\cos\theta\sin\theta$
9. $2\cos^2\theta - 1$
10. $\dfrac{2\cos\theta\sin\theta}{2\cos^2\theta - 1}$

1. $= 1 + \dfrac{\cos^2\theta}{\sin^2\theta}$

$= \dfrac{\sin^2\theta}{\sin^2\theta} + \dfrac{\cos^2\theta}{\cos^2\theta}$

$= \dfrac{1}{\sin^2\theta}$

$= \csc^2\theta$

2. $= 1 + \dfrac{\sin^2\theta}{\cos^2\theta}$

$= \dfrac{\cos^2\theta}{\cos^2\theta} + \dfrac{\sin^2\theta}{\cos^2\theta}$

$= \dfrac{1}{\cos^2\theta}$

$= \sec^2\theta$

3. $= \dfrac{1}{\cos\theta} - \dfrac{\sin\theta}{\cos\theta} \cdot \sin\theta$

$= \dfrac{1 - \sin^2\theta}{\cos\theta}$

$= \dfrac{\cos^2\theta}{\cos\theta}$

$= \cos\theta$

4. $= \dfrac{1}{\sin\theta} - \dfrac{\cos\theta}{\cos\theta} \cdot \sin\theta$

$= \dfrac{1 - \cos^2\theta}{\sin\theta}$

$= \dfrac{\sin^2\theta}{\sin^2\theta}$

$= \sin\theta$

60

	$\sin\theta$	$\cos\theta$	$\tan\theta$	$\csc\theta$	$\sec\theta$	$\cot\theta$
1.	4/5	3/5	4/3	5/4	5/3	3/4
2.	$\sqrt{2}/2$	$\sqrt{2}/2$	1	$\sqrt{2}$	$\sqrt{2}$	1
3.	1/2	$\sqrt{3}/2$	$\sqrt{3}/3$	2	$2\sqrt{3}/3$	$\sqrt{3}$
4.	4/5	3/5	4/3	5/4	5/3	3/4
5.	4/5	3/5	4/3	5/4	5/3	3/4
6.	5/13	12/13	5/12	13/5	13/12	12/5
7.	$\sqrt{2}/2$	$\sqrt{2}/2$	1	$\sqrt{2}$	$\sqrt{2}$	1
8.	15/17	8/17	15/8	17/15	17/8	8/15
9.	3/5	4/5	3/4	5/3	5/4	4/3
10.	15/17	8/17	15/8	17/15	17/8	8/15
11.	12/13	5/13	12/5	13/12	13/5	5/12

61

Use the law of cosines to state an equation to find the missing part, x.

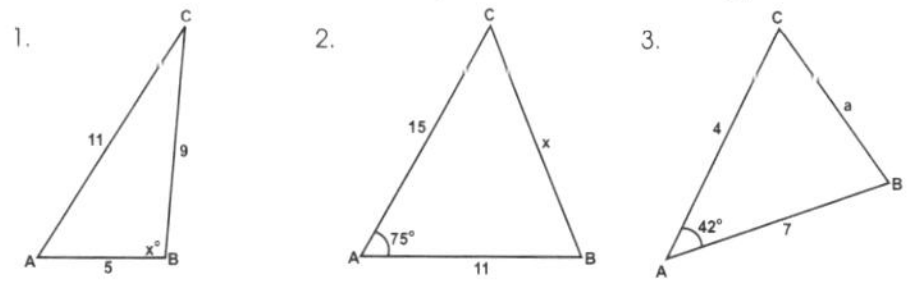

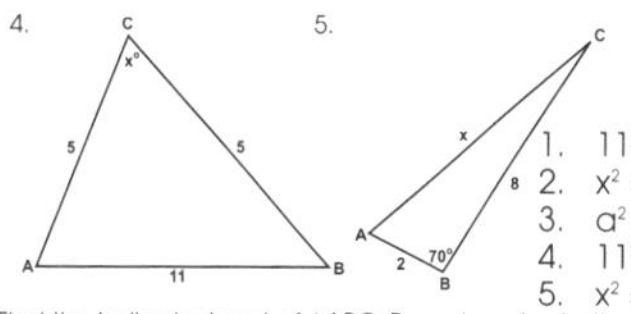

1. $11^2 = 9^2 + 5^2 - 2(9)(5) \cos B$
2. $x^2 = 15^2 + 11^2 - 2(15)(11) \cos 75°$
3. $a^2 = 4^2 + 7^2 - 2(4)(7) \cos 42°$
4. $11^2 = 5^2 + 5^2 - 2(5)(5) \cos x$
5. $x^2 = 8^2 + 2^2 - 2(8)(2) \cos 70°$

Find the indicated part of ΔABC. Round angles to the nearest tenth and lengths to three significant digits.

6. $b = 12, c = 10, \angle A = 38°, a =$ ______
7. $a = 14, b = 15, c = 18, \angle A =$ ______
8. $a = 12, c = 11, \angle B = 81°, b =$ ______
9. $a = 8, b = 9, c = 15, \angle C =$ ______
10. $a = 5, b = 7, \angle C = 40°, c =$ ______
11. $c = 20, b = 30, \angle A = 140°, a =$ ______
12. $b = 2, a = 4, \angle C = 20°, c =$ ______
13. $a = 5, b = 9, c = 11, \angle C =$ ______
14. $a = 1.5, b = 3, c = 2, \angle B =$ ______
15. $a = .6, b = .8, c = 1.2, \angle A =$ ______

6. 7.41
7. 49.2°
8. 15.0
9. 123.7°
10. 4.51
11. 47.1
12. 2.23
13. 99.6°
14. 117.3°
15. 26.4°

62

Answer Key

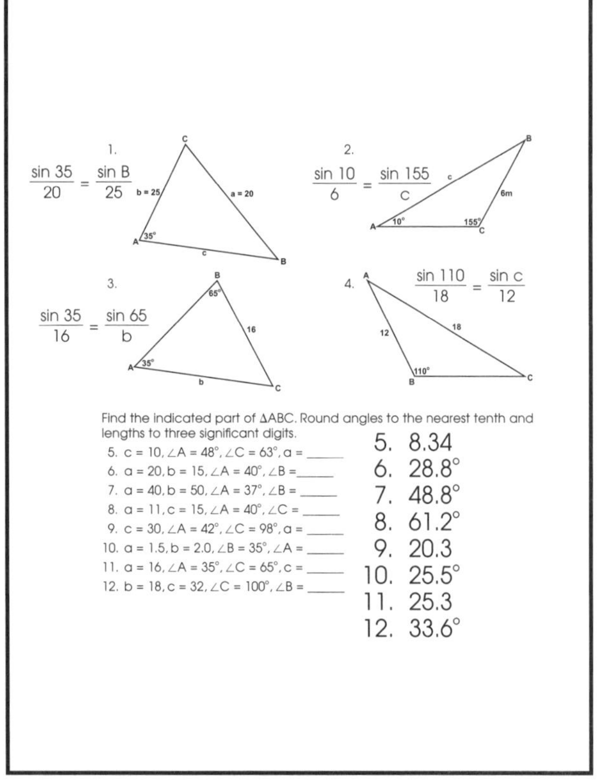

1. $\frac{\sin 35}{20} = \frac{\sin B}{25}$

2. $\frac{\sin 10}{6} = \frac{\sin 155}{c}$

3. $\frac{\sin 35}{16} = \frac{\sin 65}{b}$

4. $\frac{\sin 110}{18} = \frac{\sin c}{12}$

Find the indicated part of ΔABC. Round angles to the nearest tenth and lengths to three significant digits.

5. c = 10, ∠A = 48°, ∠C = 63°, a = ____ 5. 8.34
6. a = 20, b = 15, ∠A = 40°, ∠B = ____ 6. 28.8°
7. a = 40, b = 50, ∠A = 37°, ∠B = ____ 7. 48.8°
8. a = 11, c = 15, ∠A = 40°, ∠C = ____ 8. 61.2°
9. c = 30, ∠A = 42°, ∠C = 98°, a = ____ 9. 20.3
10. a = 1.5, b = 2.0, ∠B = 35°, ∠A = ____ 10. 25.5°
11. a = 16, ∠A = 35°, ∠C = 65°, c = ____ 11. 25.3
12. b = 18, c = 32, ∠C = 100°, ∠B = ____ 12. 33.6°

63

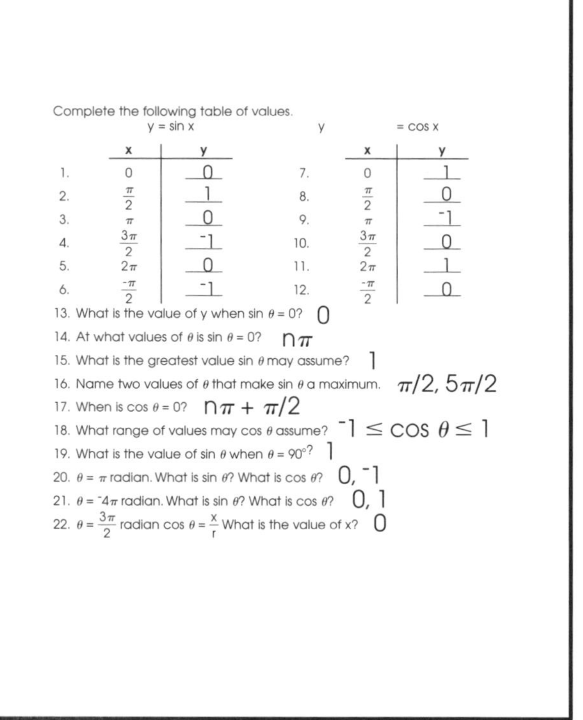

Complete the following table of values.

y = sin x | y = cos x

	x	y
1.	0	0
2.	$\frac{\pi}{2}$	1
3.	π	0
4.	$\frac{3\pi}{2}$	⁻1
5.	2π	0
6.	$\frac{-\pi}{2}$	⁻1

	x	y
7.	0	1
8.	$\frac{\pi}{2}$	0
9.	π	⁻1
10.	$\frac{3\pi}{2}$	0
11.	2π	1
12.	$\frac{-\pi}{2}$	0

13. What is the value of y when sin θ = 0? 0
14. At what values of θ is sin θ = 0? $n\pi$
15. What is the greatest value sin θ may assume? 1
16. Name two values of θ that make sin θ a maximum. $\pi/2, 5\pi/2$
17. When is cos θ = 0? $n\pi + \pi/2$
18. What range of values may cos θ assume? $^-1 \le \cos\theta \le 1$
19. What is the value of sin θ when θ = 90°? 1
20. θ = π radian. What is sin θ? What is cos θ? 0, ⁻1
21. θ = ⁻4π radian. What is sin θ? What is cos θ? 0, 1
22. $\theta = \frac{3\pi}{2}$ radian cos $\theta = \frac{x}{r}$ What is the value of x? 0

64

Consider the graph y = 2 sin x.

1. Where will it cross the x-axis?
2. Graph y = 2 sin x. $n\pi$
3. What is the range of y = 2 sin x?
$^-2 \le y \le 2$

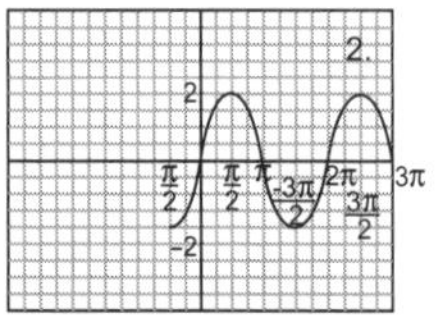

In the equations y = a sin x and y = a cos x, |a| is called the **amplitude** of the curve.

4. What is the amplitude of y = 3 cos x? 3
5. Graph y = 3 cos x.
6. Where does y = 3 cos x cross the x-axis?
$\pi/2 + n\pi$
7. How does y = 3 cos x compare to y = cos x?
taller and crosses x-axis at same places

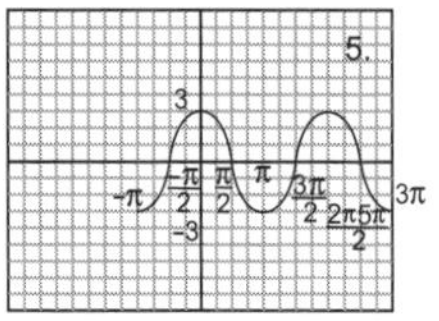

y = sin x and y = cos x ange over the same values each time x runs through an interval of 2π units. The curve is said to be periodic and its period is 2π.

8. Graph y = sin 2x. (Include at least x = 0, $x = \frac{\pi}{4}$, $x = \frac{\pi}{2}$.) 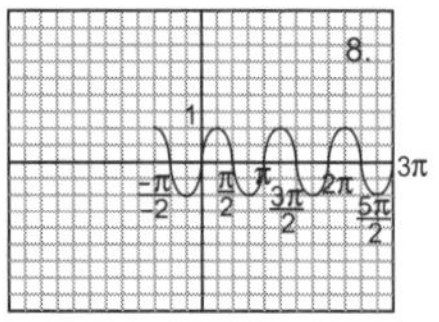
9. Where does y = sin 2x cross the x-axis?
$n\pi/2$
10. What is the period of y = sin 2x?
π

2. (graph) 2, −2, $\frac{\pi}{2}$, 3π

5. (graph) 3, −3, −π, π, 3π

8. (graph) 1, 3π

65

Identify each segment as a radius, chord or diameter.

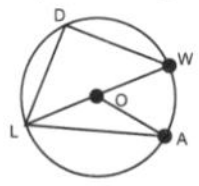

1. $\overline{OA}$ R
2. $\overline{DL}$ C
3. $\overline{LW}$ D
4. $\overline{WO}$ R
5. $\overline{DW}$ C
6. $\overline{LA}$ C
7. $\overline{AW}$ C
8. $\overline{OL}$ R

Name all examples of each term shown.

Figure 1

9. Radius $\overline{BA}, \overline{BK}, \overline{BT}, \overline{BH}, \overline{BY}$
10. Chord $\overline{AK}, \overline{AT}, \overline{TH}, \overline{KT}, \overline{AH}$
11. Center B
12. Diameter $\overline{KT}, \overline{AH}$

Figure 2

13. Name all the points in the interior of the circle. S, P, O
14. Name all the points in the exterior of the circle. M, N
15. Name all the points on the circle. Z, T, R, U, X, Q
16. Name the center of the circle. P
17. Name all segments that are a radius of the circle. $\overline{PZ}, \overline{PX}, \overline{PU}, \overline{PT}$
18. Name all segments that are a diameter of the circle. $\overline{ZU}, \overline{TX}$
19. Name all segments that are neither a radius nor a diameter of the circle. $\overline{ZR}, \overline{XU}$, PO, OX
20. Name all segments that are chords of the circle. $\overline{ZR}, \overline{XU}$

66

Answer Key

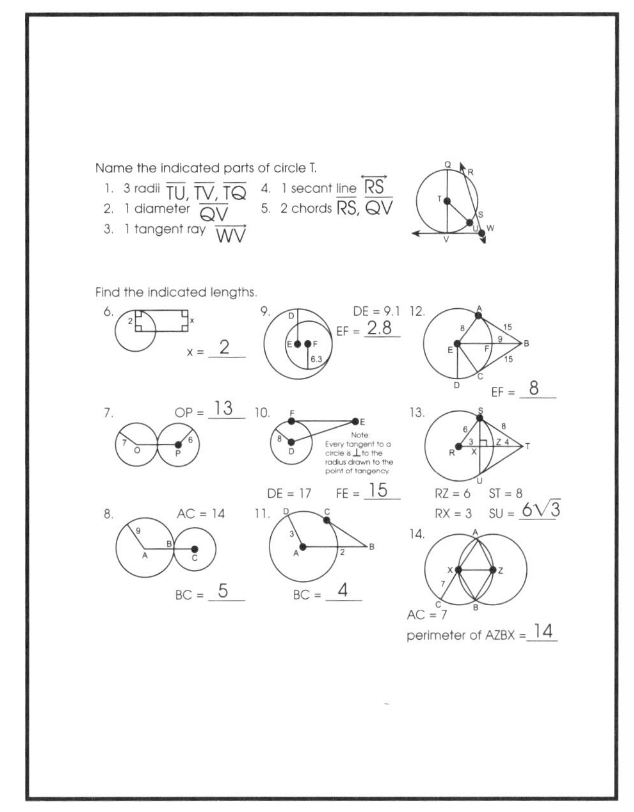

Name the indicated parts of circle T.

1. 3 radii $\overline{TU}, \overline{TV}, \overline{TQ}$
2. 1 diameter $\overline{QV}$
3. 1 tangent ray $\overrightarrow{WV}$
4. 1 secant line $\overleftrightarrow{RS}$
5. 2 chords $\overline{RS}, \overline{QV}$

Find the indicated lengths.

6. x = 2
7. OP = 13
8. AC = 14, BC = 5
9. DE = 9.1, EF = 2.8
10. DE = 17, FE = 15 (Note: Every tangent to a circle is ⊥ to the radius drawn to the point of tangency.)
11. BC = 4
12. EF = 8
13. RZ = 6, ST = 8, RX = 3, SU = $6\sqrt{3}$
14. AC = 7, perimeter of AZBX = 14

67

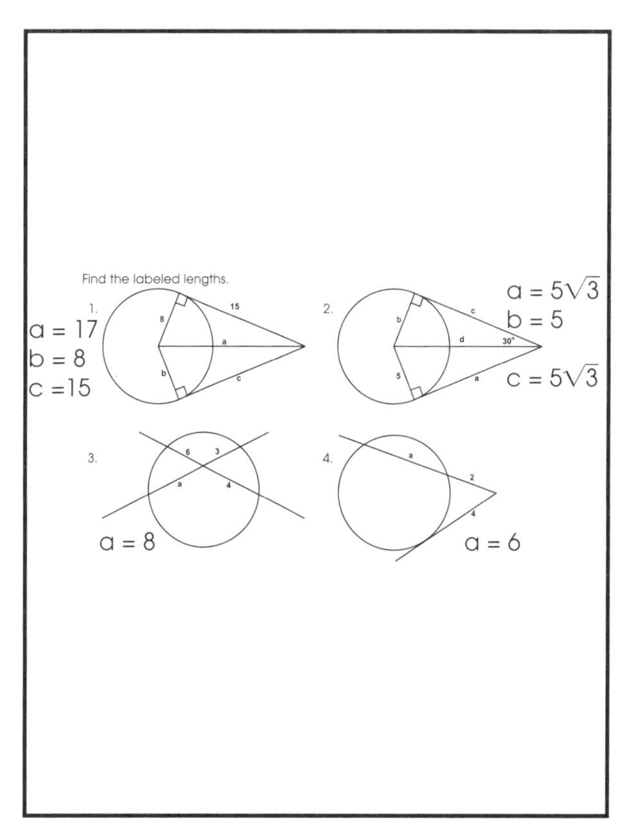

Find the labeled lengths.

1. a = 17, b = 8, c = 15
2. a = $5\sqrt{3}$, b = 5, c = $5\sqrt{3}$
3. a = 8
4. a = 6

68

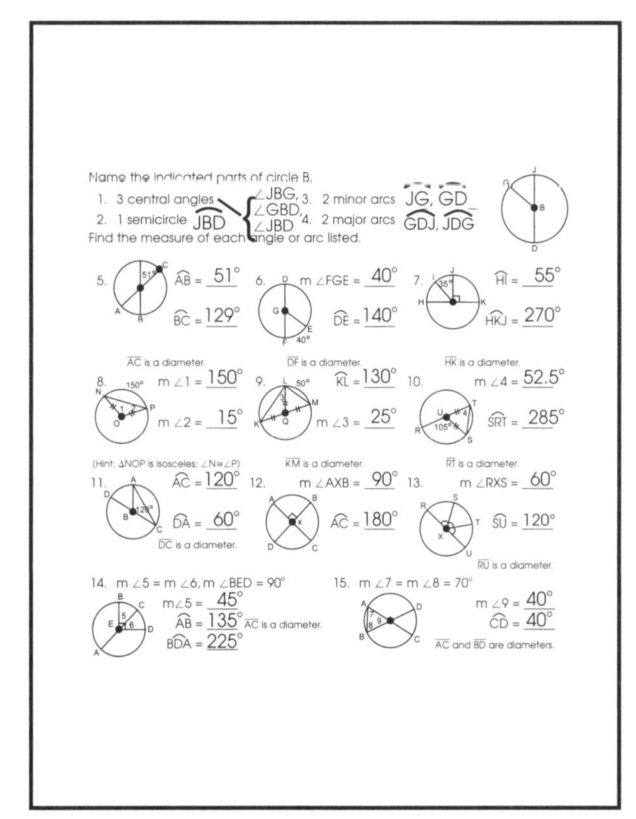

Name the indicated parts of circle B.

1. 3 central angles ∠JBG, ∠GBD, ∠JBD
2. 1 semicircle $\widehat{JBD}$
3. 2 minor arcs $\widehat{JG}, \widehat{GD}$
4. 2 major arcs $\widehat{GDJ}, \widehat{JDG}$

Find the measure of each angle or arc listed.

5. $\widehat{AB}$ = 51°, $\widehat{BC}$ = 129° — $\overline{AC}$ is a diameter.
6. m∠FGE = 40°, $\widehat{DE}$ = 140° — $\overline{DF}$ is a diameter.
7. $\widehat{HI}$ = 55°, $\widehat{HKJ}$ = 270° — $\overline{HK}$ is a diameter.
8. m∠1 = 150°, m∠2 = 15° (Hint: ΔNOP is isosceles; ∠N≅∠P)
9. $\widehat{KL}$ = 130°, m∠3 = 25° — $\overline{KM}$ is a diameter.
10. m∠4 = 52.5°, $\widehat{SRT}$ = 285° — $\overline{RT}$ is a diameter.
11. $\widehat{AC}$ = 120°, $\widehat{DA}$ = 60° — $\overline{DC}$ is a diameter.
12. m∠AXB = 90°, $\widehat{AC}$ = 180°
13. m∠RXS = 60°, $\widehat{SU}$ = 120° — $\overline{RU}$ is a diameter.
14. m∠5 = m∠6, m∠BED = 90°; m∠5 = 45°, $\widehat{AB}$ = 135°, $\widehat{BDA}$ = 225° — $\overline{AC}$ is a diameter.
15. m∠7 = m∠8 = 70°; m∠9 = 40°, $\widehat{CD}$ = 40° — $\overline{AC}$ and $\overline{BD}$ are diameters.

69

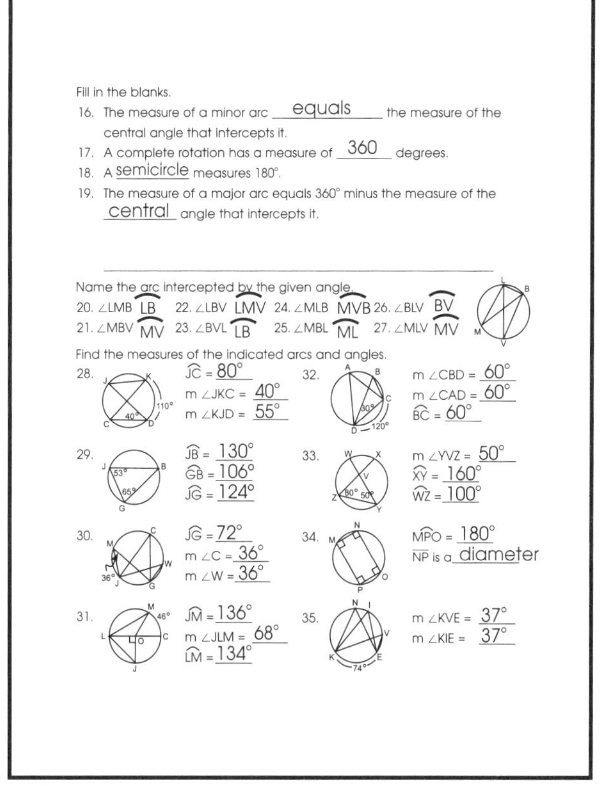

Fill in the blanks.

16. The measure of a minor arc equals the measure of the central angle that intercepts it.
17. A complete rotation has a measure of 360 degrees.
18. A semicircle measures 180°.
19. The measure of a major arc equals 360° minus the measure of the central angle that intercepts it.

Name the arc intercepted by the given angle.

20. ∠LMB $\widehat{LB}$
21. ∠MBV $\widehat{MV}$
22. ∠LBV $\widehat{LMV}$
23. ∠BVL $\widehat{LB}$
24. ∠MLB $\widehat{MVB}$
25. ∠MBL $\widehat{ML}$
26. ∠BLV $\widehat{BV}$
27. ∠MLV $\widehat{MV}$

Find the measures of the indicated arcs and angles.

28. $\widehat{JC}$ = 80°, m∠JKC = 40°, m∠KJD = 55°
29. $\widehat{JB}$ = 130°, $\widehat{GB}$ = 106°, $\widehat{JG}$ = 124°
30. $\widehat{JG}$ = 72°, m∠C = 36°, m∠W = 36°
31. $\widehat{JM}$ = 136°, m∠JLM = 68°, $\widehat{LM}$ = 134°
32. m∠CBD = 60°, m∠CAD = 60°, $\widehat{BC}$ = 60°
33. m∠YVZ = 50°, $\widehat{XY}$ = 160°, $\widehat{WZ}$ = 100°
34. $\widehat{MPO}$ = 180°, $\overline{NP}$ is a diameter
35. m∠KVE = 37°, m∠KIE = 37°

70

Answer Key

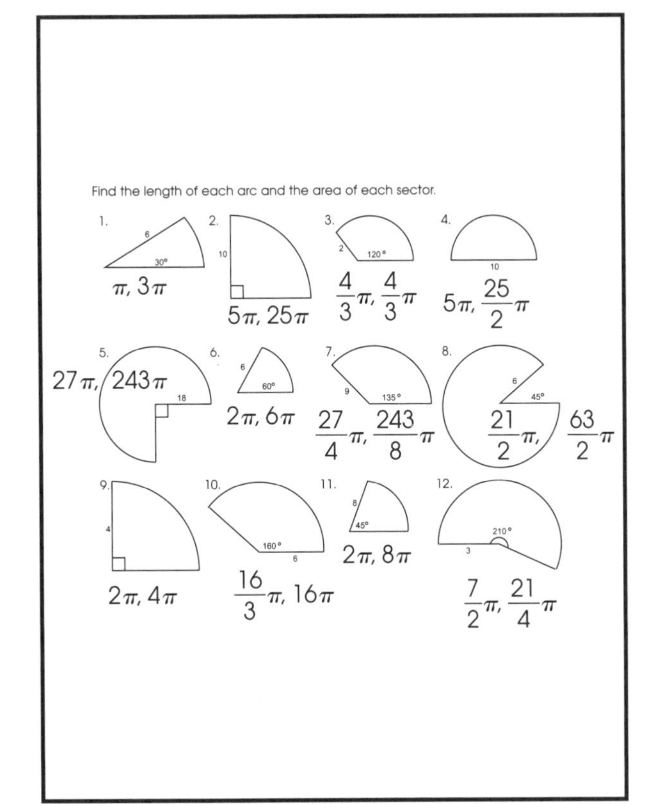
Find the length of each arc and the area of each sector.

1. $\pi, 3\pi$
2. $5\pi, 25\pi$
3. $\frac{4}{3}\pi, \frac{4}{3}\pi$
4. $5\pi, \frac{25}{2}\pi$
5. $27\pi, 243\pi$
6. $2\pi, 6\pi$
7. $\frac{27}{4}\pi, \frac{243}{8}\pi$
8. $\frac{21}{2}\pi, \frac{63}{2}\pi$
9. $2\pi, 4\pi$
10. $\frac{16}{3}\pi, 16\pi$
11. $2\pi, 8\pi$
12. $\frac{7}{2}\pi, \frac{21}{4}\pi$

71

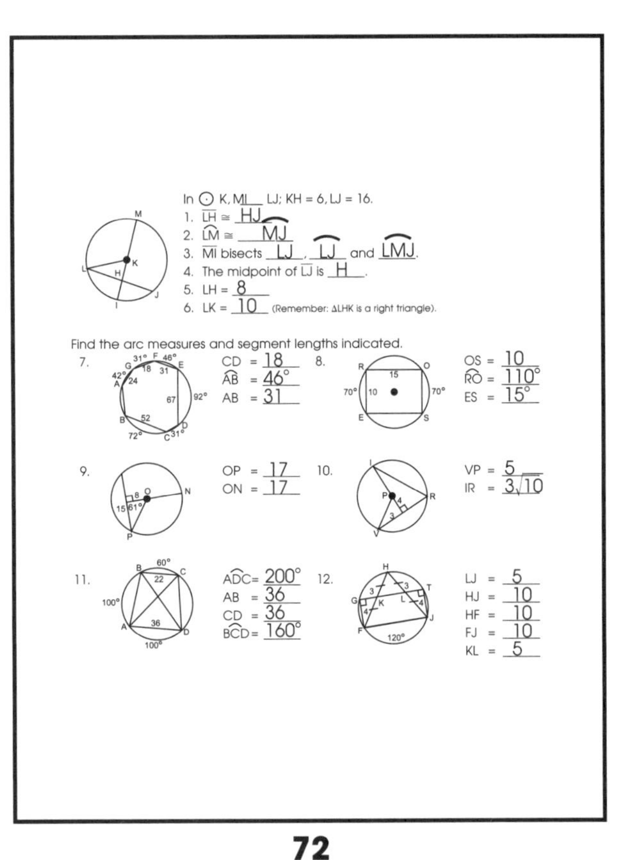
In ⊙ K, $\overline{MI} \perp \overline{LJ}$; KH = 6, LJ = 16.

1. $\overline{LH} \cong$ HJ
2. $\widehat{LM} \cong$ $\widehat{MJ}$
3. $\overline{MI}$ bisects $\overline{LJ}$, $\widehat{LJ}$ and $\widehat{LMJ}$.
4. The midpoint of $\overline{LJ}$ is H.
5. LH = 8
6. LK = 10 (Remember: △LHK is a right triangle).

Find the arc measures and segment lengths indicated.

7. CD = 18, $\widehat{AB}$ = 46°, AB = 31
8. OS = 10, $\widehat{RO}$ = 110°, ES = 15°
9. OP = 17, ON = 17
10. VP = 5, IR = $3\sqrt{10}$
11. $\widehat{ADC}$ = 200°, AB = 36, CD = 36, $\widehat{BCD}$ = 160°
12. LJ = 5, HJ = 10, HF = 10, FJ = 10, KL = 5

72

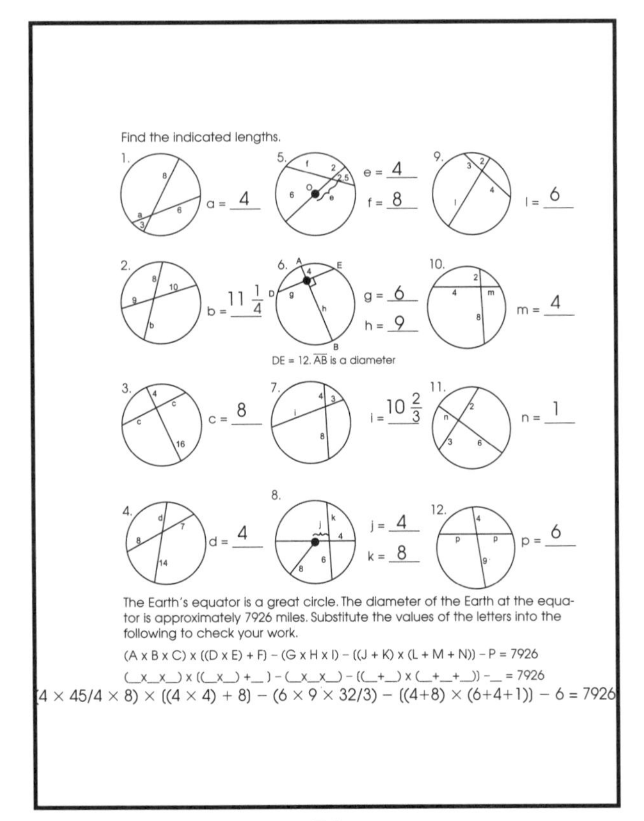
Find the indicated lengths.

1. a = 4
2. b = $11\frac{1}{4}$
3. c = 8
4. d = 4
5. e = 4, f = 8
6. g = 6, h = 9 (DE = 12. $\overline{AB}$ is a diameter)
7. i = $10\frac{2}{3}$
8. j = 4, k = 8
9. l = 6
10. m = 4
11. n = 1
12. p = 6

The Earth's equator is a great circle. The diameter of the Earth at the equator is approximately 7926 miles. Substitute the values of the letters into the following to check your work.

(A x B x C) x ((D x E) + F) – (G x H x I) – ((J + K) x (L + M + N)) – P = 7926

(_x_x_) x ((_x_) +_) – (_x_x_) – ((_+_) x (_+_+_)) –_ = 7926

$(4 \times 45/4 \times 8) \times ((4 \times 4) + 8) - (6 \times 9 \times 32/3) - ((4+8) \times (6+4+1)) - 6 = 7926$

73

Find the indicated parts.

1. $\widehat{ABC}$ = 180°, m∠AOB = 125°, $\widehat{AB}$ = 125°
2. $\widehat{EF}$ = 40°, m∠EOF = 40°, m∠CFO = 90°
3. $\widehat{IH}$ = 85°, $\widehat{GH}$ = 138°, m∠GOI = 137°
4. $\widehat{JK}$ = 44°, m∠JOK = 44°
5. $\widehat{MQ}$ = 116°, $\widehat{QP}$ = 64°, m∠PNQ = 32°, m∠MRP = 90°
6. $\widehat{BC}$ = 100°, m∠BEC = 100°, m∠CED = 80°, m∠ECD = 50°
7. $\widehat{TV}$ = 80°, m∠UOV = 100°, m∠SUT = 33°
8. $\widehat{XYZ}$ = 186°, m∠Y = 87°, $\widehat{WZY}$ = 156°
9. $\widehat{IP}$ = 76°, IP = 276°
10. x = 8, y = 11
11. x = 6
12. AT = 7, IT = 3, $\widehat{AI}$ = 180°

74

Answer Key

Convert the following angle measures from degrees to radians or from radians to degrees.

degrees $\times \frac{\pi}{180}$ = radians

radians $\times \frac{180}{\pi}$ = degrees

1. 180° 1π
2. $\frac{\pi}{2}$ radians $90°$
3. 27° 0.15π
4. 45° $\pi/4$
5. 6.2832 radians $360°$
6. 4.7 radians $269.4°$
7. 2 radians 114.6π
8. 90° $\pi/2$
9. 0.05235 radians $3°$
10. $\frac{\pi}{3}$ radians $60°$
11. 1.0472 radians $60°$
12. 36° $\pi/5$

75

The length of an arc is directly proportional to the size of the central angle. In other words, the greater the angle, the greater the arc. Since there are 2π radians in the complete circle, then the length of the arc can be expressed as $S = \theta r$, where θ is the measure of the central angle and S is the length of the arc.

Complete the following table.

	θ	r	S
1.	180°	1	π
2.	3π rad.	1	3π
3.	$\frac{\pi}{4}$ rad.	5	$5\pi/4$
4.	2π rad.	2	4π
5.	2 rad.	3	6
6.	$\frac{\pi}{2}$ rad.	4	2π
7.	45°	4	π
8.	3 rad.	4/3	4
9.	1°	1	$\pi/180$
10.	270°	6	9π

76

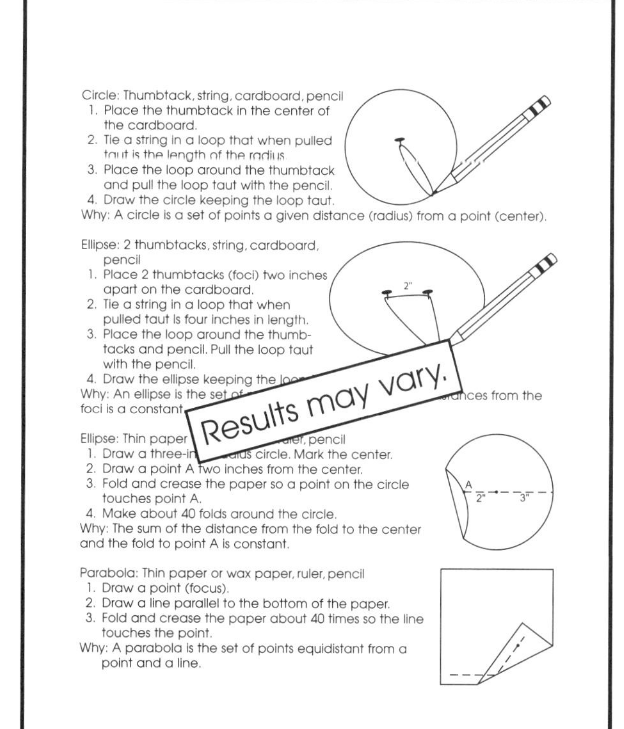

Circle: Thumbtack, string, cardboard, pencil
1. Place the thumbtack in the center of the cardboard.
2. Tie a string in a loop that when pulled taut is the length of the radius.
3. Place the loop around the thumbtack and pull the loop taut with the pencil.
4. Draw the circle keeping the loop taut.

Why: A circle is a set of points a given distance (radius) from a point (center).

Ellipse: 2 thumbtacks, string, cardboard, pencil
1. Place 2 thumbtacks (foci) two inches apart on the cardboard.
2. Tie a string in a loop that when pulled taut is four inches in length.
3. Place the loop around the thumb-tacks and pencil. Pull the loop taut with the pencil.
4. Draw the ellipse keeping the loo[illegible]

Why: An ellipse is the set of [illegible] ances from the foci is a constant[illegible]

Results may vary.

Ellipse: Thin paper [illegible] uler, pencil
1. Draw a three-in[illegible] adius circle. Mark the center.
2. Draw a point A two inches from the center.
3. Fold and crease the paper so a point on the circle touches point A.
4. Make about 40 folds around the circle.

Why: The sum of the distance from the fold to the center and the fold to point A is constant.

Parabola: Thin paper or wax paper, ruler, pencil
1. Draw a point (focus).
2. Draw a line parallel to the bottom of the paper.
3. Fold and crease the paper about 40 times so the line touches the point.

Why: A parabola is the set of points equidistant from a point and a line.

77

Give the coordinates of the following points.

C (0, 1) G (⁻5, ⁻5) P (⁻5, ⁻1) U (⁻1, 3) V (1, ⁻2) Z (⁻3, 3)

Use the coordinates to locate the correct letter on the graph.

1. Where is Rutherford B. Hayes buried?

F R E M O N T O H I O
(3, ⁻2) (3, 4) (0, ⁻4) (5, ⁻4) (⁻5, 4) (3, ⁻5) (4, 0) (⁻5, 4) (⁻5, 1) (⁻3, 0) (⁻5, 4)

2. Darwin, MN claims to have the largest what?

B A L L O F T W I N E
(1, 3) (⁻3, 5) (5, 1) (5, 1) (⁻5, 4) (3, ⁻2) (4, 0) (4, 2) (⁻3, 0) (3. ⁻5) (0, ⁻4)

3. Who was the tenth president of the United States?

J O H N T Y L E R
(5, 5) (⁻5, 4) (⁻5, 1) (3, ⁻5) (4, 0) (⁻2, ⁻5) (5, 1) (0, ⁻4) (3, 4)

4. A line is a simple figure in the coordinate plane. Name three points on the line.
⁻2, 0 -1, 1 0, 2
5. The line passes through which quadrants?
I, II, III
6. Give the location by quadrant(s) of the following points.
(⁻2, ⁻5) III (3, ⁻1) IV

Equal x- and y-coordinates. I, III
Opposite x- and y-coordinates. II, IV

78

Answer Key

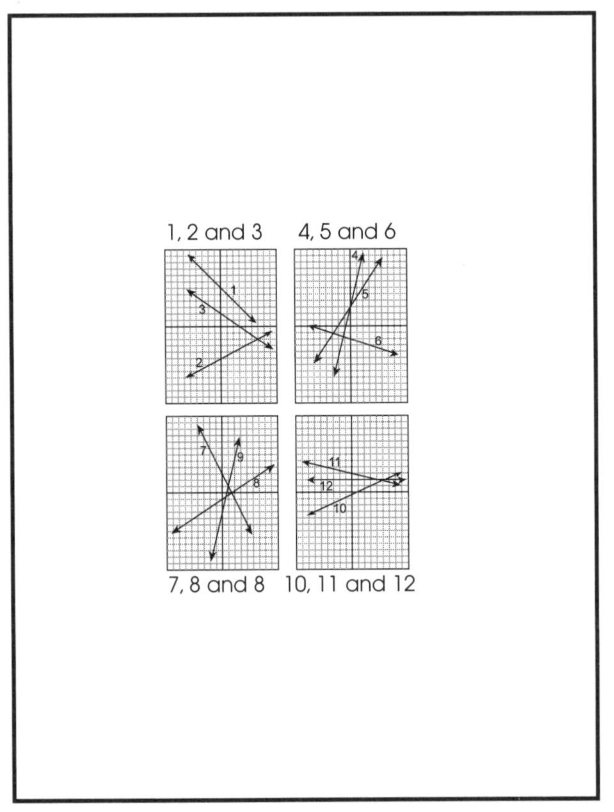

79

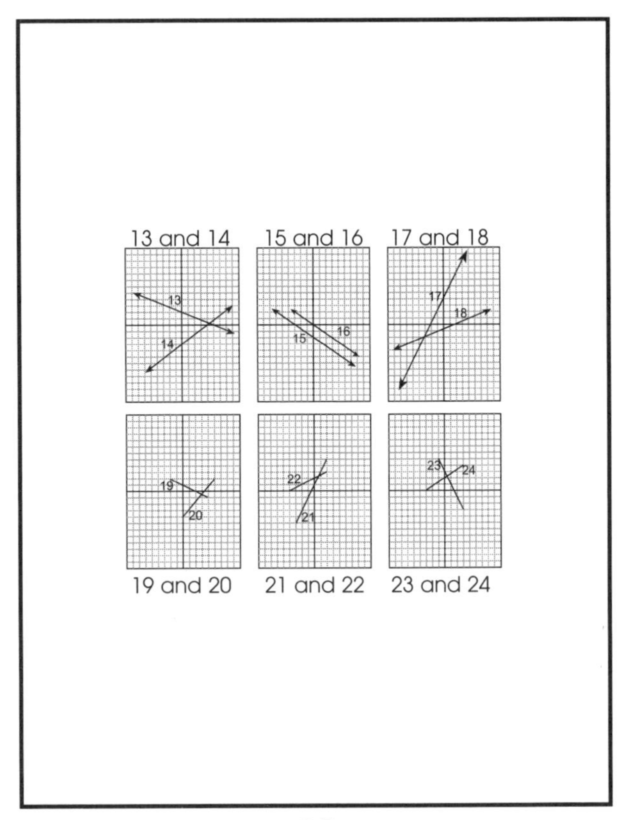

80

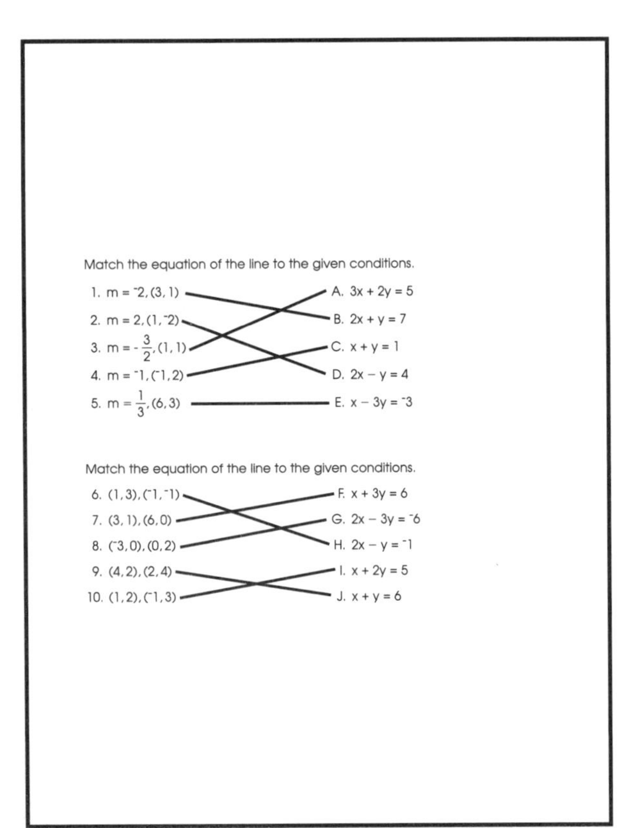

Match the equation of the line to the given conditions.

1. $m = {}^{-}2, (3, 1)$
2. $m = 2, (1, {}^{-}2)$
3. $m = -\frac{3}{2}, (1, 1)$
4. $m = {}^{-}1, ({}^{-}1, 2)$
5. $m = \frac{1}{3}, (6, 3)$

A. $3x + 2y = 5$
B. $2x + y = 7$
C. $x + y = 1$
D. $2x - y = 4$
E. $x - 3y = {}^{-}3$

Match the equation of the line to the given conditions.

6. $(1, 3), ({}^{-}1, {}^{-}1)$
7. $(3, 1), (6, 0)$
8. $({}^{-}3, 0), (0, 2)$
9. $(4, 2), (2, 4)$
10. $(1, 2), ({}^{-}1, 3)$

F. $x + 3y = 6$
G. $2x - 3y = {}^{-}6$
H. $2x - y = {}^{-}1$
I. $x + 2y = 5$
J. $x + y = 6$

81

Match the equation of the line to the given conditions.

11. Parallel to $y = 3x + 4$ through $(0, -2)$
12. Parallel to $y = \frac{1}{2}x - 3$ through $(4, 2)$
13. Parallel to $2x + 3y = 6$ through $(4, -3)$
14. Parallel to $3x - 4y = 1$ through $(2, 2)$
15. Parallel to $3x + 2y = 10$ through $(1, \frac{1}{2})$

K. $3x + 2y = 4$
L. $2x + 3y = {}^{-}1$
M. $3x - 4y = {}^{-}2$
N. $3x - y = 2$
O. $x - 2y = 0$

Match the equation of the line to the given conditions.

16. Perpendicular to $y = 3x + 4$ through $(0, {}^{-}2)$
17. Perpendicular to $y = \frac{1}{2}x - 3$ through $(1, 4)$
18. Perpendicular to $3x + 2y = 6$ through $(1, 1)$
19. Perpendicular to $2x - 5y = 2$ through $(2, 3)$
20. Perpendicular to $x + y = 4$ through $({}^{-}3, {}^{-}3)$

P. $x + 3y = {}^{-}6$
Q. $2x - 3y = {}^{-}1$
R. $2x + y = 6$
S. $x - y = 0$
T. $5x + 2y = 16$

V. Summary

Write the equation of the line with the following conditions.

21. $m = 4$ through $(\frac{1}{2}, {}^{-}2)$ $4x - y = 4$
22. Through $(2, {}^{-}1)$ and $(8, 1)$ $x - 3y = 5$
23. Parallel to $2x - y = -3$ through $(2, {}^{-}1)$ $2x - y = 5$
24. Perpendicular to $3x + 6y = 5$ through $(4, 1)$ $2x - y = 7$

82

Answer Key

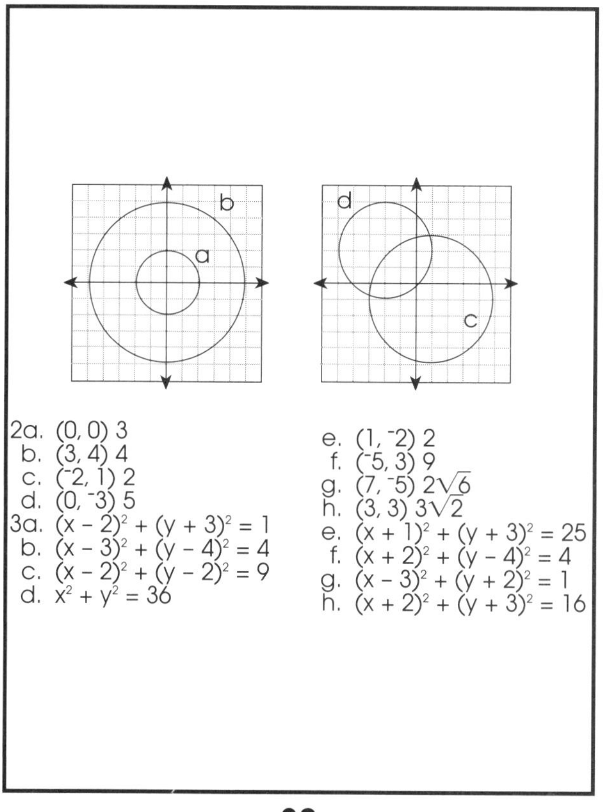

83

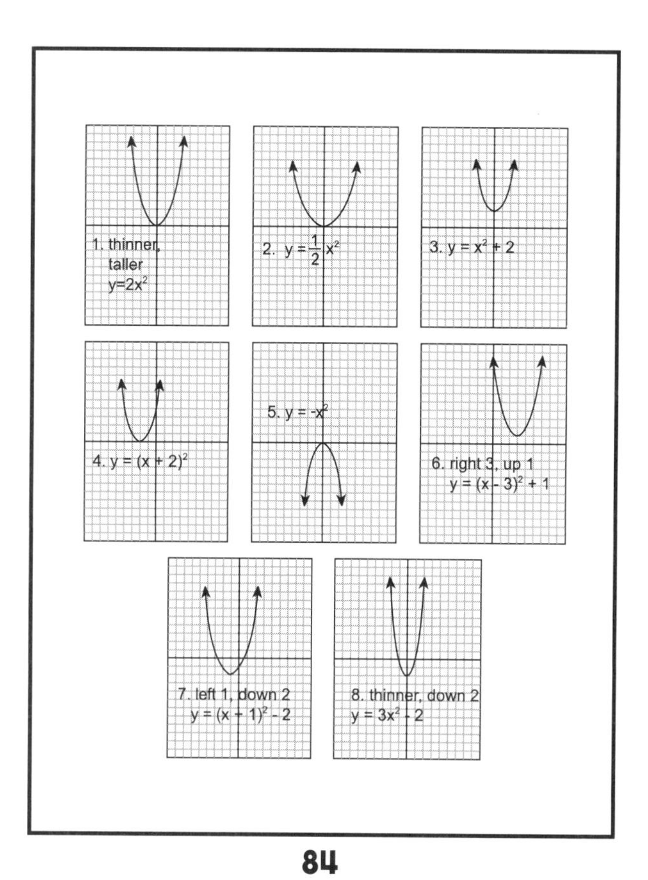

84

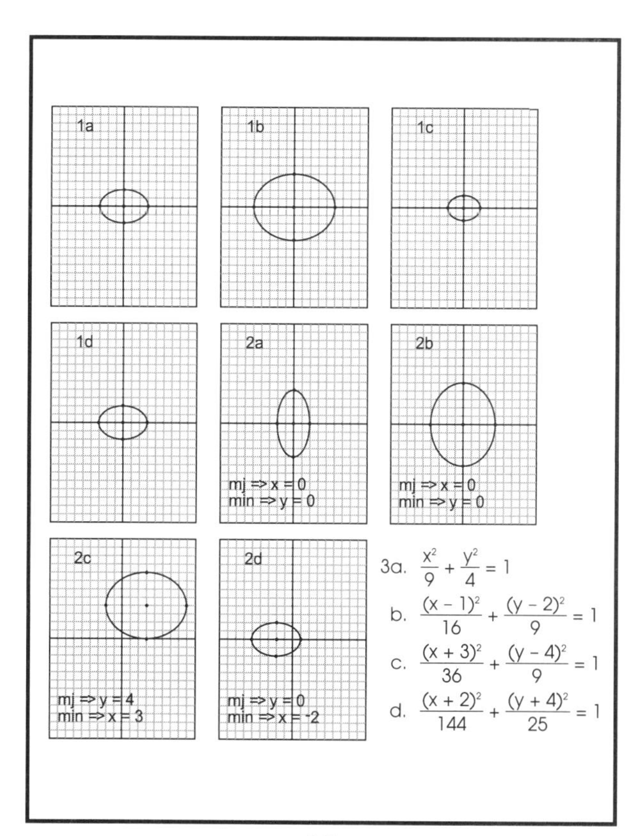

85

1a

1b

1c

1d

2a $y = \pm\frac{1}{2}x$

2b $y = \pm\frac{2}{3}x$

2c $y-1= \pm\frac{(x+1)}{3}$

2d $y-2 = \pm(x-1)$

3a. $\frac{(x - 2)^2}{4} - \frac{(y + 1)^2}{16} = 1$

b. $\frac{(x - 3)^2}{9} - \frac{(y - 2)^2}{25} = 1$

c. $\frac{(x + 2)^2}{1} - \frac{(y - 4)^2}{9} = 1$

d. $\frac{(x + 1)^2}{16} - \frac{(y + 3)^2}{4} = 1$

86

Answer Key

Graph the following polar coordinates.

1. $(2, \frac{\pi}{2})$
2. $(3, \pi)$
3. $(1, 45°)$
4. $(5, \frac{-\pi}{2})$
5. $(4, 270°)$
6. $(2, -\frac{3}{2}\pi)$
7. $(2, 540°)$
8. $(2, 2\pi)$

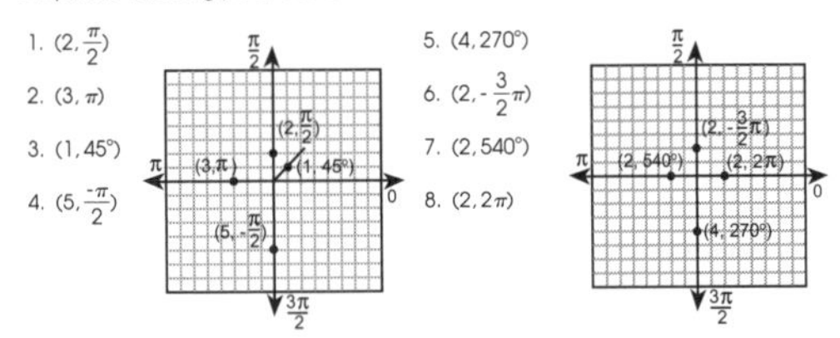

In which quadrant would you expect to find the following points?

9. $(2, \frac{\pi}{4})$ I
10. $(4, \frac{3}{4}\pi)$ II
11. $(3, \frac{-\pi}{3})$ IV
12. $(1, \pi)$ quadrantal
13. $(2, \frac{15}{3}\pi)$ quadrantal
14. $(4, -\frac{4\pi}{3})$ II
15. $(3, {}^-700°)$ I
16. $({}^-1, 45°)$ III

87

Find the polar coordinates corresponding to the following rectangular coordinates.

1. $(2, 0)$ $(2, 0)$
2. $(3, 4)$ $(5, 53.13°)$
3. $(0, 5)$ $(5, 90°)$
4. $(4, 4)$ $(4\sqrt{2}, 45°)$
5. $({}^-2, {}^-2)$ $(2\sqrt{2}, 225°)$
6. $({}^-6, 8)$ $(10, 126.87°)$

Find the rectangular coordinates corresponding to the following polar coordinates.

7. $(2, \frac{\pi}{2})$ $(0, 2)$
8. $(3, \frac{-\pi}{4})$ $(\frac{3\sqrt{2}}{2}, \frac{{}^-3\sqrt{2}}{2})$
9. $(4, \pi)$ $({}^-4, 0)$
10. $(2, 0)$ $(2, 0)$
11. $(5, \frac{\pi}{3})$ $(\frac{5}{2}, \frac{5\sqrt{3}}{2})$
12. $(\sqrt{2}, \frac{\pi}{4})$ $(1, 1)$

Express the following equations in polar notation.

13. $x = 2$
14. $x^2 + y^2 = 4$
15. $y = 2x + 1$
16. $y = 4$
17. $x^2 + y^2 - 3 = 13$
18. $y = 3x$

Express the following equations in rectangular notation.

19. $r = 5$
20. $r = 3 \csc \theta$
21. $\tan \theta = 3$
22. $r = 2 \sec \theta$
23. $r = 7$
24. $\tan \theta = 2$

13. $r = 2 \sec \theta$
14. $r = 2$
15. $\sin \theta = 2 \cos \theta \frac{+1}{r}$
16. $r = 4 \csc \theta$
17. $r = 4$
18. $\tan \theta = 3$
19. $x^2 + y^2 = 25$
20. $y = 3$
21. $\frac{y}{x} = 3$
22. $x = 2$
23. $x^2 + y^2 = 49$
24. $\frac{y}{x} = 2$

88

Find the distance between these pairs of points.

1. $(6, 4)$ and $(2, 1)$ 5
2. $({}^-2, {}^-4)$ and $(3, 8)$ 13
3. $(0, 0)$ and $(5, 10)$ $5\sqrt{5}$
4. $({}^-5, 2)$ and $(7, {}^-7)$ 15
5. $(0, {}^-8)$ and $(8, 7)$ 17
6. $({}^-2, 11)$ and $(4, 3)$ 10
7. $(2, 1)$ and $(4, 0)$ $\sqrt{5}$
8. $(6, 4)$ and $(6, {}^-2)$ 6
9. $({}^-2, 2)$ and $(4, {}^-1)$ $3\sqrt{5}$
10. $({}^-3, {}^-5)$ and $(2, 5)$ $5\sqrt{5}$
11. $({}^-4, 0)$ and $(2, 3)$ $3\sqrt{5}$
12. $({}^-1, 5)$ and $(3, {}^-3)$ $4\sqrt{5}$
13. $(0, 0)$ and $(3, 4)$ 5
14. $(1, 8)$ and $(3, 10)$ $2\sqrt{2}$
15. $(9, 8)$ and $({}^-3, 4)$ $4\sqrt{10}$
16. $(2, 2)$ and $(2, 4)$ 2

89

Find the distance and the midpoint between the given points. Cross out the correct answers below. Use the remaining letters to complete the statement.

	Distance	Midpoint
1. $({}^-2, 2)$ and $(4, {}^-1)$	$3\sqrt{5}$	$(1, 1/2)$
2. $({}^-3, {}^-5)$ and $(2, 5)$	$5\sqrt{5}$	$({}^-1/2, 0)$
3. $({}^-1, 5)$ and $(3, {}^-3)$	$4\sqrt{5}$	$(1, 1)$
4. $(0, 0)$ and $(3, 4)$	5	$(1\ 1/2, 2)$
5. $(1, 2)$ and $(4, 7)$	$\sqrt{34}$	$(5/2, 9/2)$
6. $({}^-2, 4)$ and $(3, {}^-5)$	$\sqrt{106}$	$(1/2, {}^-1/2)$
7. $(2, 2)$ and $(6, 6)$	$4\sqrt{2}$	$(4, 4)$
8. $(3, 6)$ and $(5, {}^-2)$	$2\sqrt{17}$	$(4, 2)$
9. $({}^-1, {}^-4)$ and $(3, 5)$	$\sqrt{97}$	$(0, 1/2)$

~~5~~ S	~~$(-\frac{1}{2}, 0)$~~ Q	$(0, -\frac{1}{2})$ P	~~$4\sqrt{5}$~~ U	$\sqrt{5}$ Y	10 T	~~$(1, \frac{1}{2})$~~ M	$9\sqrt{7}$ H	~~$(1, 5)$~~ G
~~$\sqrt{97}$~~ B	$3\sqrt{21}$ A	~~$(\frac{1}{2}, -\frac{1}{2})$~~ H	~~$(5/2, 9/2)$~~ R	21 G	~~$3\sqrt{5}$~~ W	$(2, 4)$ O	~~$(4, 2)$~~ E	~~$\sqrt{34}$~~ U
0 R	$(-1, -1)$ E	~~$5\sqrt{5}$~~ S	$2\sqrt{53}$ A	1 N	~~$(1\frac{1}{2}, 2)$~~ V	$(2, 1\frac{1}{2})$ T	$4\sqrt{17}$ H	~~$2\sqrt{17}$~~ S
~~$(1, 1)$~~ R	$(0, 0)$ E	34 O	~~$\sqrt{106}$~~ Y	32 R	~~$4\sqrt{2}$~~ L	$(2, 2)$ E	~~$(4, 4)$~~ T	25 M

10. This Distance Formula is based on the
P Y T H A G O R E A N T H E O R E M.

90

Answer Key

1. If I own a horse, then I own an animal.
   Converse If I own an animal, then I own a horse.
   Inverse If I do not own a horse, then I do not own an animal.
   Contrapositive If I do not own an animal, I do not own a horse.
2. If I study, then I do well in school.
   Converse If I do well in school, then I studied.
   Inverse If I do not study, then I do not do well in school.
   Contrapositive If I do not do well in school, then I did not study.
3. If today is Monday, then yesterday was Sunday.
   Converse If yesterday was Sunday, then today is Monday.
   Inverse If today is not Monday, then yesterday was not Sunday.
   Contrapositive If yesterday was not Sunday, then today is not Monday.
4. If it is Saturday, I do not go to school.
   Converse If I do not go to school, then it is Saturday.
   Inverse If it is not Saturday, I go to school.
   Contrapositive If I go to school, then it is not Saturday.
5. If I do not go to bed early, I do not sleep well.
   Converse If I do not sleep well, then I did not go to bed early.
   Inverse If I go to bed early, I sleep well.
   Contrapositive If I sleep well, then I went to bed early.
6. If $6x = 18$, then $x = 3$.
   Converse If $x = 3$, then $6x = 18$.
   Inverse If $6x \neq 18$, then $x \neq 3$.
   Contrapositive If $x \neq 3$, then $6x \neq 18$.
7. If $AB + BC = AC$, then B is between A and C.
   Converse If B is between A and C, then $AB + BC = AC$.
   Inverse If $AB + BC \neq AC$, then B is not between A and C.
   Contrapositive If B is not between A and C, then $AB + BC \neq AC$.
8. Given a conditional statement, the contrapositive statement is **always** true.
   Choose from converse, inverse, or contrapositive.)

91

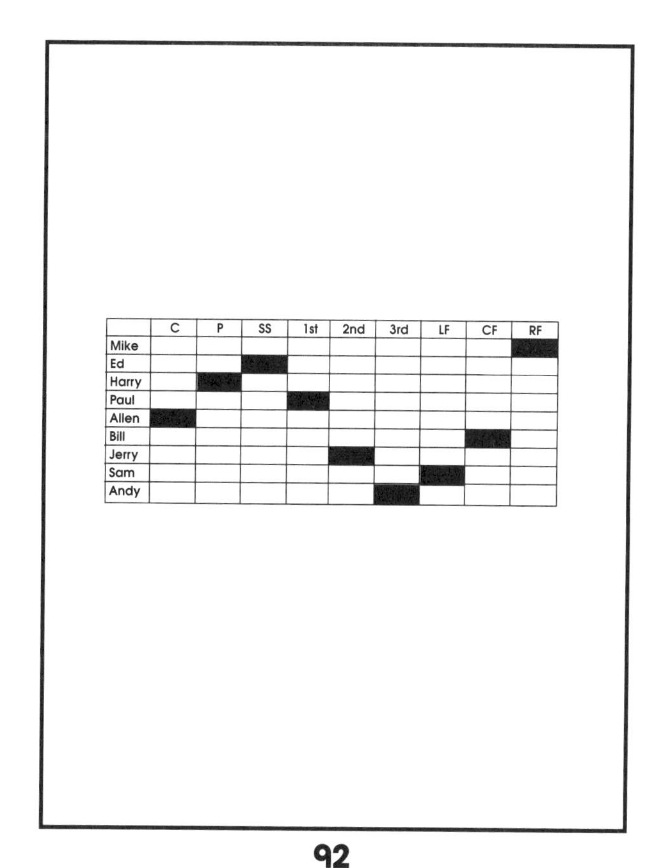

	C	P	SS	1st	2nd	3rd	LF	CF	RF
Mike									■
Ed			■						
Harry		■							
Paul				■					
Allen	■								
Bill								■	
Jerry					■				
Sam							■		
Andy						■			

92

Find the circumference and area of each circle.

1. 20π, 100π (10)
2. π, $\frac{\pi}{4}$ (1/2)
3. 24π, 144π (12)
4. 10π, 25π (10)
5. 5π, $\frac{25\pi}{4}$ (5)
6. 8π, 16π ($4\sqrt{2}$)
7. 14π, 49π (7, 60°)
8. 5π, $\frac{25\pi}{4}$ (3, 4)
9. 6π, 9π ($3\sqrt{2}$)
10. $\frac{1}{2}\pi$, $\frac{1}{16}\pi$ ($\frac{1}{4}$)
11. 3π, $\frac{9}{4}\pi$ (3)
12. $6\sqrt{2}\pi$, 18π (6, 6)
13. $\sqrt{65}\pi$, $\frac{65}{4}\pi$ (4, 7)
14. 7π, $\frac{49}{4}\pi$ (7)
15. $3\sqrt{2}\pi$, $\frac{9}{2}\pi$ (3, 3)
16. 10π, 25π (5)

93

Find the volume, lateral area, and total area of the following prisms.

1. 30, 42, 62
2. 12, 24, 36
3. 16, $16 + 8\sqrt{2}$, $32 + 8\sqrt{2}$
4. 64, 48, 112
5. 160, 112, 192
6. 480, 320, 440
7. 45, $60 + 30\sqrt{2}$, $69 + 30\sqrt{2}$
8. 27, 36, 54
9. 36, 36, 72
10. 108, 108, 144
11. 248, 176, 238
12. 90, $66 + 6\sqrt{37}$, $96 + 6\sqrt{37}$

94

Answer Key

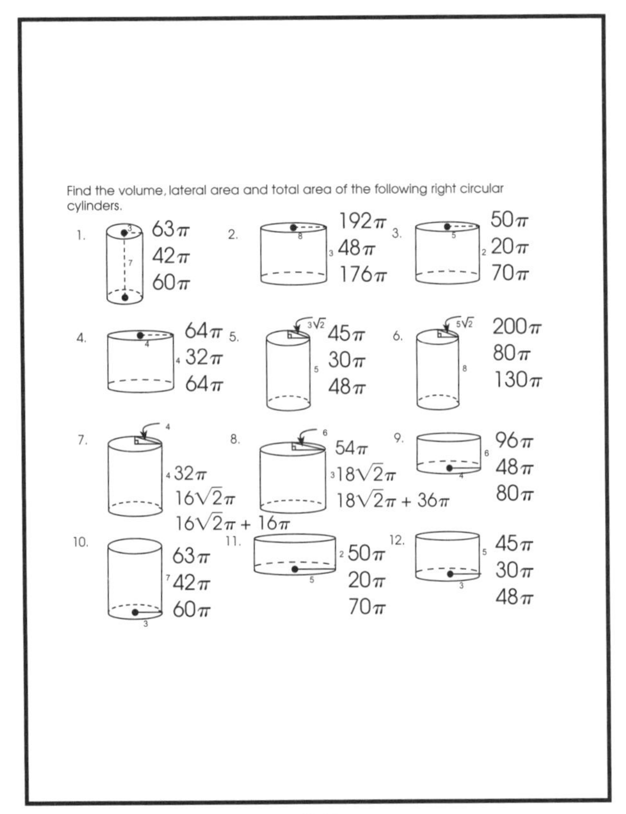

95

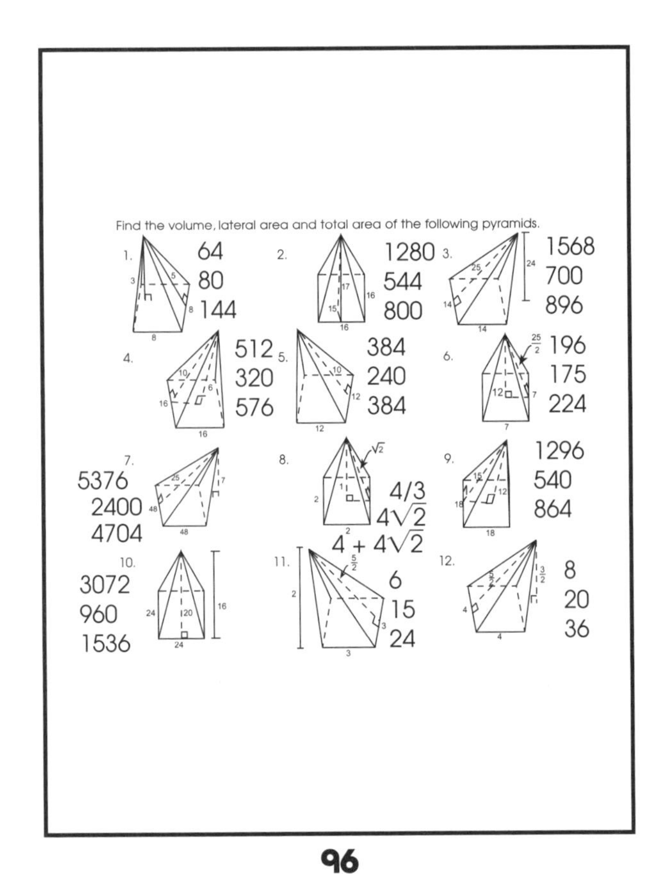

96

Find the volume, lateral area and total area of the following right circular cones.

1. 320π, 136π, 200π
2. 16π, 20π, 36π
3. $2/3\pi$, $\sqrt{5}\pi$, $(1 + \sqrt{5})\pi$
4. $8/3\pi$, $4\sqrt{2}\pi$, $4\pi + 4\sqrt{2}\pi$
5. 96π, 60π, 96π
6. 392π, 175π, 224π
7. 9π, $9\sqrt{2}\pi$, $9\pi + 9\sqrt{2}\pi$
8. 324π, 135π, 216π
9. 18π, $9\sqrt{5}\pi$, $9\pi + 9\sqrt{5}\pi$
10. 2560π, 544π, 800π
11. $\frac{1}{3}\pi$, $\sqrt{2}\pi$, $1\pi + \sqrt{2}\pi$
12. 128π, 80π, 144π

97

Determine if each region below is polygonal by breaking it into triangular regions.

1.
2.
3.
4.
5. NO
6.
7.
8. NO
9.
10.
11. NO
12. NO

98

Answer Key

Find the following for each polygon: a) The sum of the measures of the interior angles, b) The sum of the measures of the exterior angles.

1. A 32-sided polygon 5400, 360
2. A 12-sided polygon 1800, 360
3. A 6-sided polygon 720, 360
4. An 8-sided polygon 1080, 360
5. A 3-sided polygon 180, 360
6. A 5-sided polygon 540, 360

Find the following for each regular polygon: a) The measure of each exterior angle, b) the measure of each interior angle.

7. A 6-sided polygon 120, 60
8. A 5-sided polygon 108, 72
9. A 3-sided polygon 60, 120
10. An 8-sided polygon 135, 45
11. A 4-sided polygon 90, 90
12. A 10-sided polygon 144, 36
13. A regular polygon has an exterior angle with a measure of 20°. Find the number of sides. 18
14. A regular polygon has an interior angle with a measure of 120°. Find the number of sides. 6
15. A regular polygon has 20 sides. Find the measure of each exterior angle. 18°
16. A regular polygon has 10 sides. Find the measure of each interior angle. 144°

99

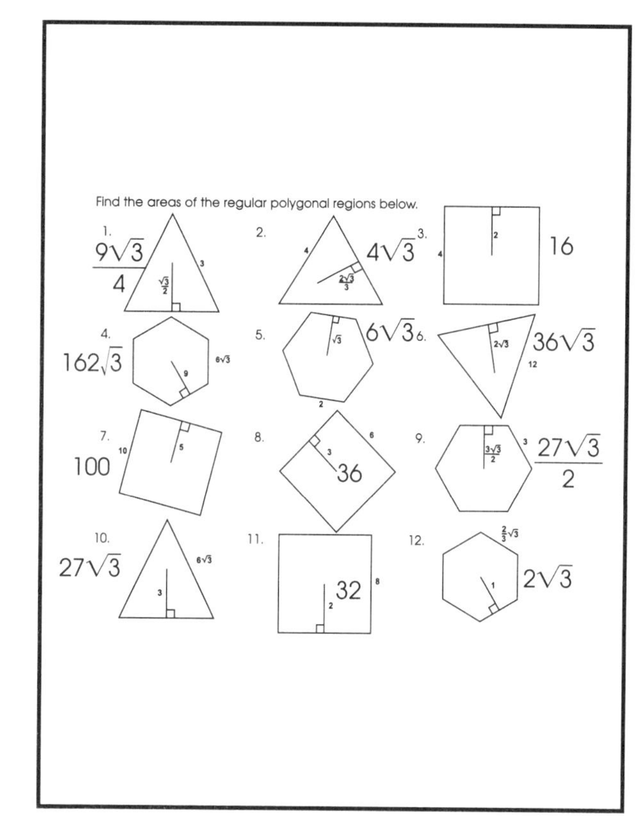

100

Use the following steps to make a Platonic Solid.

1. Cut out the circle to use as a pattern.
2. Cut out the number of circles equal to the number of faces of the Platonic Solid you are making: Tetrahedron, 4; octahedron, 8; icosahedron, 20; cube, 6; dodecahedron, 12.
3. Trace the inscribed polygon onto stiff paper to use as a folding template. Use the polygon corresponding to the shape of the face of the Platonic Solid. Triangle-tetrahedron, octahedron, icosahedron. Square-cube Pentagon-dodecahedron
4. Place the template on each circle and fold back the flaps.
5. Glue or staple flaps of the faces at each vertex to make the Platonic Solid.

Historical Comment: The Platonic Solids are named after Plato, a Greek philosopher and mathematician (427-347 B.C.). In ancient Greece, the four basic elements were identified with four of the Platonic Solids: tetrahedron, fire; cube, earth; octahedron, air; and icosahedron, water. The dodecahedron with its 12 faces was related to the universe (12 signs of the zodiac).

101

Notes